Imaging in Cellular and Tissue Engineering

Series in Cellular and Clinical Imaging

Series Editor
Ammasi Periasamy

SERIES IN CELLULAR AND CLINICAL IMAGING
Ammasi Periasamy, series editor

Imaging in Cellular and Tissue Engineering

Edited by

Hanry Yu
Nur Aida Abdul Rahim

CRC Press
Taylor & Francis Group
Boca Raton   London   New York

CRC Press is an imprint of the
Taylor & Francis Group, an **informa** business

CRC Press
Taylor & Francis Group
6000 Broken Sound Parkway NW, Suite 300
Boca Raton, FL 33487-2742

© 2013 by Taylor & Francis Group, LLC
CRC Press is an imprint of Taylor & Francis Group, an Informa business

First issued in paperback 2019

No claim to original U.S. Government works

ISBN 13: 978-0-367-44586-7 (pbk)
ISBN 13: 978-1-4398-4803-6 (hbk)

Library of Congress Cataloging-in-Publication Data

Imaging in cellular and tissue engineering / editors, Hanry Yu, Nur Aida Abdul Rahim.
    p. ; cm. -- (Series in cellular and clinical imaging ; 2)
  Includes bibliographical references and index.
  Summary: "This book covers the full range of available imaging modalities and optical methods used to help evaluate material and biological behavior. It also highlights a wide range of optical and biological applications. Each chapter in the text describes a specific application and discusses relevant instrumentation, governing physical principles, data processing procedures, as well as advantages and disadvantages of each modality. Following a broad introduction to key topics, the main chapters are divided between in vitro and in vivo applications. The final section focuses on methods for data processing and analysis"--Provided by publisher.
  ISBN 978-1-4398-4803-6 (hardback : alk. paper)
  I. Yu, Hanry, editor of compilation. II. Abdul Rahim, Nur Aida, editor of compilation. III. Series: Series in cellular and clinical imaging ; 2.
  [DNLM: 1. Cell Engineering. 2. Microscopy--methods. 3. Tissue Engineering. QU 300]

R855.3
610.28--dc23
                                                        2013008810

Visit the Taylor & Francis Web site at
http://www.taylorandfrancis.com

and the CRC Press Web site at
http://www.crcpress.com

To God who called and kept me going;

To my wife, Dr. Wing Chan, and daughter,
Isabel Yu, for enduring my neglect;

To my parents and mother-in-law who care for us;

To my role models and mentors: Sunney I. Chan for his passion for
education, Mike Sheetz for his creativity and single-minded focus on
scientific inquiry, Choon-Fong Shi for his "no wall" vision for great
universities, and Bob Nerem for his selfless support for young people;

To students and staff from whom I continue to learn, and
with whom I explore together in scientific adventure!

—Hanry Yu

Contents

PART I Overview

PART II *In Vitro* Applications

PART III *In Vivo* Applications

PART IV Data Analysis

Abbreviations

3D	Three-dimensional
AF	Autofluorescence
AFM	Atomic force microscope
ALS	Administer antilymphocyte serum
AMFPI	Active matrix flat panel imagers
AMI	Acute myocardial infarction
ANN	Artificial neural networks
ART	Algebraic reconstruction technique
BDL	Bile duct ligation
BFP	Blue fluorescent protein
BLI	Bioluminescent imaging
BMI	Body mass index
BMS	Bare metal stents
BMSCs	Bone marrow–derived stem cells
BRET	Bioluminescence resonance energy transfer
CAD	Coronary artery disease
CBRed	Click beetle luciferase red
CCD	Charge-coupled device
CEST	Chemical exchange saturation transfer
CFD	Computational fluid dynamics
CFP	Cyan fluorescent protein
CLIO	Cross-linked iron oxide
CMV	Cytomegalovirus
CNS	Central nervous system
CT	Computed tomography
CTE	Cellular and tissue engineering
DAPI	4′,6-Diamidino-2-phenylindole
DCF	Double clad fiber
DES	Drug-eluting stents
DEXA	Dual-energy x-ray absorptiometry
DO	Dissolved oxygen
DOFLA	Diversity-oriented fluorescence library approach
DSA	Digital subtraction angiography
DsRed	*Discosoma* red fluorescent protein
DTBM	Demineralized trabecular bone matrix
DTBZ	Dihydrotetrabenazine
DTTCI	3,3-Diethylthiatricarbocyanine iodide

ECM	Extracellular matrix
EEL	External elastic lamina
ELISA	Enzyme-linked immuno-sorbent assay
EM	Expectation-maximization
ES	Embryonic stem
FBP	Filtered back-projection
FDA	US Food and Drug Administration
FDG-PET	2-Deoxy-2-[^{18}F]-fluoro-D-glucose-PET
FD-OCT	Fourier-domain OCT
FEM	Finite element modeling
FID	Free induction decay
FITC	Fluorescein isothiocyanate
Fluc	Firefly luciferase
FMT	Fluorescence-mediated tomography
FT	Fourier transform
GFP	Green fluorescent protein
GLCM	Gray-level co-occurrence matrix
GLP-1R	Glucagon-like peptide-1 receptor
GVD	Group velocity dispersion
HcRed	*Heteractis crispa* red fluorescent protein
HCS	High-content screening
hfMSC	Human fetal mesenchymal stem cells
HIV-1	Human immunodeficiency virus type 1
HSCs	Hepatic stellate cells
HSV-1	Herpes simplex virus type 1
HSV-TK	Herpes simplex virus thymidine kinase
IDF	International Diabetes Federation
IEL	Internal elastic lamina
IL	Interleukin
IMCL	Intramyocellular lipids
IO	Iron-oxide
IVUS	Intravascular ultrasound
KNN	*k*-nearest neighbor
LED	Light-emitting diodes
LFM	Lateral force microscopy
LR	Logistic regression
LSECs	Liver sinusoidal endothelial cells
MEMRI	Manganese-enhanced magnetic resonance imaging
MHP36	Maudsley Hippocampal Clone 36
MIL	Mean intercept length
MIP	Mouse insulin I promoter
MMM	Multifocal multiphoton microscope
MMP	Matrix metalloproteinase
MN	Magnetic nanoparticles
MPC	Myogenic precursor cells
MPM	Multiphoton microscopy
MR	Magnetic resonance
MRI	Magnetic resonance imaging
MRM	Magnetic resonance microscopy
MRMR	Minimum redundancy maximum relevancy

MSCs	Mesenchymal stem cells
MT	Magnetization transfer
NA	Numerical aperture
NCEP	National Cholesterol Education Program
NHS	*N*-hydroxysuccinimidyl
NIR	Near-infrared
NIRF	Near-infrared fluorescent
NK	Natural killer
NMR	Nuclear magnetic resonance
NPCs	Neural progenitor cells
NTA	Ni-nitriloacetate
NTPs	Nucleotide triphosphates
OCT	Optical coherence tomography
ODD	Object-detector distance
OFDI	Optical frequency domain imaging
OPCs	Oligodendrocyte progenitor cells
OPLA	Open-cell poly(lactic acid)
OSEM	Order subset expectation maximization
PC	Phosphocreatine
PCA	Principal component analysis
PCI	Percutaneous coronary intervention
PCL	Polycaprolactone
PDP	2-Pyridyldithiopropionyl residue
PEG	Poly (ethylene glycol)
PET	Positron emission tomography
PFC	Perfluorocarbons
PGA	Polyglycolic acid
PGK	Phosphoglycerate kinase
PLS	Partial least squares
PM	Pulse modulation
PMT	Photomultiplier tubes
PRP	Platelet-rich plasma
PS-OCT	Polarization-sensitive OCT
rf	Radiofrequency
RF	Random forests
RF–RFE	Random forest–recursive feature elimination
RIA	Radioimmunoassay
RIP	Rat insulin I promoter
Rluc	Renilla luciferase
SDD	Source-detector distance
SEM	Scanning electron microscopy
SERS	Surface-enhanced Raman spectroscopy
SHG	Second-harmonic generation
SHIM	Second-harmonic imaging microscope
SNR	Signal-to-noise ratio
SOD	Source-object distance
SOFM	Self-organizing feature map
SPECT	Single-photon emission computed tomography
SPIO	Superparamagnetic iron oxide
SRB	Sulforhodamine B

SRES	Surface-enhanced Raman spectroscopy
STED	Stimulated-emission depletion
sTRAIL	Secreted form of tumor necrosis factor-related apoptosis-inducing ligand
SUR1	Sulfonylurea receptor 1
SVM	Support vector machines
SVM-RFE	SVM-recursive feature elimination
SVZ	Subventricular zone
TAA	Thioacetomide
TCFA	Thin cap fibroatheroma
Tcho	Total choline
TD-OCT	Time-domain OCT
TE	Tissue engineering
TFT	Thin film transistor
TIMP	Tissue inhibitors of metalloproteinase
TIRF	Total internal fluorescence microscopy
Ti-sa	Titanium-sapphire
TK	Thymidine kinase
TNF-α	Tumor necrosis factor-α
TPEF	Two-photon excited fluorescence
USPIO	Ultrasmall superparamagnetic iron oxide
UV	Ultraviolet
VEGF	Vascular endothelial growth factor
VMAT2	Vesicular monoamine transporter 2
WAT	White adipose tissue
WC	Waist circumference
YFP	Yellow fluorescent protein

Series Preface

A picture is worth a thousand words.

This proverb says everything. Imaging began in 1021 with the use of a pinhole lens in a camera in Iraq; later in 1550, the pinhole was replaced by a biconvex lens developed in Italy. This mechanical imaging technology migrated to chemical-based photography in 1826 with the first successful sunlight-picture made in France. Today, digital technology counts the number of light photons falling directly on a chip to produce an image at the focal plane; this image may then be manipulated in countless ways using additional algorithms and software. The process of taking pictures ("imaging") now includes a multitude of options—it may be either invasive or noninvasive, and the target and details may include monitoring signals in two, three, or four dimensions.

Microscopes are an essential tool in imaging used to observe and describe protozoa, bacteria, spermatozoa, and any kind of cell, tissue, or whole organism. Pioneered by Antoni van Leeuwenhoek in the 1670s and later commercialized by Carl Zeiss in 1846 in Jena, Germany, microscopes have enabled scientists to better grasp the often misunderstood relationship between microscopic and macroscopic behavior, by allowing for study of the development, organization, and function of unicellular and higher organisms, as well as structures and mechanisms at the microscopic level. Further, the imaging function preserves temporal and spatial relationships that are frequently lost in traditional biochemical techniques and gives two- or three-dimensional resolution that other laboratory methods cannot. For example, the inherent specificity and sensitivity of fluorescence, the high temporal, spatial, and three-dimensional resolution that is possible, and the enhancement of contrast resulting from detection of an absolute rather than relative signal (i.e., unlabeled features do not emit) are several advantages of fluorescence techniques. Additionally, the plethora of well-described spectroscopic techniques providing different types of information, and the commercial availability of fluorescent probes such as visible fluorescent proteins (many of which exhibit an environment- or analytic-sensitive response), increase the range of possible applications, such as development of biosensors for basic and clinical research. Recent advancements in optics, light sources, digital imaging systems, data acquisition methods, and image enhancement, analysis and display methods have further broadened the applications in which fluorescence microscopy can be applied successfully.

Another development has been the establishment of multiphoton microscopy as a three-dimensional imaging method of choice for studying biomedical specimens from single cells to whole animals with sub-micron resolution. Multiphoton microscopy methods utilize naturally available endogenous fluorophores—including NADH, TRP, FAD, and so on—whose autofluorescent properties provide a label-free approach. Researchers may then image various functions and organelles at molecular levels using two-photon and fluorescence lifetime imaging (FLIM) microscopy to distinguish normal from cancerous conditions. Other widely used nonlabeled imaging methods are coherent anti-Stokes Raman scattering spectroscopy (CARS) and stimulated Raman scattering (SRS) microscopy, which allow imaging of molecular function using the molecular vibrations in cells, tissues, and whole organisms. These techniques have been widely used in gene therapy, single molecule imaging, tissue engineering, and stem cell research. Another nonlabeled method is harmonic generation (SHG and THG), which is also widely used in clinical imaging, tissue engineering, and stem cell research. There are many more advanced technologies developed for cellular and clinical imaging, including multiphoton tomography, thermal imaging in animals, ion imaging (calcium, pH) in cells, and so on.

The goal of this series is to highlight these seminal advances and the wide range of approaches currently used in cellular and clinical imaging. Its purpose is to promote education and new research across a broad spectrum of disciplines. The series emphasizes practical aspects, with each volume focusing on a particular theme that may cross various imaging modalities. Each title covers basic to advanced imaging methods, as well as detailed discussions dealing with interpretations of these studies. The series also provides cohesive, complete state-of-the-art, cross-modality overviews of the most important and timely areas within cellular and clinical imaging.

Since my graduate student days, I have been involved and interested in multimodal imaging techniques applied to cellular and clinical imaging. I have pioneered and developed many imaging modalities throughout my research career. The series manager, Luna Han, recognized my genuine enthusiasm and interest to develop a new book series on Cellular and Clinical Imaging. This project would not have been possible without the support of Luna. I am sure that all the volume editors, chapter authors, and myself have benefited greatly from her continuous input and guidance to make this series a success.

Equally important, I personally would like to thank the volume editors and the chapter authors. This has been an incredible experience working with colleagues who demonstrate such a high level of interest in educational projects, even though they are all fully occupied with their own academic activities. Their work and intellectual contributions based on their deep knowledge of the subject matter will be appreciated by everyone who reads this book series.

Ammasi Periasamy, PhD
Series Editor
Professor and Center Director
W.M. Keck Center for Cellular Imaging
University of Virginia
Charlottesville, Virginia

Preface

Cellular and tissue engineering (CTE) encompasses a wide scope of effort that combines the engineering of suitable materials, along with biochemical and biomechanical factors, as well as with cells, for the purpose of improving or indeed replacing biological functions. The specificity of application requires optimized functionality. Thus, CTE requires precise control of all processes related to it, from inception of idea, to design and fabrication, and monitoring over time; for both *in vitro* maintenance as well as *in vivo* implantation. Quantitative methods are needed to measure structural integrity and functional changes in constructs over time. This is where various imaging modalities enter the picture. Imaging is a problem of scale. For small tissue pieces *in vitro*, high precision instrumentation can be used to understand the construct of both material and biological behavior. However, transferring these modalities to *in vivo* applications is a complex process, rendering them less useful. Although the current clinical imaging modalities allow us to look at specimens on a larger scale, data obtained provide only an idea of general patterns and do not allow us to understand what happens on a smaller scale. Optical methods, while providing great detail, have poor penetration depth. There is always a compromise. Although some of these issues can be addressed, others await a technological breakthrough that will propel the field forward. All of these applications and issues will be discussed in this book.

Various articles and chapters exist that address specific imaging modalities for different CTE applications. This book, however, seeks to gather these scattered pieces of information into an all-encompassing tome. Each chapter describes a specific CTE application and discusses the relevant optical instrumentation as well as physical principles governing the imaging method. Authors consider advantages and disadvantages as well as strengths and weaknesses. They present data processing procedures that have been and are being developed. Therefore, the book covers both the fundamentals and state-of-the-art applications.

The scope of CTE is broadening at an accelerated rate. While traditionally known to include applications for repair or replacement of whole tissues (i.e., bone, cartilage, blood vessels, bladder, etc.), the field now also encompasses more novel artificially created support systems such as an artificial pancreas and bioartificial liver. The aim of this book is to showcase the wide range of optical and biological applications, for the purpose of assisting investigators who wish to assess the suitability of specific imaging modalities for specific applications with certain functional requirements.

We hope that this book will capture the interest of experts and newcomers, both from the imaging community as well as from the CTE community. It is therefore geared toward scientists, researchers, professionals, and students across various fields of bioengineering with an interest in biomedical imaging.

Acknowledgments

We would like to express our gratitude to many people who have helped us with this book and made this book possible. We thank Dr. Ammasi Periasamy for his invitation to write this book and CRC Press/Taylor & Francis Group for enabling us to publish it.

We offer our appreciation to those who have contributed to the book contents. These include Abhishek Ananthanarayanan, Anju Raja, Athanassios Sambanis, Baixue Zheng, Chen-Yuan Dong, George K. Radda, Hsuan-Shu Lee, Huipeng Li, Jagath C. Rajapakse, Jerry K.Y. Chan, Jiakai Lin, Jinling Lu, Kai-Hsiang Chuang, Kishore Kumar Bhakoo, Liu Linbo, Nicholas E. Simpson, Philip W. Kuchel, Ping-Chin Cheng, Piyushkumar A. Mundra, Roy E. Welsch, Sheng-Xiang Xie, Shiwen Zhu, Shu Wang, Shuoyu Xu, Teoh Swee Hin, Wanxin Sun, Wei-Liang Chen, Weiping Han, Ying Zhao, and Zhang Zhiyong.

We are grateful to current and past staff and students of the Laboratory of Cellular and Tissue Engineering for their scientific contributions to this book. Special thanks to Koh Kang Sheing Phoebe, Li Huipeng, Peng Qiwen, and Wang Junjie for their remarkable effort in working on the layout, formatting, and references.

We thank our family members and friends for their support and encouragement throughout the time we worked on this book.

Editors

Hanry Yu is a professor in the Department of Physiology, Yong Loo Lin School of Medicine, National University Health System, Singapore. He concurrently holds the positions of group leader at the Institute of Bioengineering and Nanotechnology, A*STAR, principal investigator in the Mechanobiology Institute, Singapore, and visiting professor at the Department of Biological Engineering in the Massachusetts Institute of Technology (MIT). He received his PhD from Duke University, North Carolina, as a cell biologist. This was followed by a postdoctoral research position at the European Molecular Biology Laboratory (EMBL), Germany, as a Human Frontier Science Program (HFSP) fellow. He organized the first conference interfacing imaging and biomaterials/tissue engineering 10 years ago, and since then has established major microscopy facilities in Asia. Since joining the National University of Singapore (NUS) in 1997, he has established and managed major cross-disciplinary graduate education and research programs at university and institute levels in Singapore. His research activities over the past decade involve interfacing biomaterials with tissue engineering, bioimaging, computation and systems biology, mechanobiology, and he takes pride in the many PhD students and postdoctoral researchers trained in his laboratory in the art and science of carrying out integrative cross-disciplinary scientific studies, and who then went on to outstanding careers in academia and industry.

Nur Aida Abdul Rahim was a postdoctoral researcher in the Laboratory of Cellular and Tissue Engineering, Department of Physiology, Yong Loo Lin School of Medicine, National University of Singapore (NUS). This position is funded by the Global Enterprise for Micro-Mechanics and Molecular Medicine (GEM4). Her undergraduate and master's degrees were from the University of Cambridge, England, while her PhD was from the Massachusetts Institute of Technology (MIT), Massachusetts, all in mechanical engineering. Her research interests range from imaging intracellular fluorescent fusion protein interaction kinetics to stem cell migration and differentiation in microfluidic devices.

Contributors

Abhishek Ananthanarayanan is a PhD candidate and a laboratory executive at the Institute of Bioengineering and Nanotechnology, Singapore. He has published book chapters and reviews and coauthored peer-reviewed publications in the field of liver tissue engineering. His main interests include developing liver tissue models to understand the mechanisms of hepatotoxicity, the pathophysiology of pathogens affecting the liver and to develop more personalized *in vitro* screens to better equip the pharmaceutical industry in early drug discovery.

Kishore Kumar Bhakoo, BSc (Hons), PhD, graduated in medical biochemistry from the University of Kent at Canterbury in 1978. He received his PhD from the Institute of Neurology, University of London in 1983, and completed his postdoctoral training at the Ludwig Institute for Cancer Research. In 1986, he was appointed a Wellcome Research Associate at the Royal College of Surgeons, Institute for Child Health, London.

Professor Bhakoo was appointed as a staff scientist and university research lecturer at the MRC Magnetic Resonance Spectroscopy Unit, Department of Biochemistry, University of Oxford in 1996.

Professor Bhakoo returned to London as a MRC group head and senior lecturer at the MRC Clinical Sciences Center, Hammersmith Hospital, Imperial College London (2002), to lead the Stem Cell Imaging Group. The group developed preclinical models directed toward improving our understanding of stem cell behavior posttransplantation, in different tissues, including the brain, spinal cord, and the heart.

In 2009, Professor Bhakoo was appointed as the head of the Translational Molecular Imaging Group, based within the Singapore Bioimaging Consortium (SBIC), where he is developing multimodal imaging technologies to monitor stem cell–based therapies, inflammatory processes and chemotherapeutic responses in oncology. In 2011, Professor Bhakoo was appointed director of the newly established facility at A*STAR, the Translational Imaging Industrial Laboratory, which works directly with the industry to develop new drugs through imaging technologies.

Jerry K. Y. Chan, consultant at the KK Women's and Children's Hospital and associate professor at Duke-NUS Graduate Medical School, completed his PhD doctoral thesis at the Imperial College London, UK in 2006. His research interests include fetal stem cell biology, intrauterine stem cell and gene therapy, tissue engineering and regenerative medicine, developmental programming in nonhuman primates and reproductive medicine. Dr. Chan has published over 90 papers and book chapters and is the recipient of many awards, including Young Investigator Award from the International Society of Prenatal Diagnosis and Therapy, Miami Beach, USA, 2012; Young Scientist Award from Bone-Tec, Hannover, 2011; Best Poster Award from the International Bone Tissue Engineering Congress, Hannover, Germany, 2010; Clinician Scientist Award (Investigator) from NMRC 2009 and the Vandenburg-Storz Prize for Best Young Investigator from the 27th International Fetal Medicine and Surgery Society (IFMSS), Athens, Greece, 2008.

Wei-Liang Chen received PhD in physics in 1998 from the Johns Hopkins University, modeling molecular line emissions from the outflow of late-type stars. From 1998 to 2004, he taught physics and chemistry at the Milken Community High School in Los Angeles. Since 2004, he has been a postdoctoral researcher at the National Taiwan University where he works on the application of multiphoton and second-harmonic generation imaging for biomedical research, and on the application of spectroscopy to novel nanomaterials.

Ping-Chin Cheng obtained his PhD in anatomy from the University of Illinois Medical Center where he began his research career using x-ray microscopy. He went on to hold assistant professor positions at the State University of New York (SUNY) at Buffalo within several departments, including the Department of Anatomical Sciences, the Department of Biological Sciences, and finally the Department of Electrical and Computer Engineering where he became professor.

During his career Professor Cheng has taken several sabbatical posts at the Center for Nano-photonics at Swinburne University of Technology, Australia and the Institute of Plant and Microbial Biology, Academia Sinica, Taiwan. He has also held the position of professor at the National University of Singapore's Departments of Diagnostic Radiology, Bioengineering, Electrical and Computer Engineering and also the Department of Biological Sciences. After his stay in Singapore, Professor Cheng returned to SUNY, where he now resides, to continue his service as professor in the Department of Electrical Engineering.

Kai-Hsiang Chuang is the head of the Magnetic Resonance Imaging Group in the Singapore Bioimaging Consortium and research scientist at the Clinical Imaging Research Centre, National University of Singapore. He received PhD in electrical engineering from the National Taiwan University, Taipei, Taiwan in 2001, where he studied methods for detecting human brain function using magnetic resonance imaging (MRI). He worked as a postdoc with Dr. Alan Koretsky at the National Institutes of Health, USA till 2007, where he focused on contrast-enhanced methods especially using manganese as a functional contrast agent in animals. He joined the Singapore Bioimaging Consortium in 2008. He has published more than 40 papers on MRI. His research interest is in the development of functional imaging methods, including pulse sequences and contrast agents, for early detection of neurodegenerative diseases and cancers.

Chen-Yuan Dong received a PhD in physics from the University of Illinois at Urbana under the guidance of Enrico Gratton in 1998. His doctoral thesis was on the development and application of pump-probe and two-photon fluorescence microscopy for biological applications. From 1998 to 2000, he received an NIH postdoctoral fellowship to work with Peter So at MIT's Department of Mechanical Engineering where he gained additional experience in the development and applications of multiphoton imaging. In 2001, Dr. Dong joined the Department of Physics at the National Taiwan University where he developed a biophysics and biophotonics research program. Dr. Dong's group pioneered the extension of nonlinear optical microscopy in many areas of tissue applications as he is interested in both answering fundamental questions of physiological phenomena *in vivo* and the potential extension of optical imaging in clinical applications. Dr. Dong has published more than 90 journal papers and is on the editorial board of the *Journal of Biomedical Optics*.

Weiping Han received his BS in medical biology from Nankai University, and PhD in physiology from Cornell University. He worked as a postdoctoral researcher at the University of Pittsburgh School of Medicine, where he studied cellular mechanisms of hormone secretion regulation using advanced fluorescence imaging techniques. He then moved to Dallas and worked as an HHMI (Howard Hughes Medical Institute) associate at the UT (University of Texas) Southwestern Medical Center, where he studied molecular mechanisms of neurotransmitter release using molecular genetic and cell biological methods. In 2003, he was promoted to research assistant professor in the Center for Basic Neuroscience at UT Southwestern Medical Center. Since then, his research has been focused on the molecular mechanisms of hormone secretion and signaling. In December 2005, he moved to Singapore to set up a research program in the Laboratory of Metabolic Medicine (LMM) at SBIC. His current research focuses on understanding the biological basis of diabetes and its link to cognitive impairment.

Teoh Swee Hin is currently professor, director for the Renaissance Engineering Programme and head, Division of Bioengineering, School of Chemical and Biomedical Engineering, Nanyang Technological University (NTU). He is also a Fellow of the Academy of Engineers, Singapore. He received his BEng (1st Hons) and PhD from the Materials Engineering Department, Monash University, Australia in 1978 and 1982, respectively. He was the director at the Centre of Biomedical Materials & Applications (BIOMAT) in the Department of Mechanical Engineering and executive member of the National University of Singapore (NUS) Tissue Engineering Program, National University of Singapore. He was the founder chairman of the NUS Graduate Bioengineering Programme (GPBE) from 1999 to 2005. GPBE attracted one of the best cohort and most talented group of graduate students in bioengineering in Singapore who eventually established themselves as postdocs in prestigious universities such as MIT, Johns Hopkins, Yale, and Columbia. Prior to these appointments he was the founder deputy director of the Institute of Materials Research and Engineering (IMRE) for 2 years (1997–1998).

The entrepreneurial spirit of Professor Teoh led him to spin off more than three companies, among the more notable is Osteopore International. It is here that he guided the company to obtain FDA approval and CE mark for bone scaffolds in the craniofacial indications. The scaffolds have now been implanted successfully in more than 1500 patients. He received the prestigious Golden Innovation Award, Far East Economic Review, and the Institute of Engineers Prestigious Engineering Achievement Award in 2004, for development of the platform technology for scaffolds in bone tissue engineering. He is one of the few experts who teach regulatory affairs in medical devices and cGMP manufacturing.

Professor Teoh's main field of research is in biomaterials and tissue engineering. He is well known for his outstanding contribution to the 3D bioresorbable scaffolds for bone. His research focused on the study of mechanisms that promote cells proliferation and differentiation as a result of mechno induction through load-bearing scaffolds for tissue regeneration and remodeling. He did pioneering work in fracture-wear-resistant biomaterials. His group was ranked first in bone tissue engineering scaffolds in World Web of *Science*, 2010. He has supervised more than 60 graduate students, filed 6 patents, given more than 45 keynote/invited lectures, and published more than 320 technical papers. Professor Teoh also contributes significantly to professional bodies. He is presently the chairman, Singapore Academy, Asia Regulatory Professional Association (ARPA). He is one among the board of editors for *Tissue Engineering, Journal of Tissue Engineering and Regenerative Medicine, Journal of Mechanical Behaviour of Biomedical Materials, Journal of Oral & Maxillofacial Research*, and *Proceedings of the Institution of Mechanical Engineers Part H: Journal of Engineering in Medicine*.

Huipeng Li is currently a PhD student in computational and systems biology under the program Singapore–MIT Alliance. He obtained his bachelor's in physics from Fudan University in China in 2008. His main research interest includes computational modeling of biological systems and bio-image analysis. He has been working on computational analysis of a bi-stable model of TGF-b1 in liver fibrosis, looking for effective combination therapies for liver fibrosis. He has also been working on image analysis of liver surface images from BDL rat model to look for regional difference of vascular structure, in order to correlate regional vascular change with the progression of liver disease. In addition, he has also been working on TSP1 protein cleavage by plasmin in calcium free and replete cell culture condition, to find the effect of calcium in the regulation of active TGF-b1.

Jiakai Lin received his PhD in life sciences from the National University of Singapore (NUS) in 2012. He is currently a postdoctoral fellow at the Institute of Bioengineering and Nanotechnology. His research interests include microRNAs in cancer therapy and tissue engineering.

Liu Linbo received a BEng in precision instrument in 2001 and MEng in optical engineering in 2004 from Tianjin University, China. He received his PhD in bioengineering in 2008 from the National University of Singapore (NUS) before his postdoctoral training at the Wellman Center in Photomedicine, Harvard Medical School (HMS) and Massachusetts General Hospital (MGH) from 2008 to 2011. He was promoted as an instructor in dermatology at HMS. Currently, Dr. Liu is with School of Electrical and Electronic Engineering and School of Chemical and Biomedical Engineering as a Nanyang assistant professor since September, 2012.

Jinling Lu received a BS in biotechnology, MS and PhD in biomedical engineering from Huazhong University of Science and Technology, China in 2000, 2003, and 2007, respectively. She carried out her postdoctoral research in the Laboratory of Metabolic Medicine in the Singapore Bioimaging Consortium of Agency for Science, Technology and Research (A*STAR), Singapore from 2008 to 2011. After completing her postdoctoral research, she joined the Britton Chance Center for Biomedical Photonics, Wuhan National Laboratory for Optoelectronics, Huazhong University of Science and Technology, China. Her research interests focus on molecular imaging and optical imaging of protein interactions.

Piyushkumar A. Mundra is a research fellow at the BioInformatics Research Center, School of Computer Engineering, Nanyang Technological University (NTU), Singapore. Dr. Mundra received PhD in computer engineering from NTU (Singapore), a master of technology in bioprocess technology from the Institute of Chemical Technology, Mumbai, India, and a bachelor of engineering in chemical engineering from the Nirma Institute of Technology, India. He was a research project assistant for more than a year at the National Chemical Laboratory, India. Dr. Mundra's current research focuses on the analysis of microarray gene-expression data for classification, gene network inference, and biological image analysis. His research interests include gene selection, methods to infer gene regulatory network inferences, image feature selection, high-dimensional regression techniques, and classification algorithms.

Sir George K. Radda went to Oxford in 1956 from Hungary. He obtained his MA in chemistry and DPhil in physical organic chemistry at Merton College, Oxford. Following a year of postdoctoral work with Melvin Calvin in Berkeley, California (1962–1963), he returned to Oxford. Since then he has held the following posts: Junior Research Fellow, Merton College (1961–1964); lecturer in organic chemistry, St. John's College (1963–1964); departmental demonstrator in biochemistry (1964–1966); fellow and tutor in organic chemistry, Merton College (1964–1984); university lecturer in biochemistry (1966–1984); British Heart Foundation professor of molecular cardiology (1984–2003); Professorial Fellow, Merton College (1984–2003); honorary director, MRC Biochemical and Clinical Magnetic Resonance Unit (1988–1996); head, Department of Biochemistry (1991–1996); chief executive, Medical Research Council (on secondment) (1996–2003); chairman, National Cancer Research Institute (2001–2003); and emeritus professor of molecular cardiology and Fellow of Merton College, Oxford (2003–2006). In 2006, Oxford University brought him out of retirement and appointed him as professor and head of the Department of Physiology, Anatomy & Genetics, University of Oxford (2006–2008) to bring about the merger of two departments. This new department was judged the top pre-clinical department in the UK Research Selectivity Exercise in 2008. Being professorial Fellow, Merton College, Oxford (2006–2008), he returned to his positions as emeritus professor in the University and emeritus Fellow of Merton College (November 2008); and scientific director and chairman of the Singapore Bioimaging Consortium (2005–2010). He was a scientific advisor to the dean, National University of Singapore Medical School (2005–2011); director, Functional Metabolism Research Group (2010–to date) and chairman, Biomedical Research Council of A*-STAR (Agency for Science, Technology and Research), Singapore (2008–to date).

Anju Mythreyi Raja is a lecturer at the Ngee Ann Polytechnic in Singapore. She earned her PhD from the National University of Singapore (NUS) in the Graduate Programme in BioEngineering. Her research interests are breast cancer, animal models for cancer studies, and bioimaging. She has worked on developing nonlinear imaging methods for studying breast cancer and liver fibrosis first at the Institute of Bioengineering and Nanotechnology and later at HistoIndex Pte. Ltd. Anju completed her undergraduate studies at the Birla Institute of Technology and Science in India, where she received her BE Hons in electronics and instrumentation.

Jagath C. Rajapakse received an MSc and PhD in electrical and computer engineering from the University at Buffalo, Buffalo, New York. He is currently a professor of computer engineering and the director of the BioInformatics Research Center, Nanyang Technological University (NTU), Singapore. He is also a visiting professor in the Department of Biological Engineering, MIT, Cambridge. He was a postdoctoral researcher at the Max-Planck Institute of Cognitive and Brain Sciences, Germany, and the National Institute of Mental Health, USA. His current research interests include neuroinformatics and bioinformatics. He has authored or coauthored over 210 research papers in refereed journals, books, and conference proceedings. He was listed among the most cited scientists of all fields.

Professor Rajapakse is an associate editor of the *IEEE Transactions on Medical Imaging*, the *IEEE Transactions on Computational Biology and Bioinformatics*, and the *IEEE Transactions on Neural Networks*.

Professor Rajapakse is a fellow of American Institute of Medical and Biological Engineering (AIMBE) and Institute of Electrical and Electronics Engineering (IEEE).

Athanassios Sambanis received his PhD in chemical engineering from the University of Minnesota. Following his postdoctoral appointment at the Massachusetts Institute of Technology Biotechnology Process Engineering Center, he joined the Georgia Institute of Technology in 1989, where he is currently professor in the School of Chemical & Biomolecular Engineering and in the Emory/Georgia Tech Department of Biomedical Engineering. His research interests are in CTE, and specifically in developing cell- and tissue-based therapies for diabetes, cell technologies for tissue engineering applications, cell encapsulation, monitoring of tissue constructs *in vitro* and postimplantation *in vivo*, cell and tissue cryopreservation, cell and tissue functional evaluation, and mathematical modeling of biological systems. He has authored or coauthored more than 85 book chapters and journal publications. He is a fellow of the American Institute for Medical and Biological Engineering.

Nicholas E. Simpson received his PhD in cancer biology from Wayne State University, Detroit, Michigan. He carried out his postdoctoral work at Emory University, Atlanta, Georgia and joined the faculty of the Department of Medicine at the University of Florida in 2002. Currently, he directs research in the Endocrinology Division's Laboratory for Tissue Engineering. His research interests include developing a tissue construct comprised of insulin-secreting cells entrapped in a matrix (an aptly named "bioartificial pancreas") as a replacement for the insulin-secreting cells that have failed in persons having diabetes (particularly Type 1). Dr. Simpson also uses NMR approaches to study mitochondrial processes in diabetes, cancer, and rare inborn errors of metabolism through NMR spectroscopic techniques and computer modeling analysis. He is an author or coauthor of over 45 journal articles and book chapters.

Wanxin Sun is an application scientist in the Bruker Nano Surface Division, responsible for application and engineering developments for surface metrology technologies, ranging from scan probe microscopy, optical interferometry, to mechanical property testing. Dr. Sun received bachelor's and master's in analytical instruments from Tianjin University and a PhD in physics from the National University of Singapore (NUS). His research interests include tip-enhanced Raman scattering (TERS), nonlinear optical microscopy, material-related properties characterization at the nanometer scale, including mechanical, electrical, magnetic and thermal properties. He also applied these technologies to the regimen of imaging-based bioinformatics, for example, cell/tissue dynamics under different mechanical and chemical stimuli. He has published more than 40 articles in international journals and delivered invited talks at international conferences, ranging from material, nanotechnology to biological imaging. He is also an owner of several US patents. Besides his own research and development work, he takes the responsibility to chair sessions and co-organize symposiums at different international conferences.

Atsushi Tanaka received both an MD and PhD from Osaka City University in 1990 and 2003, respectively. He joined the Wellman Center of Photomedicine at Massachusetts General Hospital. He is an associate professor at Wakayama Medical University and serves as a percutaneous coronary intervention cardiologist. His research interest is coronary artery imaging, including intravascular ultrasound, multidetector computed tomography, and optical coherence tomography. He has published more than 70 papers in this field and received awards, especially the clinical research award from the Japanese College of Cardiology.

Shu Wang received his PhD in cell and molecular biology from Gothenburg University, Sweden, in 1993. He was a postdoctoral fellow at Harvard Medical School (HMS) and University of California, San Francisco. He currently holds a joint appointment as a group leader at the Institute of Bioengineering and Nanotechnology and an associate professor at the Department of Biological Sciences, NUS. His research interests include gene and protein delivery, tissue engineering, stem cell biology, and cancer therapy.

Roy E. Welsch received his AB in mathematics from Princeton and his MS and PhD from Stanford also in mathematics. He joined the MIT faculty in 1969 as an assistant professor, became an associate professor 4 years later and was promoted to professor in 1979. He held positions such as Leaders for Manufacturing (now called Leaders for Global Operations) chair from 1988 until 1993. From 1973 until 1979 he was also a senior research associate at the National Bureau of Economic Research where he participated in the development of the Troll econometric, financial, and statistical modeling system. He was co-director or director of the MIT Statistics Center from 1981 until 1989.

Professor Welsch is widely recognized for his book (with Edwin Kuh and David Belsley) on regression diagnostics and for his work on robust estimation, multiple comparison procedures, nonlinear modeling, and statistical computing. He is currently involved with research on robust process control and experimental design, credit scoring models and risk assessment, diagnostics for checking model and design assumptions, reliability measurement in electronic commerce, and volatility modeling in financial markets.

Professor Welsch is a fellow of the Institute of Mathematical Statistics, the American Statistical Association, and the American Association for the Advancement of Science.

Shuoyu Xu studied electronics engineering at Peking University in Beijing, P.R. China. After graduating in 2007, he received a scholarship from the Singapore–MIT Alliance to carry out his PhD study under the Computation and System Biology (CSB) program. His studies dealt with the application of nonlinear optics microscopy in life sciences and medicine, particularly focusing on liver fibrosis. He is interested in developing image processing, image analysis, and pattern recognition tools and their applications in biological and biomedical research. He has worked at Biosystems and Micromechanics (BioSyM) IRG, Singapore–MIT Alliance for Research and Technology (SMART) Center as a research assistant since July 2011, where he continues his work on the computer-aided diagnosis and digital pathology of liver fibrosis.

Xiaochun Xu is currently a research associate at the Mechanobiology Institute, Singapore. The Mechanobiology Institute (MBI) is one of the Research Centres of Excellence in the National University of Singapore (NUS), whose mission is to develop a new paradigm of biomedical research by focusing on the quantitative and systematic understanding of dynamic functional processes. Dr. Xu received a bachelor's and master's in biology from the Huazhong University of Science and Technology (HUST), China in 2008 and 2011, respectively. His research interests include bio-photonics, bioinformatics, and biomedical system, including x-ray cone-beam microtomography, single molecular tracking, cubic membrane, and ophthalmic device. He has published several journal papers in these areas.

Ying Zhao is a manager at the Preclinical Cancer Test Facility, Karolinska Institutet, Sweden. She received a BS in pharmacy and an MS in pharmaceutics from Peking University, China in 2001 and 2004, respectively. She earned her PhD in bioengineering from the National University of Singapore (NUS) and the Institute of Bioengineering and Nanotechnology, Singapore in 2009 and continued her postdoctoral research at the Institute of Bioengineering and Nanotechnology until 2012. In 2012, she joined Karolinska Institutet as the manager of Preclinical Cancer Test Facility. She has published 15 papers in peer-reviewed journals. Her research interest includes cellular and gene therapy, cancer therapy, and molecular imaging.

Baixue Zheng received a BS in life science and computational biology from the National University of Singapore (NUS) in 2006. She received a PhD in computational and systems biology in 2011 from Singapore–MIT Alliance, a global partnership in graduate education between MIT, NUS, and the Nanyang Technological University. After working as an image processing specialist in the Mechanobiology Institute Singapore from 2010 to 2012, she joined Genometry Inc. in Boston, Massachusetts as director of data analysis. Her research interests includes high-throughput screening, biostatistics and algorithm development for image processing, gene sequencing, and data analysis.

Zhang Zhiyong is currently an associate professor in Shanghai 9th People's Hospital, School of Medicine, Shanghai Jiao Tong University and a principal investigator at the National Tissue Engineering Center of China and Shanghai Key Laboratory of Tissue Engineering. Trained as a bioengineer in multidisciplinary interfaces, Dr. Zhang Zhiyong holds great passion for clinical translation and commercialization of biomedical research. After completing a BSc in biology from Xiamen University, China in 2004, he joined NUS for his PhD study, where he was awarded PhD in 2009. During his PhD and postdoctoral studies, he received a thorough and comprehensive academic training across different disciplines, including biomaterials science, stem cell and molecular cell biology research, bioreactor development, bioimaging, animal model, and bioprocessing technology. He has filed 2 US patents, published 14 academic papers in international top-tiered journals, including *Stem Cells, Biomaterials, Cell Transplantation*, and *Tissue Engineering* with a accumulative impact factor of 89.4 (average IF: 6.4 per paper), and has authored 4 book chapters. He has presented his research work at more than 20 international conferences and was bestowed 8 prestigious international awards, including the young scientist awards, best oral presentation awards, and so on. His research effort has led to success in securing more than 3 million Singapore Dollar research grants and a successful commercialization of a unique bioreactor device. His main research interest lies in developing a clinically and commercially feasible bone tissue engineering strategy for large defect treatment.

Shiwen Zhu received her BS in electronic and information science from Peking University, P.R. China in 2007, and continued her research as a PhD candidate in computation and system biology, Singapore–MIT Alliance, Singapore. Her study mainly focused on the quantification and prediction of protein subcellular localization from high-content two-dimensional images and high-resolution three-dimensional images. She is interested in developing computational methods, including image pre-processing, image segmentation, feature extraction, and statistical analysis, to solve biological problems. Currently she continues her study on colocalization analysis of transcriptional factors in mESCs as a research assistant at the Centre for BioImaging Sciences (CBIS), NUS since December 2011.

I

Overview

1

Hanry Yu
*National University Health
System*

Mechanobiology Institute

*Institute of Bioengineering
and Nanotechnology*

*Massachusetts Institute of
Technology*

Abhishek
Ananthanarayanan
*National University of
Singapore*

*Institute of Bioengineering
and Nanotechnology*

Introduction to Cellular and Tissue Engineering

1.1 Introduction

Regenerative medicine to facilitate the regeneration and repair of the damaged tissues and organs due to diseases or injuries has evolved over the past few centuries in different forms (Mironov et al., 2004). Prostheses made with biocompatible materials, drugs to stimulate regeneration, cells transplanted to lesions, and devices to support or replace the functions of the damaged organs, have been developed (Field et al., 1998; Humes, 2000; Sato et al., 2008). Recent developments to incorporate cutting-edge biomaterials and biotechnologies in developing tissue constructs for improved regenerative performance have increased in sophistication (Uygun et al., 2010). While the traditional approach of implanting materials and devices or transplanting cells *in vivo* have persisted in healing tissue damages in simpler organs such as skin, cartilage, bone, or muscle (Groeber et al., 2012; Salgado et al., 2004), various *in vitro* methodologies to build structural and functional mimics of native biological tissues especially for structurally complicated internal organs (e.g., liver, kidney, bladder, lung) have been attempted with increasing degrees of success (Baker, 2011). Cell and tissue-engineered constructs are being developed from these *in vitro* methodologies to better control the tissue structures and functions prior to implantation into the human body; with the hope that further tissue remodeling will occur *in vivo* (Kedem et al., 2005). In some cases, this assumption works and in others, it is not true that the *in vivo* remodeling will correct the design flaw of the *in vitro* engineered tissue constructs.

The basic paradigm of cell and tissue engineering is to develop proper cell sources; design and manufacture proper biomaterial scaffolds; put them together *in vitro* into tissue constructs that resemble the native biological tissues/organ both structurally and functionally; and integrate them into the living hosts (Griffith and Swartz, 2006). Enabling accurate engineering of tissue constructs for *in vivo* repair and regeneration would require proper understanding of how these constructs are formed *in vitro*, and possible ways in which they would be remodeled for integration into the living hosts *in vivo* (Williams, 2009). Thus, various biological imaging tools for acquiring images through turbid media or highly scattering dense tissues, image processing and analysis algorithms that can quantify structures and functions at various spatial and temporal scales are needed for effective cell and tissue engineering in biomedical applications.

1.2 Historical Perspectives

Intervention to the process of tissue repair and regeneration coincides with the practice of modern surgery (Garner, 2004). Even in the earliest surgical practices, materials either from the patients themselves (autologous), other patients (allogeneic), animals (xenogeneic), or nonliving materials such as prosthesis, tissue fixture like nails, glues, antiseptics, or sutures have been utilized. Such materials and devices became more sophisticated over centuries until more conscious engineering efforts in recent decades were made to control the process of regeneration (Kneser et al., 2006). Following an unsuccessful proposal to the National Science Foundation by YC Fung, the term "Tissue Engineering" was officially defined in the Granlibakken Conference at Lake Tahoe in 1987 as "application of the principles and methods of engineering and life sciences toward fundamental understanding of structure–function relationship in normal and pathological mammalian tissues and the development of biological substitutes for the repair or regeneration of tissue or organ function" (Skalak and Fox Proceeding). There has been little documented history on how the term "tissue engineering" was conceived and coined. Interviewing earlier participants of the events suggests a few people such as YC Fung, Shu Chien, Robert Nerem, Savio L-Y Woo, Van C Mow, Fred Fox, and Dick Skalak were principal proponents. Later in 1993, Langer and Vacanti summarized it in a *Science* article (Langer and Vacanti, 1993, pp. 920–926) as "an interdisciplinary field that applies the principles of engineering and life sciences toward the development of biological substitutes that restore, maintain, or improve tissue or organ function" that expanded the coverage and impact of the field. There are now wide areas of overlap between tissue engineering research and regenerative medicine, stem cell research, drug and gene delivery, and biomaterials research (Williams, 2009). Cellular engineering arises from the technical improvement of cellular functions through genetic manipulation, immortalization, and clonal selection often employed in the biochemical engineering field to generate cells with desirable phenotypes, for example, to increase the production of antibodies/proteins (Omasa et al., 2010). As the tissue engineering research grows with increasing sophistication, cellular (or cell) and tissue engineering often refers to a growing approach in tissue engineering and regenerative medicine research that pays attention to structure and function controls at cellular resolution, particularly via control of the extracellular microenvironments (Atala et al., 2006; Toh et al., 2007). Imaging sciences have played a supporting role in tissue engineering research but have become more critical as the cell and tissue engineering field evolves (REFon microCT, confocal handbook and other chapters in this book).

1.3 *In Vivo* versus *In Vitro* Approaches

Damaged or injured tissues can be repaired or regenerated with different capacities and effectiveness *in vivo* depending on location and the function that the tissue serves. For example, the liver repairs and regenerates routinely (Fausto, 2004). Skin repairs readily but forms scars (Beanes et al., 2003). Likewise, peripheral nerve sprouts readily regenerate as long as there is no scar formation blocking the path while the central nervous system inhibits the nerve sprouting processes to avoid misconnection that can be fatal (Yiu and He, 2006). Therefore, engineering different tissues would require different approaches to facilitate the repair and regeneration of tissue function. At one end of the spectrum where the native regenerative capability of the tissue is high, all we need to do is to induce and stimulate the body's native ability to regenerate the tissue. For example, liver tissue engineers aim to develop life support system to support patients' lives long enough (2–4 weeks, transplant hepatocytes/stem cells (Espejel et al., 2010) or introduce drugs/genes into the patients to stimulate the patients' own liver to regenerate (Ueki et al., 1999)) and do so only in severe cases of hepatic failure or cirrhosis. In such tissues, it is reasonable to assume that our body is an active and good bioreactor to remodel tissues regardless of structure and form of the engineered tissue implanted in the host. All we need to do is to induce the body to regenerate instead of worrying about how to control the process of repair and regeneration. Therefore, it is surprising that liver tissue engineers are the first to embark on precision engineering of tissue constructs *in vitro*. They do so because the microarchitecture of the liver is extremely critical in the maintenance of viability and liver-specific function but focused on engineering small pieces of liver tissues for drug-testing applications *in vitro* (Khetani and Bhatia, 2008). Other examples are peripheral nerve and skin. Much attention has been paid to prevent or inhibit scar formation so that the innate regenerative capabilities of these tissues can yield the desirable outcome (Shaunak et al., 2004).

At the other end of the spectrum, we have tissues that cannot fully regenerate themselves either due to the large size of the injury (e.g., bone; Mandal et al., 2012), critical function requiring precise tissue structures (central nervous system and many epithelial tissues such as intestine, lung, and kidney; Petersen et al., 2010), or lack of vasculature to supply the nutrients, oxygen, and other essential support for rapid repair and regeneration (Arkudas et al., 2007). In these tissues, it is not correct to assume that the body is an ideal bioreactor for regeneration. One marker of these tissues is that the cells constituting these tissues are often terminally differentiated cells with complex cellular functions that have lost their ability to proliferate (Khetani and Bhatia, 2008). The microenvironments in these tissues have been changed to support the terminally differentiated function of the cells in these tissues (Khademhosseini et al., 2006), and are greatly different from the microenvironments that are conducive to cell proliferation, migration, and other dynamic behaviors in development when these tissues were originally formed (on developing tissues with high cell proliferation and dynamics). These tissues retain very limited repair capabilities but not regenerative capabilities in adult and thus cannot remodel engineered tissue constructs unless we gain better and quantitative understanding of the limit of regeneration/remodeling in these tissues; especially on how the body responds to engineered tissue constructs that are implanted *in vivo*. This is where systems tissue engineering and biomedical imaging would greatly facilitate the success of such an *in vitro* approach. The *in vitro* biomaterials scaffolds, bioreactors, and tissue construct technologies have been rapidly evolving with many components actively tested in clinical trials, especially for skin, bone, and cartilage (Ikada, 2006). With proper systems level and quantitative understanding of the cell and tissue responses to biomaterials or tissue-engineered constructs, tissue engineers can rationally design biomaterial scaffolds, bioreactors, microenvironments, and tissue constructs at the desirable level of precision and scales such that we have a certain degree of control on the facilitated repair and regeneration process of more sophisticated tissues once the engineered tissue constructs are integrated *in vivo*.

1.4 Top-Down versus Bottom-Up Approaches

Two approaches emerge over the past two decades on how to construct tissues *in vitro* that to some degree recapitulate the tissue structure and function relationship *in vivo* (Ananthanarayanan et al., 2011; Khademhosseini et al., 2006). One is the top-down approach that fabricates the final desirable shape of the tissue constructs with suitable biomaterials. Upon cell seeding in the appropriate bioreactor, scaffold surface modification, scaffold properties such as porosity and stiffness, or upon implantation *in vivo*, tissue engineers rely upon cellular dynamics and remodeling to form the desired tissue constructs. A classic example is the tissue-engineered human ear where a biodegradable polymer scaffold was made into the shape of the ear, and cells were seeded into the scaffold; cultured in a bioreactor for a period of time such that the cells adhere to surfaces, establish cell–cell junctions, and cover the entire scaffold (Langer and Vacanti, 1993). Such an approach has been successfully used in bladder (Atala et al., 2006), and more recently, utilizing decellularized matrices from organs for heart, liver, and other more sophisticated tissues (Du et al., 2008; Uygun et al., 2010). The major limitation of the constructs created with top-down approach is that they often lack the detailed tissue structures that might be critical for certain tissue functions. For example, culturing hepatocytes in microgels or scaffolds helps the maintenance of the differentiated status of hepatocytes for various applications, however, they lack the intricate structural features which are representative of the liver. Likewise, though tissue-engineered hearts can beat, they function poorly to respond to endocrine regulation of the blood circulation needs. This top-down approach has worked well for skin, bone, and cartilage where some key functions can be restored without the reconstruction of fine structure details (MacNeil, 2007). Thus, the field has evolved to develop technologies using a bottom-up approach. A comprehensive bottom-up review can be found elsewhere (Ananthanarayanan et al., 2011; Novo and Parola, 2008). In short, the focus here is on constructing the fine tissue structures at the scale of tens or hundreds of cells with precision control of microenvironments such that the microscale pieces of tissues are engineered *in vitro*. This can be either by putting a group of cells together, or culturing cells on a microengineered substrate or in 3D hydrogel droplets. Technologies have also been developed to glue these precision-engineered constructs together into larger and higher order tissue structures and larger constructs (Friedman, 2008; Novo and Parola, 2008). This approach has yielded tissue constructs that are used in *in vitro* applications and are yet to prove their utility in the *in vivo* applications. With the integration of systems biology and imaging technologies into the bottom-up approach, in future, there will be a complement of bottom-up approach and the top-down approach to result in truly functional tissue constructs for regenerative medicine and other biomedical applications.

1.5 Imaging Technologies in Cell and Tissue Engineering

Much of the imaging technologies applied in cell and tissue engineering has been adopted from those developed for either clinical imaging or biological imaging. Clinical imaging modalities such as ultrasound, x-ray, MRI, CT, PET, and so on have been miniaturized for animal studies (Badea et al., 2012; Greco et al., 2012; Langham et al., 2009; Manook et al., 2012) or extended with higher resolution to observe more details down to millimeter scales (Badea et al., 2008) with more recent instruments pushing down to hundreds of micrometers with limited temporal resolution (Lau et al., 2007). In the case of micro-CT, there have been efforts to improve the imaging contrasts of soft-tissue features and biological specificity with the proper development of contrast agents (Metscher, 2009). They focus on imaging and quantifying the gross structure and function features of large tissues and organs *in vivo*. These modalities can image through dense tissues and even the entire human body but their spatial and temporal resolution limits along with the lack of biological specificity for large and complex tissue constructs restricted their ability to connect with the vast amount of molecular and cellular information available in the literature that cell and tissue engineering (especially the bottom-up approach) is increasingly built upon (Ananthanarayanan et al., 2011).

At smaller scales, biological optical microscopy has played significant roles to image and quantify features at cellular or molecular resolution. Recent developments on super-resolution imaging at single molecule resolution with high temporal resolution coupled to the rapid developments in mechanobiology to monitor mechanical microenvironments and cellular responses start defining the causal relationships and physiological relevance of many molecules in the context of tissue functions and disease processes (http://mechanobiology.info/; Gauthier et al., 2011; Thomas et al., 2004). Cellular and tissue niche is a term that is often used to describe microenvironments surrounding stem cells as they play a role in development and regeneration (Morrison and Spradling, 2008). Niche has been extended to study cancer stem cell and cancer microenvironments that regulate cancer cell behaviors (Borovski et al., 2011). Now it occupies an important place in defining and understanding tissue dynamics and integration of tissue-engineered constructs in living hosts at high spatial and temporal resolution (Nuttelman et al., 2008). High throughput and high content imaging modalities have now been extended to image tissues and tissue constructs at cellular or molecular resolution to reach and bridge the size and depth limits of clinical imaging modalities (Bahlmann et al., 2007). Multiphoton imaging, optical coherence tomography, optoacoustic microscopy, endomicroscopy, and so on have been developed over the decades and innovations on deeper imaging (beyond sub-millimeter) through dense tissues remain a challenge, which are being actively tackled via hardware and algorithm developments (Singh et al., 2012). Finally, we see the extension of imaging modalities traditionally reserved for pure materials science and engineering research now being used to image biological cells and tissues. For imaging tissue constructs that contain both living and nonliving materials, some of these modalities have already found great use in top-down tissue engineering endeavors as this book illustrates while challenges remain to truly have imaging modalities developed and adopted to bridge and satisfy multitudes of requirements in cell and tissue engineering.

References

Lau, S. H., Feng, W., and Yu, H (eds.). 2007. Virtual noninvasive 3D imaging of biomaterials and soft tissue with a novel high contrast CT, with resolution from mm to sub-30 nm. *International Conference on Materials for Advanced Technologies (ICMAT)*. Singapore.

Ananthanarayanan, A., Narmada, B. C., Mo, X., Mcmillian, M., and Yu, H. 2011. Purpose-driven biomaterials research in liver-tissue engineering. *Trends Biotechnol*, 29, 110–18.

Arkudas, A., Tjiawi, J., Bleiziffer, O. et al. 2007. Fibrin gel-immobilized Vegf and bfgf efficiently stimulate angiogenesis in the AV loop model. *Mol Med*, 13, 480–7.

Atala, A., Bauer, S. B., Soker, S., Yoo, J. J., and Retik, A. B. 2006. Tissue-engineered autologous bladders for patients needing cystoplasty. *Lancet*, 367, 1241–6.

Badea, C. T., Drangova, M., Holdsworth, D. W., and Johnson, G. A. 2008. *In vivo* small-animal imaging using micro-CT and digital subtraction angiography. *Phys Med Biol*, 53, R319–R350.

Badea, C. T., Stanton, I. N., Johnston, S. M., Johnson, G. A., and Therien, M. J. 2012. Investigations on x-ray luminescence CT for small animal imaging. *Proceedings of SPIE*, 8313, 83130T.

Bahlmann, K., So, P. T., Kirber, M. et al. 2007. Multifocal multiphoton microscopy (MMM) at a frame rate beyond 600 Hz. *Opt Express*, 15, 10991–8.

Baker, M. 2011. Tissue models: A living system on a chip. *Nature*, 471, 661–5.

Beanes, S. R., Dang, C., Soo, C., and Ting, K. 2003. Skin repair and scar formation: The central role of TGF-beta. *Expert Rev Mol Med*, 5, 1–22.

Borovski, T., Melo, F. D. E., Vermeulen, L., and Medema, J. P. 2011. Cancer stem cell niche: The place to be. *Cancer Res*, 71, 634–9.

Du, Y., Lo, E., Vidula, M. K., Khabiry, M., and Khademhosseini, A. 2008. Method of bottom-up directed assembly of cell-laden microgels. *Cell Mol Bioeng*, 1, 157–62.

Espejel, S., Roll, G. R., Mclaughlin, K. J. et al. 2010. Induced pluripotent stem cell-derived hepatocytes have the functional and proliferative capabilities needed for liver regeneration in mice. *J Clin Invest*, 120, 3120–6.

Fausto, N. 2004. Liver regeneration and repair: Hepatocytes, progenitor cells, and stem cells. *Hepatology*, 39, 1477–87.

Field, C., Frazier, J. A., Wong, M. L. et al. 1998. Polymerization of purified yeast septins: Evidence that organized filament arrays may not be required for septin function. *J Cell Biol*, 143, 737–749.

Friedman, S. L. 2008. Mechanisms of hepatic fibrogenesis. *Gastroenterology*, 134, 1655–69.

Garner, J. P. 2004. Tissue engineering in surgery. *Surgeon*, 2, 70–8.

Gauthier, N. C., Fardin, M. A., Roca-Cusachs, P., and Sheetz, M. P. 2011. Temporary increase in plasma membrane tension coordinates the activation of exocytosis and contraction during cell spreading. *Proc Natl Acad Sci USA*, 108, 14467–72.

Greco, A., Mancini, M., Gargiulo, S. et al. 2012. Ultrasound biomicroscopy in small animal research: Applications in molecular and preclinical imaging. *J Biomed Biotechnol*, 2012, 519238.

Griffith, L. G. and Swartz, M. A. 2006. Capturing complex 3D tissue physiology in vitro. *Nat Rev Mol Cell Biol*, 7, 211–24.

Groeber, F., Holeiter, M., Hampel, M., Hinderer, S., and Schenke-Layland, K. 2012. Skin tissue engineering— *In vivo* and *in vitro* applications. *Adv Drug Deliv Rev*, 63, 352–66.

Humes, H. D. 2000. Bioartificial kidney for full renal replacement therapy. *Semin Nephrol*, 20, 71–82.

Ikada, Y. 2006. Challenges in tissue engineering. *J R Soc Interface*, 3, 589–601.

Kedem, A., Perets, A., Gamlieli-Bonshtein, I., Dvir-Ginzberg, M., Mizrahi, S., and Cohen, S. 2005. Vascular endothelial growth factor-releasing scaffolds enhance vascularization and engraftment of hepatocytes transplanted on liver lobes. *Tissue Eng*, 11, 715–22.

Khademhosseini, A., Langer, R., Borenstein, J., and Vacanti, J. P. 2006. Microscale technologies for tissue engineering and biology. *Proc Natl Acad Sci USA*, 103, 2480–7.

Khetani, S. R. and Bhatia, S. N. 2008. Microscale culture of human liver cells for drug development. *Nat Biotechnol*, 26, 120–6.

Kneser, U., Polykandriotis, E., Ohnolz, J. et al. 2006. Engineering of vascularized transplantable bone tissues: Induction of axial vascularization in an osteoconductive matrix using an arteriovenous loop. *Tissue Eng,* 12, 1721–31.

Langer, R. and Vacanti, J. P. 1993. Tissue engineering. *Science,* 260, 920–6.

Langham, M. C., Magland, J. F., Epstein, C. L., Floyd, T. F., and Wehrli, F. W. 2009. Accuracy and precision of MR blood oximetry based on the long paramagnetic cylinder approximation of large vessels. *Magn Reson Med,* 62, 333–40.

Macneil, S. 2007. Progress and opportunities for tissue-engineered skin. *Nature,* 445, 874–80.

Mandal, B. B., Grinberg, A., Gil, E. S., Panilaitis, B., and Kaplan, D. L. 2012. High-strength silk protein scaffolds for bone repair. *Proc Natl Acad Sci USA,* 109, 7699–704.

Manook, A., Yousefi, B. H., Willuweit, A. et al. 2012. Small-animal PET imaging of amyloid-beta plaques with [C-11]PiB and its multi-modal validation in an APP/PS1 mouse model of Alzheimer's disease. *Plos One,* 7, e31310.

Metscher, B. D. 2009. MicroCT for comparative morphology: Simple staining methods allow high-contrast 3D imaging of diverse non-mineralized animal tissues. *BMC Physiol,* 9, 11.

Mironov, V., Visconti, R. P., and Markwald, R. R. 2004. What is regenerative medicine? Emergence of applied stem cell and developmental biology. *Expert Opin Biol Ther,* 4, 773–81.

Morrison, S. J. and Spradling, A. C. 2008. Stem cells and niches: Mechanisms that promote stem cell maintenance throughout life. *Cell,* 132, 598–611.

Novo, E. and Parola, M. 2008. Redox mechanisms in hepatic chronic wound healing and fibrogenesis. *Fibrogenesis Tissue Repair,* 1, 5.

Nuttelman, C. R., Rice, M. A., Rydholm, A. E., Salinas, C. N., Shah, D. N., and Anseth, K. S. 2008. Macromolecular monomers for the synthesis of hydrogel niches and their application in cell encapsulation and tissue engineering. *Progr Polymer Sci,* 33, 167–179.

Omasa, T., Onitsuka, M., and Kim, W. D. 2010. Cell engineering and cultivation of Chinese hamster ovary (Cho) cells. *Curr Pharm Biotechnol,* 11, 233–40.

Petersen, T. H., Calle, E. A., Zhao, L. et al. 2010. Tissue-engineered lungs for *in vivo* implantation. *Science,* 329, 538–41.

Salgado, A. J., Coutinho, O. P., and Reis, R. L. 2004. Bone tissue engineering: State of the art and future trends. *Macromol Biosci,* 4, 743–65.

Sato, Y., Murase, K., Kato, J. et al. 2008. Resolution of liver cirrhosis using vitamin A-coupled liposomes to deliver siRNA against a collagen-specific chaperone. *Nat Biotechnol,* 26, 431–42.

Shaunak, S., Thomas, S., Gianasi, E. et al. 2004. Polyvalent dendrimer glucosamine conjugates prevent scar tissue formation. *Nat Biotechnol,* 22, 977–84.

Singh, V. R., Cha, J. W., Nedivi, E., and So, P. T. C. 2012. Photon reassignment of scattered emission photons in multifocal multiphoton microscope (MMM). *Proc SPIE,* 8588, 8588–54.

Thomas, J. D., Lee, T., and Suh, N. P. 2004. A function-based framework for understanding biological systems. *Annu Rev Biophys Biomol Struct,* 33, 75–93.

Toh, Y. C., Zhang, C., Zhang, J. et al. 2007. A novel 3D mammalian cell perfusion-culture system in microfluidic channels. *Lab Chip,* 7, 302–9.

Ueki, T., Kaneda, Y., Tsutsui, H. et al. 1999. Hepatocyte growth factor gene therapy of liver cirrhosis in rats. *Nat Med,* 5, 226–30.

Uygun, B. E., Soto-Gutierrez, A., Yagi, H. et al. 2010. Organ reengineering through development of a transplantable recellularized liver graft using decellularized liver matrix. *Nat Med,* 16, 814–20.

Williams, D. F. 2009. On the nature of biomaterials. *Biomaterials,* 30, 5897–909.

Yiu, G. and He, Z. 2006. Glial inhibition of CNS axon regeneration. *Nat Rev Neurosci,* 7, 617–27.

II

In Vitro Applications

2

Confocal Microscopy for Cellular Imaging: High-Content Screening

Baixue Zheng
National University of Singapore

Abhishek Ananthanarayanan
National University of Singapore

Institute of Bioengineering and Nanotechnology

2.1 Introduction

The ultimate goal of tissue engineering is to find effective and economical solutions to improve lives and replace vital organs, which are lost or diseased owing to a variety of reasons. In recent years, efforts have been devoted to develop better *in vitro* constructs mimicking *in vivo* phenotypes. These advances are important for better fundamental understanding of the various life processes and facilitating screening of compounds earlier in the drug-discovery pipeline. The need to obtain predictive responses in a highly reproducible manner imposes constraints when designing such higher-order systems. It has necessitated the development of new technologies and adaptation of expertise from a diverse field in order to recapitulate the intricate features of tissues. These include the use of proper matrices to control the orientation of tissue constructs, substrates with desirable material characteristics, the use of appropriate soluble factors, and the precise control of the cellular microenvironment. It is crucial to provide an *in vivo*-like microenvironment, so as to achieve the natural orientation, biological, chemical, and mechanical cues to a cell. The complexity of biological systems requires the variations of a variety of parameters to obtain the optimal responses. Often, these parameters are important factors that decide or predict the authenticity and predictability of responses obtained from a system. As the number of experiments increase exponentially with the number of parameters, rapid, cheap, and accurate methods are needed to test multiple conditions in parallel experiments.

In this chapter, we focus on the principle and applications of high-content screening (HCS) technology, which is an image-based integrated platform for large-scale high-throughput studies. The chapter includes an overview of the various HCS systems, their components, applications in cell and tissue engineering, and future perspective of HCS.

2.2 Overview of High-Content Screening

2.2.1 Imaging Prospective in Cell and Tissue Engineering Applications

Complex biological interactions involving multiple pathways or multiple subcellular compartments should be studied in the cellular environment, which preserves the native spatial and temporal information. Over the last few decades, the development of various imaging modalities has provided vital insights into the functions of various organs especially the brain where little was known about before the advent of imaging technologies like functional magnetic resonance imaging (fMRI). Imaging techniques are noninvasive and provide the means to visualize cells, tissue constructs, and their behaviors in their appropriate spatial and temporal domains. The single cell level information gathered from microscopic images allows for identification and screening of subclusters of rare events in the population. Spatial information in the cell samples is useful for colocalization studies, such as studying of focal adhesion complex components (Zhang et al., 2010). Imaging method is also useful for studying individual cell types in a coculture system (Wang et al., 2011). Time-lapse studies allow continuous observation of the cellular behavior over time and migration patterns (Beerling et al., 2011). Quantitative information can also be extracted from carefully designed imaging experiment. Recent technology advance such as multifocal multiphoton microscopy (Bahlmann et al., 2007) further improves resolution and speed by parallel illumination. HCS, which is at the other end of the spectrum, helps integrating and automating microtiter plates, liquid handler, and microscopes with increased computational power to streamline experimental procedures, hence free humans from repeated manual labor, minimize human errors, and increase throughput.

2.2.2 Why High-Content Screening?

Earlier tissue engineering techniques have various levels of success in recreating various organs and tissues. Organs where structure–function relationships are not tightly coupled such as bone, bladder, and skin have been reconstructed with relative success. However, owing to the poor understanding of architectural and compositional requirements of various other tissues, organs such as the heart, liver, and kidney where structure–function relationships are tightly coupled has had limited success.

The evolution of various tools such as microfabrication technology, microfluidic chambers, and well-controlled synthetic microenvironments and its application to the field of tissue engineering has led to the recreation of various tissues or functional entities of complex organs. The best examples of which are the microscale primary coculture of hepatocytes, microfluidic chambers containing hepatocytes. The tissue constructs maintain the functional phenotype of liver cells and open up entirely new avenues for drug testing, drug interactions, and importantly evaluating the mechanisms of toxicity. These techniques have also been extrapolated to the application of culturing cardiomyocytes, where they are used to manipulate the aspect ratio of cells to create and maintain cardiomyocyte phenotype. These systems help us study various processes such as propagation and perturbation of electrical impulses and cardiotoxicity.

These systems recapitulate the *in vivo* physiological phenotypes and are also amenable to manipulations and thereby of great value to the pharmaceutical industry and basic sciences to study various disease processes and mechanisms of toxicity if performed in a large scale and with higher throughput and reproducibility.

Driven by the need of a high-throughput system for large-scale studies and screenings, Cellomics Inc. created the first HCS machine in 1996, known as Cellomics ArrayScan VTi, which offered an integrated platform including a wide-field image acquisition module, as well as image processing and informatics tools for automated image acquisition and data analysis of arrayed samples. The HCS technology has evolved further over years. Current HCS platform can be either wide-field or confocal, to image fixed or live samples. With multicolor fluorescent techniques, automated liquid handler, fast image acquisition platform, efficient data storage, and management system, HCS is capable of rapidly generating terabytes of data at subcellular resolution for large-scale genetics, biochemical, biomedical, and pharmacological studies. HCS systems are used both in academia and in the pharmaceutical companies to study the function and behavior in cells and target identification and validation in drug screening (Rauwerda et al., 2006).

The merit of HCS is its streamlined flow from sample to biological meaningful information in an automated fashion, which could free researchers from labor intensity work of sample preparation, image acquisition, and data mining. More samples could be tested in a relatively shorter period of time as a result and the system can be used to study larger cell populations in order to achieve statistical significance. In addition, the online feedback mechanism provided by many of the commercial HCS systems actively interacts with image-acquisition modules and image-processing software for active feedbacks and result-dependent intelligent acquisition. Image acquired can be immediately quantified and the results are used to control the image-acquisition procedure. For example, it is possible to configure an HCS system to capture a predetermined number of cells from a variable number of frames before moving to the next well. This function can, on one hand, ensure that a statistical significant number of cells are collected regardless of the cellular density; and avoid oversampling in wells with high cellular density. For a multiwell plate with great variation in cellular densities, such features can significantly shorten the experimental duration.

2.3 System Requirements

HCS is an integrated system including both hardware and software for cell culture preparation, image processing, and data analysis. In this chapter, we will discuss the detailed aspects of each of these components.

2.3.1 Reagents

A cell can be made visible under a microscope by numerous physical or chemical meanings. Current HCS system typically uses fluorescence technology for such purpose. Chemical dyes or transfected fluorescent peptides are tagged to proteins of interest and imaged by an optical microscope system. Such approach has relatively higher consistency, reproducibility, and flexibility in terms of the choices of targets and dyes. In addition, the immuno-fluorescent staining procedure can be easily scaled up with the use of a liquid handler. In a typical HCS study, samples are simultaneous prepared by an automated liquid handler in 96- or 384-well plates. Thousands of perturbations can be tested in a single day.

Multiplexing of fluorophores need to be optimized prior to the actual experiments. Several criteria need to be noted when designing an experiment: (1) High quantum yield of the fluorophore (the ratio of the number of photons emitted to the number of photons absorbed) is desirable. Quantum yield depends on the chemical property of the fluorophore, solvent polarity, ion concentration as well as pH. The optimal solvent condition will help to ensure high quantum yield. If multiple dyes are used, the fluorophores need to have similar quantum yield and binding affinity to further minimize crosstalk. (If one fluorophore A has significantly higher quantum yield than the other fluorophore B, A appears much brighter than B. To detect signals from B, the detector gain needs to be turned up and as a result, bleed through signal from A in channel B also increases. It is possible to get a false-positive signal in channel B that actually bleeds through signal from channel A.) (2) When multiple tagging is required, fluorophores should be selected to have well-separated excitation and emission wavelengths so as to minimize excitation crosstalk and emission bleed-through among channels. The species specificity of the primary and secondary antibodies should be checked to avoid possible cross reactivity. (3) A photostable dye is

TABLE 2.1 List of Companies with HSC Reagent Kits

Cell-Signaling Technology (CST)
Invitrogen
GE Healthcare
ThermoFisher
MDS Analytical Technologies

desirable and it should be stable in the mounting media. (4) For live cell imaging, the molecular dynamics need to be checked by control experiments to ensure that the addition of fluorescent tags does not significantly alter the native dynamics of the protein. Fluorescent tags should not induce toxicity to the cells. Phototoxicity and photobleaching should be minimized during image acquisition.

To save time and effort from optimizing experimental conditions, several companies have offered HCS reagent kits for common applications such as cell proliferation and apoptosis (Table 2.1). Nuclear dyes such as DAPI or Hoechst are typically included in the kits. The nuclear location identified by DAPI/Hoechst stain is commonly used in the cell segmentation algorithm during image processing. Besides the nuclear dye, one or two additional dyes (normally green and red fluorescent probes) are used to tag the protein(s) of interest.

2.3.2 Hardware Aspect of HCS: Automated Image Acquisition and Integration for High-Throughput Research

Speed, resolution, and sensitivity are the three important parameters for HCS. Ideally, the experiment should be designed to use the minimal time to achieve the highest resolution and sensitivity. Unfortunately, there is a trade-off of the three parameters. Before a breaking through technology is available, increasing one aspect will unavoidably lower the other one or two parameters. Hence, choice of the best combination of the three parameters is highly dependent on its application.

2.3.2.1 Automated Image-Acquisition Modules

Automated image-acquisition modules and the subsequent image-analysis capabilities are the core technologies of HCS. The first HCS machine was created by Cellomics Inc in 1996. Since then, both hardware and software of the platform technology has evolved tremendously. A list of automated HCS systems can be found in Table 2.2. This list includes both microscope and cytometry-based systems from major HCS companies.

A wide-field fluorescence image system collects fluorescence signals from a thick optical section. A confocal system excites and collects fluorescence signals from only a small volume of sample, as most of the out-of-focus light from outside the focal plane is blocked by the use of a pinhole. Newly emerged confocal technology also uses optical interference (pseudo-confocal) (Zeiss Apotome, ArrayScan) or Nipkow spinning disk (spinning confocal) (BD Biosciences Pathway) to achieve confocal effect. By reducing the background noise from the focal plane, a confocal microscope can achieve much higher resolution in both axial and lateral directions than a comparable wide-field microscope. However, a confocal microscope has a smaller field of view and hence it captures fewer cells per image. This means that more images and longer time need to be taken by a confocal microscope in order to achieve a statistical significant population of cells for downstream analysis. The compromise between resolution and time is an important consideration when choosing the right HCS system to image 2D cell monocultures on optical plates. Although compromise between image resolution and acquisition time is unavoidable, images obtained should be of the best possible quality in order to generate meaningful conclusions. Subsequent image-processing steps such as background correction and deconvolution can improve signal to noise ratio, however, information content will not be increased. The confocal system is a more suitable choice for thick substrates, 3D-cell or tissue cultures.

TABLE 2.2   List of HCS Image Acquisition Modules

Model	Company
Wide-Field System	
ImageXpress Micro	Molecular Devices
ArrayScan	ThermoFisher Scientific
KineticScan	ThermoFisher Scientific
In Cell Analyzer 2000	GE Healthcare
Operetta	Perkin Elmer
Cell-IQ	The Automation Partnership
Confocal System	
ImageXpress Ultra	Molecular Devices
In Cell Analyzer 3000	GE Healthcare
Opera	Perkin Elmer
BD Pathway 435	BD Biosciences Pathway
BD Pathway 855	BD Biosciences Pathway
Cytometer	
IsoCyte	Molecular Devices
Acumen eX3	TTP LabTech Acumen
iCyte	CompuCyte
iCys	CompuCyte
iColor	CompuCyte

Current commercial high-content systems commonly use an inverted microscope with a long-range air objective. The numerical aperture of the objective should be chosen as high as possible, to maximize the amount of light collected and image resolution. Besides the commonly used air objective in HCS systems, water objective is also available for OPERA system by Perkin Elmer. The magnification of the objective should be chosen based on the cellular details that of interest to the researchers. For example, a 5× or 10× air objective is suitable for cell counting and cell proliferation analysis. A 20× or 40× objective is more suitable for protein colocalization studies. More images are required to reach a statistical significant number of cells at higher magnifications as each image captures less number of cells.

Wavelength and intensity are two considerations when choosing a light source (Table 2.3). The wavelength of a light source should match with the excitation spectrum of the fluorophore. A narrower bandwidth (wavelength range) should be used for fluorophores with small Stroke shift. Higher light intensity helps to shorten the exposure time and minimize photobleaching of biological sample. There are three types of light source commonly used in microscopy: lamp, light-emitting diodes (LED), and laser. Lamp covers a broad range of wavelength in the visible spectrum, but intensity at a particular wavelength is low; on the other hand, laser source has limited wavelengths, but the intensity is high. LED has intermediate properties compared with lamp and laser. In addition, laser light is coherent. It can be transmitted more efficiently than the other two types of light in the optical system, hence achieving higher intensity.

Current HCS systems can typically acquire 20,000 high-resolution images per day (Imagexpress Micro). The throughput for cytometry-based system is higher, reaching approximately 150,000

TABLE 2.3   Comparison of Various Light Sources

	Bandwidth	Intensity	Cost	Coherent Light
Lamp	Broad (e.g., mercury lamp 330–760 nm)	Dim	Least expensive	No
LED	Relatively narrow (<10 nm)	Relatively bright	Moderate expensive	No
Laser	Narrow (1 nm)	Bright	Most expensive	Yes

(Molecular Devices Isocyte) to 300,000 (TTP LabTech Acumen Acumen eX3) images per day. HCS speed can be improved by multifocal parallel imaging techniques, faster stage, and more efficient auto-focus method as discussed below.

Multifocal imaging technique splits a laser beam into multiple parallel illuminations and the signals can be collected simultaneously from multiple locations by the use of a segmented high-sensitivity charge-coupled device (CCD) (Bahlmann et al., 2007). Such technique has not been integrated into the conventional HCS systems, but given its great advantage in term of speed and resolution, it certainly will place a more important role in HCS in the future, especially in tissue engineering applications, in which relatively large tissue surface needs to be studied in a speedy manner.

To image multiple positions from a microscope slide or multiwell plate, automated stage is normally used to bring different positions of the sample into the focal region. Currently, the best stage can achieve 100 nm lateral resolution (Molecular device Imagexpress Micro). Tilting can be achieved automatically to increase effective field of view and cell count as cells at image border without tilting is normally removed for analysis, however, after tilting, some of the border is no long present in the bigger image, hence more cells can be counted. BD Biosciences Pathway offers an alternative design of motionless stage and movable optics, which has 100 nm lateral and 50 nm axial resolutions. The motionless stage minimizes motion artifact and enhances image stability. Such stage is useful for imaging loosely adherent cells.

Auto-focus technique is very important to keep all images in focus in an HCS study that acquires a large number of images. Currently, there are two auto-focus techniques: image-based and laser focus. In image-based focus, fast z-stack scan is carried out and the correct focus is determined by the image sharpness and contrast. For laser focus, the bottom of the well is found instead and an offset depth for the object is added. Image-based focus is slower and generates more photo-damage than laser focusing. The two focusing method can be used simultaneously in some systems (BD Biosciences Pathway) to give more flexible control over image acquisition.

2.3.2.2 Environment Control for Live Cell Imaging

Live imaging is possible for a subset of the systems (e.g., Kinetic Scan, ThermoFisher Scientific) with integrated environmental control XE "environmental control" and fluidics modules, which maintain constant temperature, CO_2 concentration, and humidity to sustain cell growth for extended period of time in the microscope. Special attention should be given to HCS studies with live cells. The acquisition time should be kept minimal to minimize phototoxicity and photobleaching.

2.3.2.3 Liquid Handler System

Walk-away high-throughput automation is achieved when the image-acquisition module is integrated with a liquid handler (Table 2.4). A syringe pump-based liquid handler is a programmable liquid-dispensing system that can aspirate or dispense fluid through its 96- or 384-array tip head. Such system can be used for serial dilution of fluid, cell culture, and immuno-fluorescence staining of multiple samples in a parallel and automatic fashion.

Commercial liquid-handling system can automate and streamline the sample preparation steps to increase throughput and minimize sample-to-sample variations. Most of the companies offer numerous add-on modules to the liquid handler to suit the specific experimental needs and bring the pipetting precision to subnanoliter range. It is more challenging to stain cells in 3D configuration owing to penetration issue. The dimension of a single cell as well as the thickness of the sample should be taken into consideration. Optimization is needed to find the optimal dye (with good penetration) and staining protocols (dye concentration, permeabilizing chemicals, and incubation duration).

Customization is offered by most companies for the addition and integration of additional components to allow for even more flexible and robust applications. Figure 2.1 shows the layout of a customized liquid-handling system that is integrated with a barcode reader, a fluorescence and absorbance reader, a plate washer, a centrifuge, a thermal cycler, and an incubator. An independent robotic arm serves to

TABLE 2.4 List of Companies Selling Liquid Handler

Company
Agilent
Aurora Biomed
Beckman Coulter
BioTec
BioTek
Caliper Life Sciences
CyBio
Digilab
Gilson
Hamilton Robotics
Hudson Robotics
Labcyte
Protedyne
Redd & Whyte
Rigaku
Tecan Group, Ltd.
The Automation Partnership (TAP)
Titertek
Wako
Xiril
Zinsser Analytic

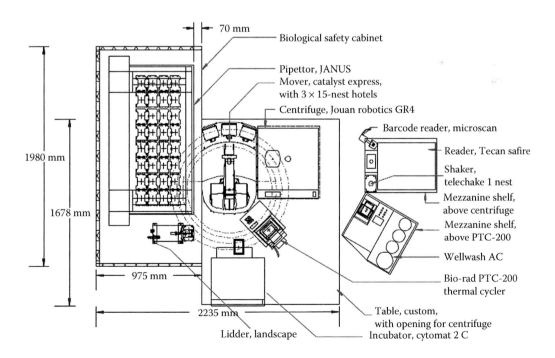

FIGURE 2.1 Schematic of RoboTox platform. (Reprinted from *Biomaterials, 32*, Zhang, S. et al., A robust high-throughput sandwich cell-based drug screening platform, 1229–41, Copyright (2011), with permission from Elsevier.)

transport multiwell plates among the instruments (Zhang et al., 2011). Besides sample preparation for HCS applications, the system can also be used to carry out ELISA and PCR experiments.

2.3.3 Image Processing and Statistical Analysis Software

Image-processing software serves to remove background noise, segment cells, and convert image into numerical values for quantitative analysis. Image-processing protocols should be robust enough to handle images with variable quality. The image quality can vary drastically owing to a variety of reasons such as imperfect focus depth, impurity in the sample, bubbles, and uneven cell density. In addition, segmentation algorithm should be optimized according to the samples as factors such as cell shape, intensity distribution, and cell density can affect the performance of the segmentation algorithm. In addition, the segmentation algorithm also depends on the labeling dyes (e.g., whether nuclear and cell membrane dye is included) and the study aim (e.g., average protein expression vs. cytoplasm to nucleus translocation). The acceptable degree of error should be taken into consideration while doing statistical analysis.

Background correction serves to flatten uneven illumination and remove noise so as to enhance the signal-to-noise ratio. Such procedure is especially important if fluorescence intensity is analyzed. There are several methods to subtract background intensity. The simplest approach is to identify an area in the image as background and subtract the mean intensity of this background area from the whole image. Such approach is very efficient for images with low cell density and relatively even background. In the case that the background intensity varies greatly owing to uneven illumination, the background intensity can be estimated by polynomial approximation from several background sample pixels. The third approach to background subtraction is called morphological top hat (Meyer and Beucher, 1990), in which the background is estimated by morphological opening of the image. The size of the filter (e.g., Gaussian filter) used for morphological opening requires the input of the object dimension, and it works well if the objects are of similar size. The image can be further processed using noise filters such as mean, median, or Gaussian filters. Periodic noise can be removed using frequency domain band pass filter. In addition, in the image preprocessing step, out of focus image should be identified and removed from further analysis. Partial objects at the image border should also be removed.

Cell segmentation is a very critical and the most challenging step in image processing. It is important to correctly identify the boundary of each cell and the subcellular compartments so that cellular features extracted subsequently can correctly represent the cellular phenomenon of interest. There are numerous segmentation methods and the correct segmentation should be chosen based on cell shape, cell density, signal-to-noise ratio, and the biological question to be studied. The most accurate segmentation is to identify the cell boundary manually by a trained researcher. Although such approach is not applicable to large-scale HCS study, it can guide the researcher to identify the most appropriate automatic segmentation method. The simplest automatic segmentation method is called thresholding, which uses gray-scale intensity information to determine a cutoff (threshold) intensity that distinguish signal from noise. Any intensity below this threshold is considered background. For example, Otsu's thresholding method automatically identifies the threshold by maximizing the variance between signal and background, and minimizing the variance within each of the two groups (Otsu, 1979). Such method works well for images with even illumination and high signal-to-noise ratio. Additional steps are needed to separate touching objects. Watershed and modified watershed algorithms are commonly used to separate touching nuclei (Higgins and Ojard, 1993). Such morphological segmentation approach is fast, but prone to under and oversegmentation, hence it is seldom used to segment cells with irregular shape. Once the nuclei are correctly segmented, it can be used to guide the segmentation of cell cytoplasm. The simple approach is based on the assumption that cell edge is at roughly equal distance to their respective nuclei. The algorithm enlarges the nuclear channel mask until the neighboring touches each other and the touching edge is defined to be the cell edge. More advanced methods utilizes (e.g., EGVD) both the morphological as well as intensity information to find cell edge (Yu et al., 2010). Such approach is more computational

TABLE 2.5   Example of Cellular Features from an HCS Study

Size	Shape	Fluorescence Intensity	Distribution
Total surface area	Circularity	Total intensity	Mean distance to other objects
Mean surface area	Geometry centroid	Average intensity	Median distance to other objects
Median surface area	Center of mass	Mean intensity	
Variance of surface area	Perimeter	Median intensity	
	Length of major axis	Variance of intensity	
	Length of minor axis		

demanding and slower, however, it generally generates better results for cells with extended morphology such as neuronal cells. Segmentation algorithms for 3D cells are also available. It is possible to segment cells with only membrane staining (e.g., Imaris 7.3.0). Another segmentation approach is edge detection. The shape gradient transition at the edge of an object is detected using algorithms such as Sobel, Prewitt, and Canny (Maini and Aggarwal, 2009).

After cells are properly segmented, cellular information such as size, shape, fluorescence intensity, and distribution can be quantified (Table 2.5). Multiple sets of phenotypic features can be generated at individual cell level or at population level. End users have the flexibility of selecting relevant features of interest that can address application-specific biological questions. Such feature selection will greatly simplify the downstream statistical analysis and save memory from uninformative or redundant data.

All current commercial HCS systems come with proprietary software, which was developed with the specific applications in mind. For example, ThermoFisher Scientific's Cellomics BioApplications Analysis Software which is packaged with ArrayScan system provides 19 application modules. Together with Cellomics HCS reagent kits, a wide range of biological and cellular processes can be studied from out-of-the-box solution packages, such as receptor activation, cell membrane receptor binding, and GPCR internalization. Commercial HCS software-packaged tools are aimed to address biological questions and hypothesis in a user-friendly and intuitive manner without tedious programming to allow new users to easily learn and use the readily available and validated solution packages. Existing statistical tools and bioapplication packages lay down the groundwork for HCS data analysis. However, there is no generic solution, customization and personalization is the key to success in HCS. For example, the same set of data could be analyzed by two different methods to generate different biological insights (Perlman et al., 2004; Loo et al., 2007). For assay development, bioapplication software packaged with HCS instrument may not be sufficient and flexible enough for exploratory analysis. It is possible to export the dataset either as raw images or numerical feature values to third-party software for processing and further analysis. Some of these softwares are listed in Table 2.6.

Statistical approach is highly dependent on the underlying questions to be addressed. Mean and median are the simple statistic parameters for analysis at the population level, while rare events or subpopulation study can be done when individual cell is the unit entity in the analysis. Temporal information of cell or subcellular events can be easily followed in a kinetic sequence of data using software such as Imaris. To convert numerical feature values to biological meaningful and statistically significant knowledge, various data mining techniques can be used, such as principal component analysis, neural network, and support vector machine.

2.3.4 Hybrid-Integrated Systems

The standard imaging features includes object shape, area, intensity level, and localization. It is possible to couple biosensors to the HCS system to measure additional parameters such as pH values, oxygen content, and electrical impedance in addition to imaging techniques on live cells (Lob et al., 2007). It is also possible to have force measurements from HCS-based system (Vandenburgh, 2010).

TABLE 2.6   List of Software Useful in HCS Study

Commercial HCS Software
ThermoFisher Scientific: HCS Discover April 2009
GE: In Cell Miner High Content Manager 2008
SpotFire
Molecular Devices: Metamorph, MetaXpress
AcuityXpress
Image Processing
ImageJ
MATLAB®
Imagepro plus
Metamorph
Cellprofiler
Definiens
BioImageXD
Imaris
Velocity
Huygens
Statistical Analysis
MATLAB
SPSS
R

2.4 Applications/Case Studies

HCS methods were developed over 10 years ago and have firmly established itself in the industrial and academic setting for various applications. One of the main applications of this technology has been in the drug discovery where it is used for primary screens and also increasingly applied to secondary screens and toxicology assessment.

HCS allows for the study of a large variety of assays, some of them being assays for morphological analysis (e.g., study of neurite outgrowth and membrane perturbation of cells). It also allows for the study of perturbations to a subpopulation of cells to study cellular differentiation, infection assays, and cell cycle analysis. HCS have been used for genome wide gene functional analysis (Pelkmans et al., 2005; Sonnichsen et al., 2005), tracking of proteome subcellular localization (Huh et al., 2003), study of protein–protein interaction, and drug screening (Perlman et al., 2004; Mitchison, 2005). To interpret information captured in images obtained from HCS, numerous efforts have been made using advance statistical methods, such as neural network (Bakal et al., 2007), support vector machine (Loo et al., 2007), and factor-analysis (Young et al., 2008).

The multiparametric nature of the system, availability of subcellular information, and the diverse response to agents obtained in a qualitative and a quantitative manner makes HCS a tool of choice for the pharmaceutical industry.

2.4.1 High-Content Screening in Primary Compound Screening

High-content cellular imaging enables disease-relevant screening earlier in the drug development process. Traditionally HCS technologies were only used in secondary screening, since it was initially believed that the throughput of this technology was too low and the methodology too expensive. Improvements in instrumentation have helped move this technology from the traditional secondary screening more to the forefront for primary screens. It is also being recognized that the multiparametric nature of this technology allows for screening multiple parameters, resulting in secondary screens happening earlier in the development process. Screening of multiple parameters of a phenotype increases accuracy, thereby reducing false positives or negatives. Imaging-based methods allow for detecting higher hit rates than biochemical screens owing to the fact that cells are analyzed for the biological responses and the systems level information obtained from HCS technique. This possibly omits the need to perform target validation, which is inherently a costly, time-consuming process without clear definitions defining a target. The compounds detected in cellular assays usually enter the cell, and therefore, satisfy the criteria for ADMET and incorporation of secondary assays into primary screens can reduce timelines and cost.

2.4.2 HCS in Lead Optimization

During the process of lead optimization compounds are improved with regard to their specificity, efficacy in iterative rounds of synthesis and biological testing. The HCS platforms report multiple endpoints, are high throughput thus decrease the cycle time and also allow the lead compounds to be tested for specificity, toxicity, efficiency, cell penetration, and stability. An important requirement for a system to be used during lead optimization is the ability to produce statistically robust data. The HCS achieves that and allows the generation of IC50 data for multiple criteria. HCS also yields more sensitive endpoints as it allows for individual cell analysis. It allows for identification of toxic effects to a subset of population, which is an important criterion for lead prioritization. It also allows for a more sensitive method to identify cellular toxicity other than measuring cell death using traditional assays. Multiparametric mode of this system provides mechanistic understanding of modes of action, which will be important to elucidate structure–activity relationships (SARs).

2.4.2.1 Elucidating SARs

SARs play a very important role in the modern drug development paradigm especially during lead optimization wherein chemical modifications to the compound are assessed for their solubility, specificity, toxicity, and preliminary ADME properties. HCS technologies have been increasingly used to establish SARs. HCS-based SARs requires quantitative compound activities. IC50 values for PI3K inhibition using chemical methods agree well with HCS-based techniques. Dose-dependent accumulation of fluorescent protein on treatment with drug candidates has also been used subsequently to support SAR process that result in PI3K inhibition.

2.4.2.2 HCS in Assessment of Toxicity

The current paradigm in drug discovery is the assessment of toxicity earlier and earlier in the drug-discovery pipeline. HCS has been utilized to improve the predictability of drug-induced responses and to determine parameters which predict toxicity before actual cell death occurs. The multiplexing ability of the system allows multiple pathways of drug responses to be studied simultaneously. The system was primarily used to assess liver toxicity, which seems to affect majority of the compounds and is a major cause of drug withdrawal. Some groups have demonstrated the usefulness of this system in elucidating the toxicity of different chemicals while understanding the pathways through which the chemicals could elicit their toxicity. These techniques could also be applied to various other cell types and can also be used to determine/assess developmental toxicity using stem cells. The proper assessment of toxicity requires the need for physiologically relevant models.

2.4.2.3 HCS and Stem Cells

The multiparametric nature of HCS devices is very useful in studying homogeneity of a cell population or occurrence of rare phenotypes in a heterogeneous population. Therefore, this system has found great applicability toward stem cell biology, which holds promise toward fields such as regenerative medicine, drug discovery, and tissue engineering. The characteristics that make stem cells so exciting is that they are capable of (1) self-renewal (pluripotency), and (2) differentiation into specific cell types. HCS can help provide analytical characterization of stem cell and quantitative information on stem cell growth and differentiation. HCS provides a good tool to monitor self-renewal and differentiation on a large scale in a scalable manner, which will enable selecting better chemical and genomic constructs from libraries for the desired application. Various groups have monitored the expression of OCT-4 GFP and a screened multiple compounds for sustaining pluripotency in the absence of feeder layers, serum. HCS has been used to identify molecules, which lead to the differentiation of ESC to specific cell types namely the differentiation of cells to neuronal, cardiovascular, and pancreatic phenotypes. This technology is increasingly being used to screen for chemicals, which can reprogram somatic cells to pluripotent cells. Imaging-based techniques is garnering increasing attention to screen for molecules that induce pluripotency, helping overcome the limitations of iPS cells such as low reprogramming efficiency and genomic instability and alterations caused by viral infection.

2.4.3 Case Study

The case study here shows an example of using the HCS approach to design an innovative drug screening platform and to derive rich biologically relevant information to help prioritize drugs for *in vivo* testing.

Currently, many antifibrotic drugs with high *in vitro* efficacies fail to produce significant effects *in vivo*, in part owing to low correlation between the *in vitro* and *in vivo* drug responses. Antifibrotic drug discovery efforts typically follow a sequential procedure. Drug candidates are first subjected to a series of *in vitro* experiments, and those satisfying a set of predefined criteria are advanced to *in vivo* animal testing. Very often, *in vitro* data have poor correlation with *in vivo* drug effects owing to the complicated pathophysiological background of hepatic fibrogenesis. As a result, a significant number of drugs fail to show desirable *in vivo* effects (Sato et al., 2008). Not only does such process have low success rate, it also prolongs the drug-discovery process. To overcome these limitations, it is important to take the *in vivo* response of a drug into consideration as early as possible in the drug development process. Here, we show that by integrating HCA and application-specific statistical analysis, a numerical predictor ($E_{predict}$) from HCS experiment can linearly correlate with *in vivo* efficacies of antifibrotic compounds.

Forty-nine compounds with 11 concentrations each were used to treat hepatic stellate cells (HSCs) LX-2, and the drug-treated cells were stained with 10 fibrotic markers. Approximately 0.3 billion feature values from more than 150,000 images were quantified to reflect the drug-induced cellular responses. Subsequently, statistical methods were used to reduce the data complexity down to a single novel score per drug, the $E_{predict}$, the magnitude of which can positively reflect the *in vitro* antifibrotic efficacy of a drug.

On the other hand, a systematic literature search was performed on the *in vivo* effects of all 49 drugs on hepatofibrotic rats, yielding 28 papers with pathologist-graded histological scores between 1986 and 2009. The data were extracted from these papers and summarized into another novel score per drug, $E_{in\ vivo}$, which reflects the *in vivo* antifibrotic efficacy of a drug.

A linear correlation was consistently observed between the $E_{predict}$ and the $E_{in\ vivo}$ in the training data set that was used to build the model as well as in two independent validation data sets (Figure 2.2). Such kind of relationship has never been reported before and we believe that it could only been achieved in a HCS system in which data from multiple fibrotic markers could be unbiasedly combined and analyzed.

By utilizing the linear relationship between the $E_{predict}$ and the $E_{in\ vivo}$, we demonstrated that it was possible to generate prediction about the likely *in vivo* outcome of a drug based on *in vitro* experimental data. In addition, seeking insight into the modes of drug response, we used $E_{predict}$ values to cluster drugs according to efficacy, and found that high-efficacy drugs tended to target proliferation, apoptosis, and contractility of HSCs. Currently, our system was optimized with data from *in vivo* studies on rats. However, we had seen evidence that the $E_{predict}$ may also have implications on human clinical data.

Our work is the first attempt to demonstrate an *in vitro–in vivo* correlation in the liver fibrosis drug discovery context, where the drug efficacy assessed *in vitro* on a multiwell dish can be directly linked to the drug efficacy *in vivo* in animal models with liver fibrosis.

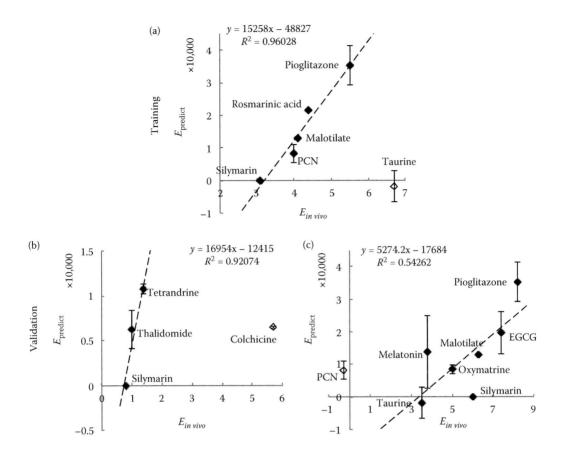

FIGURE 2.2 Correlation in training data set and two independent validation data sets in prediction of drug efficacy.

2.5 Future of High-Content Screening in Cellular and Tissue Engineering

Emerging new technologies such as light-sheet microscopy and multifocal, multiphoton microscopy may alleviate the limitations of current HCS platform by further improving image resolution and throughput at the same time.

In light sheet microscope, a specimen is illuminated by a sheet of light and the emission light orthogonal to the illumination axis is collected (Huisken et al., 2004). The concept of light sheep microscopy was first mentioned by Siedentopf and Zsigmondy in 1903. The work by Stelzer et al. in 2008 highlighted the advantage of this imaging technique: long-term noninvasive live cell imaging capability with low phototoxicity, 3D imaging with relatively deep penetration depth, fast image acquisition, and high image resolution (Keller et al., 2008).

Whole-organ screening may be feasible in the near future using the multifocal, multiphoton microscopy technology. The system can generate from whole organ to subcellular level resolution images. The potential rich information generated from such studies will provide a systematic level understanding of the tissue architecture and how organ and its various components work.

References

Bahlmann, K., So, P. T., Kirber, M. et al. 2007. Multifocal multiphoton microscopy (MMM) at a frame rate beyond 600 Hz. *Optics Express,* 15, 10991–8.

Bakal, C., Aach, J., Church, G., and Perrimon, N. 2007. Quantitative morphological signatures define local signaling networks regulating cell morphology. *Science,* 316, 1753–6.

Beerling, E., Ritsma, L., Vrisekoop, N., Derksen, P. W., and Van Rheenen, J. 2011. Intravital microscopy: New insights into metastasis of tumors. *Journal of Cell Science,* 124, 299–310.

Higgins, W. E. and Ojard, E. J. 1993. Interactive morphological watershed analysis for 3D medical images. *Computerized Medical Imaging and Graphics: The Official Journal of the Computerized Medical Imaging Society,* 17, 387–95.

Huh, W. K., Falvo, J. V., Gerke, L. C. et al. 2003. Global analysis of protein localization in budding yeast. *Nature,* 425, 686–91.

Huisken, J., Swoger, J., Del Bene, F., Wittbrodt, J., and Stelzer, E. H. 2004. Optical sectioning deep inside live embryos by selective plane illumination microscopy. *Science,* 305, 1007–9.

Keller, P. J., Schmidt, A. D., Wittbrodt, J., and Stelzer, E. H. 2008. Reconstruction of zebrafish early embryonic development by scanned light sheet microscopy. *Science,* 322, 1065–9.

Lob, V., Geisler, T., Brischwein, M., Uhl, R., and Wolf, B. 2007. Automated live cell screening system based on a 24-well-microplate with integrated micro fluidics. *Medical & Biological Engineering & Computing,* 45, 1023–8.

Loo, L. H., Wu, L. F., and Altschuler, S. J. 2007. Image-based multivariate profiling of drug responses from single cells. *Nature Methods,* 4, 445–53.

Maini, R. and Aggarwal, H. 2009. Study and comparison of various image edge detection techniques. *International Journal of Image Processing,* 3, 1–60.

Meyer, F. and Beucher, S. 1990. Morphological segmentation. *The Journal of Visual Communication and Image Representation,* 1, 21.

Mitchison, T. J. 2005. Small-molecule screening and profiling by using automated microscopy. *Chembiochem,* 6, 33–9.

Otsu, N. 1979. Threshold Selection Method from Gray-Level Histograms. *IEEE Transactions on Systems Man and Cybernetics,* 9, 62–6.

Pelkmans, L., Fava, E., Grabner, H. et al. 2005. Genome-wide analysis of human kinases in clathrin- and caveolae/raft-mediated endocytosis. *Nature,* 436, 78–86.

Perlman, Z. E., Slack, M. D., Feng, Y., Mitchison, T. J., Wu, L. F., and Altschuler, S. J. 2004. Multidimensional drug profiling by automated microscopy. *Science,* 306, 1194–8.

Rauwerda, H., Roos, M., Hertzberger, B. O., and Breit, T. M. 2006. The promise of a virtual lab in drug discovery. *Drug Discovery Today,* 11, 228–36.

Sato, Y., Murase, K., Kato, J. et al. 2008. Resolution of liver cirrhosis using vitamin A-coupled liposomes to deliver siRNA against a collagen-specific chaperone. *Nature Biotechnology,* 26, 431–42.

Sonnichsen, B., Koski, L. B., Walsh, A. et al. 2005. Full-genome RNAi profiling of early embryogenesis in *Caenorhabditis elegans*. *Nature,* 434, 462–9.

Vandenburgh, H. 2010. High-content drug screening with engineered musculoskeletal tissues. *Tissue Engineering Part B: Reviews,* 16, 55–64.

Wang, D. Y., Wu, S. C., Lin, S. P., Hsiao, S. H., Chung, T. W., and Huang, Y. Y. 2011. Evaluation of transdifferentiation from mesenchymal stem cells to neuron-like cells using microfluidic patterned co-culture system. *Biomedical Microdevices,* 13, 517–26.

Young, D. W., Bender, A., Hoyt, J. et al. 2008. Integrating high-content screening and ligand-target prediction to identify mechanism of action. *Nature Chemical Biology,* 4, 59–68.

Yu, W. M., Lee, H. K., Hariharan, S., Bu, W. Y., and Ahmed, S. 2010. Evolving generalized Voronoi diagrams for accurate cellular image segmentation. *Cytometry Part A,* 77A, 379–86.

Zhang, S., Tong, W., Zheng, B. et al. 2011. A robust high-throughput sandwich cell-based drug screening platform. *Biomaterials,* 32, 1229–41.

Zhang, X., Tee, Y. H., Heng, J. K. et al. 2010. Kinectin-mediated endoplasmic reticulum dynamics supports focal adhesion growth in the cellular lamella. *Journal of Cell Science,* 123, 3901–12.

3

Use of Multiphoton Microscopy for Tissue Engineering Applications

Wei-Liang Chen
National Taiwan University

Hsuan-Shu Lee
National Taiwan University

Chen-Yuan Dong
National Taiwan University

3.1 Introduction

In tissue engineering, a scaffold provides the necessary framework for cells to grow into the desired three-dimensional (3D) tissues or organs (Sachlos et al., 2003). The ability to capture the 3D dynamic of the growing process calls for a minimally invasive imaging modality with 3D imaging at sufficient resolution to identify the scaffold and the cell components of a tissue engineering construct. The ability of a multiphoton microscope to simultaneously register multiple spectral fluorescence signals and the second-harmonic generation (SHG) signal provides a rich amount of information for both scaffold and cell component characterization. The use of a long excitation wavelength and a no-pinhole detection scheme decreases the sample scattering effect both in the beam excitation path and in the signal detection path. The minimally invasive nature of multiphoton microscopy together with its intrinsic optical sectional capability makes it particularly suitable for capturing long-term 3D dynamics in tissue engineering.

In this chapter, we will first describe multiphoton fluorescence and SHG processes which provide the contrast mechanism in a multiphoton microscopy image. The nature of these processes is responsible for the design of a two-photon microscope and many strengths of this technique. Finally, we will describe application of multiphoton microscopy to the characterization of tissue engineering scaffold, and an example of a long-term observation using this technique.

3.2 Principles of Multiphoton Fluorescence and SHG

3.2.1 Multiphoton-Excited Fluorescence

In the multiphoton-excited fluorescence process, an electron in the ground electron state of a fluorophore is excited to a higher electronic state by the absorption of a minimum of two photons with lower energy (see Figure 3.1). The excited electron then loses energy by collisional deexcitation before decaying

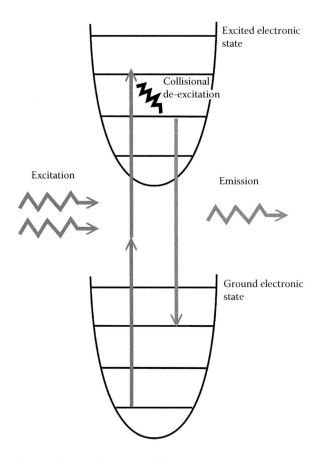

FIGURE 3.1   Diagram showing the two-photon excited fluorescence process.

back to the electronic ground state and emits a photon in the process with the emitted photon at less than the total energy of the two absorbed photon. As a higher-order process, multiphoton absorption occurs with a much lower probability than single-photon absorption. As an example, two-photon fluorescence depends quadratically on the intensity of the excitation laser, and is emitted only at the focal point of the excitation light. Confining the fluorescence emission to the focal point leads to intrinsic optical sectioning capability and a detection scheme that is less susceptible to fluorescence emission scattering by the sample.

3.2.2 Second-Harmonic Generation

In addition to fluorescence generation, the nonlinear process of SHG can also be used as a contrast mechanism for tissue engineering application. In this process, two photons of lower energy are converted to a single photon of exactly twice the energy. Like two-photon excitation, the SHG process also has a quadratic dependence on the incident laser intensity. However, since this process conserves energy, no energy is deposited on the sample. Owing to the coherent nature of SHG, the emission polarization, intensity, and direction is highly dependent on the macromolecular structure of the sample and on the polarization of the incident laser light (Roth and Freund, 1979, Mertz and Moreaux, 2001, Stoller et al., 2002, Brasselet et al., 2004, Williams et al., 2005, Nadiarnykh et al., 2007). As a result, SHG contains a rich amount of information that allows the probing of molecular structure below the optical resolution (Brasselet et al., 2004, Plotnikov et al., 2006, Su et al., 2010, 2011).

3.3 Setup of a Multiphoton Microscope

The most commonly used multiphoton microscopes are based on the single-point scanning technique (So et al., 2000). A diagram of a typical laser scanning two-photon microscope is shown in Figure 3.2. The excitation light source is provided by a femtosecond titanium–sapphire (Ti-sa) laser that is tunable from 700 to 1000 nm. For two-photon excited fluorescence in the visible spectrum, the laser wavelength is tuned close to 800 nm. The use of longer excitation wavelength leads to reduced scattering and allows greater imaging depth to be achieved. The pump source for the Ti-sa oscillator is often provided by a diode-pumped, solid-state laser.

The power and polarization of the excitation source is controlled by a set of wave plates and polarizer. It is then guided into a standard optical microscope by a pair of galvanometer-driven mirrors whose rotational axes are mutually orthogonal. The computer controls the motion of the two mirrors in order to raster scan the focused laser light over the specimen. To achieve optimal resolution, the excitation source is beam expanded to over fill the objective's back aperture, leading to a focused spot size that is diffraction limited. At the excitation wavelength of 800 nm, an oil immersion objective with a numerical aperture (NA) of 1.4 can achieve a lateral and axial resolution of 0.69 and 1.63 μm. Following excitation, two-photon fluorescence and the backward SHG signals are then collected by the focusing objective in the epi-illuminated geometry. If desired, the forward SHG signal can be simultaneously collected in the transmission mode by similar collection of objective and detector. The collected signal photons are then further separated into desired spectral bands by additional dichroic mirrors and band pass filters prior to reaching the photomultiplier tubes (PMTs). In the case that the single-photon detection mode is used, discriminators are used to process the signal photons. The computer coordinates the scanning mirror motion with the detected signals to form images. Since the generation of the multiphoton signals depends

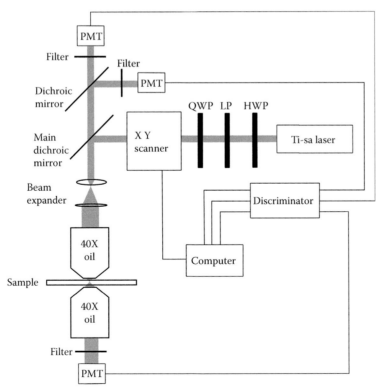

FIGURE 3.2 Diagram showing the instrumental setup of a typical two-photon microscope. QWP, quarter-wave plate; LP, linear polarizer; HWP, half-wave plater; PMT, photomultiplier tube.

nonlinearly on the excitation powers, efficient multiphoton imaging is only achievable with high incident photon flux. The required photon flux is achieved by using a femtosecond-pulsed laser source which bunches the laser photons in time and by focusing the laser source to obtain high spatial concentration.

3.4 Scaffold Imaging with Multiphoton Microscopy

To demonstrate the ability of two-photon microscopy to characterize tissue engineering scaffolds, Figure 3.3 shows large area two-photon imaging of five commercially available scaffolds that are commonly used in tissue engineering (Sun et al., 2008): open-cell poly-lactic acid (OPLA, BD Bioscience, San Jose, CA) (Vogelin et al., 2005), polyglycolic acid (PGA, Synthecon Inc, Houston, TX) (Lee et al., 2006), collagen composite scaffold (BD Bioscience, CA) (Takezawa et al., 2000), collagraft bone graft matrix strip (Zimmer, Warsaw, IN) (Takezawa et al., 2000), and nylon (Zytel, DUPONT, Wilmington, DE). The SHG image is shown in the first column followed by autofluorescence (AF) images acquired from the blue, green, and red channels, in the remaining three columns. From Figure 3.3, we see that all five scaffolds emit SHG, though they have different spectral characteristics which can be quantified by calculating the ratios of the intensities of green and red spectral bands with respect to that of the blue band (Table 3.1). As Table 3.1 shows, PGA has a stronger AF emission spectrum in the green channel;

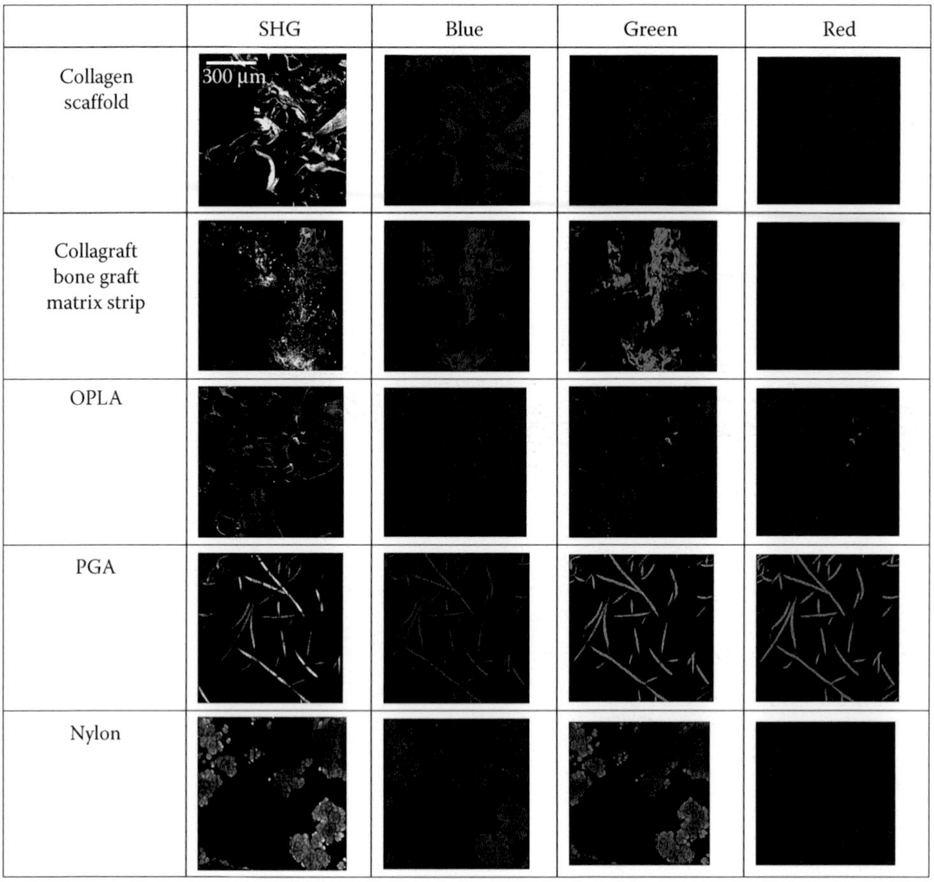

	SHG	Blue	Green	Red
Collagen scaffold	300 μm			
Collagraft bone graft matrix strip				
OPLA				
PGA				
Nylon				

FIGURE 3.3 Multiphoton AF and SHG imaging of the tissue engineering scaffolds of collagen, collagraft, OPLA, PGA, and nylon. AF is separated into the blue (435–485 nm), green (500–550 nm), and red (550–630 nm) channels. (Sun, Y. et al. Imaging tissue engineering scaffolds using multiphoton microscopy. *Microscopy Research and Technique.* 2008. 71, 140–145. Copyright Wiley-VCH Verlag GmbH & Co. KGaA. Reproduced with permission.)

TABLE 3.1 Spectral Analysis of the Scaffold Materials

Material/Ratio	Green/Blue	Red/Blue
Collagen scaffold	0.73	0.30
Collagraft bone graft matrix strip	0.80	0.26
OPLA	0.81	0.45
PGA	1.15	0.66
Nylon	0.82	0.37

Source: Sun, Y. et al. Imaging tissue engineering scaffolds using multiphoton microscopy. *Microscopy Research and Technique.* 2008. 71, 140–145. Copyright Wiley-VCH Verlag GmbH & Co. KGaA. Reproduced with permission.

Note: Blue, green, and red represent AF intensities of the blue, green, and red channels, respectively.

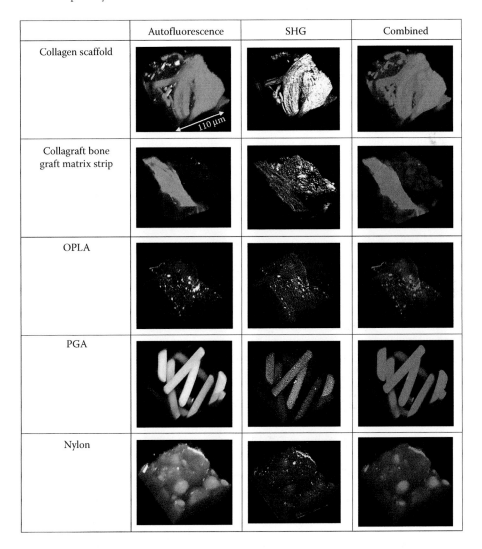

FIGURE 3.4 (**See color insert.**) 3D multiphoton images of the tissue engineering scaffolds of collagen, collagraft, OPLA, PGA, and nylon. Shown are the AF (left panel), SHG (middle panel), and combined (right panel) images. (Sun, Y. et al. Imaging tissue engineering scaffolds using multiphoton microscopy. *Microscopy Research and Technique.* 2008. 71, 140–145. Copyright Wiley-VCH Verlag GmbH & Co. KGaA. Reproduced with permission.)

while collagen, collagraft, OPLA, and nylon have similarly AF in the blue and green channels. From Figure 3.3, we also observe that PGA has a wide emission spectrum, with strong intensities in all the spectral channels. These results show that two-photon imaging can differentiate the five scaffolds by their distinct spectral signatures.

3D structures of the scaffolds can also be visualized by combining a stack of multiphoton images acquired at different depth. Figure 3.4 shows the 3D structure of the five scaffolds in AF, SHG, and combination of AF + SHG.

In scaffolds composed of blending materials such as collagraft (collagen and hydroxyl apatite), the combined AF and SHG images in Figure 3.4 show that these signals were emitted from different parts of the specimen, suggesting the presence of individual domains within the different part of blending materials. Blended materials play an important role in tissue engineering owing to the added flexibility in trying to produce scaffolds of the desired chemical and physical properties. Similar method was used in a study to visualize the phase-separated microstructure of polymeric blend of nylon and chitosan (Tan et al., 2008).

The minimally invasive nature and the intrinsic optical sectioning capability of two-photon imaging allow it to perform long-term studies of tissue engineering process. In a chondrogenic process investigation, multiphoton microscopy was used to perform long-term imaging of human mesenchymal stem cells (MSCs) cultured in chitosan scaffold (Chen et al., 2010). Figure 3.5 shows the same location and depth imaging of chitosan scaffold at different time points over a period of 49 days. The chitosan scaffold is characterized by strong AF signal (green), while the collagen produced emit strong SHG signal (blue). The linkage formed by the collagen from adjacent chitosan domain (yellow arrow) and the formation of lacunae-like structure (red arrow) are visible in the images.

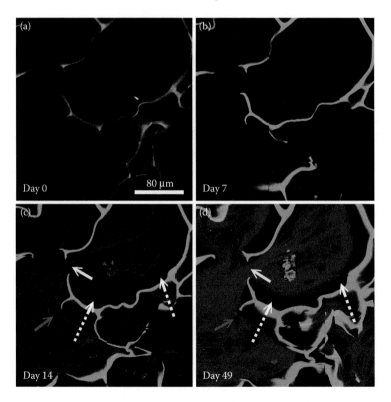

FIGURE 3.5 (**See color insert.**) Two-photon AF (green) and SHG (blue) images of engineered tissues from human MSCs cultured in chitosan scaffold. Representative images shown were acquired at the depth of 15 μm on days 0, 7, 14, and 49. (Adapted from Chen, W. L. et al. 2010. *Tissue Engineering Part C-Methods*, 16, 913–920. With permission.)

3.5 Conclusion

The main advantages of using multiphoton imaging for tissue engineering are its ability to obtain sub-micron resolution with depth discrimination images of tissue engineering construct. The availability of the fluorescent spectral information allows differentiation of scaffold types and the identification of cellular components. The minimally invasive nature of the technique and the availability of endogenous fluorescence from many scaffolds and cells allow it to perform long-term monitoring of tissue engineering processes. However, since multiphoton imaging is limited to the depths on the order of hundreds of microns, this approach is more suitable for monitoring the early stages of tissue engineering process, where the sample is still relatively thin. Even with the implementation of adaptive optics (Neil et al., 2000, Marsh et al., 2003, Schwertner et al., 2004, Rueckel et al., 2006), the improvement in the depth penetration is likely to be still too limited. Development of probe-like two-photon microscopy devices (Helmchen et al., 2001, Bird and Gu, 2003, Flusberg et al., 2005), however, has potential to be used to monitor the interaction of implanted tissue cells with the host cells. As an imaging modality, multiphoton microscopy has great potential not only to be used in the monitoring of the quality of engineered tissues, but may also be applied for studying basic physiological processes during the tissue-generation process.

References

Bird, D. and Gu, M. 2003. Two-photon fluorescence endoscopy with a micro-optic scanning head. *Optics Letters,* 28, 1552–1554.

Brasselet, S., Le Floc'h, V., Treussart, F. et al. 2004. *In situ* diagnostics of the crystalline nature of single organic nanocrystals by nonlinear microscopy. *Physical Review Letters,* 92, 207401.

Chen, W. L., Huang, C. H., Chiou, L. L. et al. 2010. Multiphoton imaging and quantitative analysis of collagen production by chondrogenic human mesenchymal stem cells cultured in chitosan scaffold. *Tissue Engineering Part C-Methods,* 16, 913–920.

Flusberg, B. A., Lung, J. C., Cocker, E. D., Anderson, E. P., and Schnitzer, M. J. 2005. *In vivo* brain imaging using a portable 3.9 gram two-photon fluorescence microendoscope. *Optics Letters,* 30, 2272–2274.

Helmchen, F., Fee, M. S., Tank, D. W., and Denk, W. 2001. A miniature head-mounted two-photon microscope: High-resolution brain imaging in freely moving animals. *Neuron,* 31, 903–912.

Lee, H. S., Teng, S. W., Chen, H. C. et al. 2006. Imaging human bone marrow stem cell morphogenesis in polyglycolic acid scaffold by multiphoton microscopy. *Tissue Engineering,* 12, 2835–2841.

Marsh, P. N., Burns, D., and Girkin, J. M. 2003. Practical implementation of adaptive optics in multiphoton microscopy. *Optics Express,* 11, 1123–1130.

Mertz, J. and Moreaux, L. 2001. Second-harmonic generation by focused excitation of inhomogeneously distributed scatterers. *Optics Communications,* 196, 325–330.

Nadiarnykh, O., Lacomb, R., Campagnola, P. J., and Mohler, W. A. 2007. Coherent and incoherent SHG in fibrillar cellulose matrices. *Optics Express,* 15, 3348–3360.

Neil, M. A. A., Juskaitis, R., Booth, M. J., Wilson, T., Tanaka, T., and Kawata, S. 2000. Adaptive aberration correction in a two-photon microscope. *Journal of Microscopy-Oxford,* 200, 105–108.

Plotnikov, S. V., Millard, A. C., Campagnola, P. J., and Mohler, W. A. 2006. Characterization of the myosin-based source for second-harmonic generation from muscle sarcomeres. *Biophysical Journal,* 90, 693–703.

Roth, S. and Freund, I. 1979. 2Nd Harmonic-generation in collagen. *Journal of Chemical Physics,* 70, 1637–1643.

Rueckel, M., Mack-Bucher, J. A., and Denk, W. 2006. Adaptive wavefront correction in two-photon microscopy using coherence-gated wavefront sensing. *Proceedings of the National Academy of Sciences of the United States of America,* 103, 17137–17142.

Sachlos, E., Reis, N., Ainsley, C., Derby, B., and Czernuszka, J. T. 2003. Novel collagen scaffolds with pre-defined internal morphology made by solid freeform fabrication. *Biomaterials,* 24, 1487–1497.

Schwertner, M., Booth, M. J., and Wilson, T. 2004. Characterizing specimen induced aberrations for high NA adaptive optical microscopy. *Optics Express,* 12, 6540–6552.

So, P. T. C., Dong, C. Y., Masters, B. R., and Berland, K. M. 2000. Two-photon excitation fluorescence microscopy. *Annual Review of Biomedical Engineering,* 2, 399–429.

Stoller, P., Kim, B. M., Rubenchik, A. M., Reiser, K. M., and Da Silva, L. B. 2002. Polarization-dependent optical second-harmonic imaging of a rat-tail tendon. *Journal of Biomedical Optics,* 7, 205–214.

Su, P. J., Chen, W. L., Chen, Y. F., and Dong, C. Y. 2011. Determination of collagen nanostructure from second-order susceptibility tensor analysis. *Biophysical Journal,* 100, 2053–2062.

Su, P. J., Chen, W. L., Li, T. H. et al. 2010. The discrimination of type I and type II collagen and the label-free imaging of engineered cartilage tissue. *Biomaterials,* 31, 9415–9421.

Sun, Y., Tan, H. Y., Lin, S. J. et al. 2008. Imaging tissue engineering scaffolds using multiphoton microscopy. *Microscopy Research and Technique,* 71, 140–145.

Takezawa, T., Inoue, M., Aoki, S. et al. 2000. Concept for organ engineering: A reconstruction method of rat liver for *in vitro* culture. *Tissue Engineering,* 6, 641–650.

Tan, H. Y., Lin, M. G., Hsiao, W. C. et al. 2008. Characterizing phase-separated microstructure of polymeric blended membrane using combined multiphoton and reflected confocal imaging. *Optics Express,* 16, 3818–3827.

Vogelin, E., Jones, N. F., Huang, J. I., Brekke, J. H., and Lieberman, J. R. 2005. Healing of a critical-sized defect in the rat femur with use of a vascularized periosteal flap, a biodegradable matrix, and bone morphogenetic protein. *Journal of Bone and Joint Surgery-American Volume,* 87A, 1323–1331.

Williams, R. M., Zipfel, W. R., and Webb, W. W. 2005. Interpreting second-harmonic generation images of collagen I fibrils. *Biophysical Journal,* 88, 1377–1386.

4

Two-Photon Microscopy for Surface Mapping and Organ Characterization

Anju Mythreyi Raja
Ngee Ann Polytechnic

Since the first demonstration of nonlinear imaging of biological samples in 1990 (Denk et al., 1990), multiphoton microscopy's (MPM) applications in the field of biomedical imaging is exponentially growing. This chapter focuses on giving an insight into the history and physical principles of nonlinear optics, distinct advantages MPM holds for biomedical application and its application in the organ-level characterization of liver fibrosis using liver biopsy and surface mapping. We envision that MPM will be a future tool for tissue engineering applications to monitor liver fibrosis progression and regression in demonstrating new therapies and regeneration of the liver.

4.1 Introduction to Two-Photon Microscopy and Second-Harmonic Generation Imaging

4.1.1 History of Nonlinear Processes

The first-known nonlinear optical effect was observed by John Kerr in 1875, when he discovered the quadratic electro-optic effect in glass. The nonlinear optical effect was theorized by Maria Goppert-Mayer in terms of two-photon absorption and two-photon emission in her doctoral dissertation in 1931 (Göppert-Mayer, 1931). Nonlinear optical phenomenon was also reported by G. N. Lewis in 1941 (Lewis et al., 1941), when fluorescein molecules embedded in boric acid rigid media responded nonlinearly to increasing excitation signal. The development of lasers in 1960 (Maiman, 1960) led to the demonstration of several nonlinear optical phenomena which were not possible with conventional light sources. In 1961, Franken et al. demonstrated second-harmonic generation (SHG) in quartz crystal (Franken et al., 1961). He demonstrated that the frequency of a ruby laser doubled when it passed through a quartz crystal. From then on, several nonlinear optical phenomena such as optical sum frequency generation, optical third-harmonic generation, optical rectification, optical parametric amplification, and oscillation were demonstrated (He and Liu, 1999).

The first nonlinear microscope was built by Sheppard and Kompfen in 1978 in Oxford, UK (Sheppard and Kompfen, 1978). The article demonstrated using resonators that can be used to improve the signal acquisition from harmonic generation of the sample. They were limited by the laser source as the energy delivered to the sample (without destroying it) is rather limited to generate harmonic signals and two-photon fluorescence. The development of pulsed laser in 1980 removed this limitation and revolutionized the field of nonlinear optical microscope. The first demonstration of two-photon-excited fluorescence (TPEF) using pulsed laser came from Denk, Strickler, and Webb in 1990 and was published in *Science* (Denk et al., 1990). They demonstrated that a large number of high-energy red photons can excite fluorophore that are normally excitable by ultraviolet (UV) laser in living cells. These pioneering works in nonlinear optics opened new doors to explore the myriads of possibility of applying them in biological systems.

4.1.2 Physical Principle of Nonlinear Optical Imaging

The physical basis of two-photon excitation and absorption was theoretically predicted by Maria Göppert–Mayer in her doctoral dissertation. The detailed quantitative explanation of this phenomenon can be found in the works of G. Baym (Baym, 1974) and J. R. Lakowicz (Callis, 1997). We briefly describe TPEF and SHG.

A fluorophore can be excited either by one or two photons. In one-photon excitation, when the energy of a single photon in the UV or visible spectrum is sufficient to overcome the energy gap between the ground state and the excited state. Once the fluorophore is excited, the fluorophore will return to the ground state emitting the acquired energy as fluorescence. As the energy of the photons are uniformly distributed in the illumination path, the fluorophores above and below the focal planes are also excited. Hence, there is a need for a pinhole to collect the fluorescence only from the focal plane.

TPEF is a result of a nonlinear interaction between the fluorophore and the excitation electromagnetic field. One photon of lower energy from the red or infrared spectrum is absorbed by the fluorophore to reach an intermediate energy state followed by the absorption of the second photon by the fluorophore to reach the final excited state. Figure 4.1 shows the Jablonski diagram for the energy transitions of one photon and TPEF. As you can see, two photons need to reach the fluorophore almost instantaneously for the excitation to occur. Thus, the probability of two-photon excitation is restricted to the focal volume where the photon concentration is sufficient. The layers of fluorophore above and below the focal volume are not excited protecting the sample from unnecessary photobleaching and thus there is no need for a pinhole. Figure 4.2 shows the difference between 1P and 2P excitation volume.

The polarization of a field is related to the electric field strength as follows:

$$P = \chi(1) \, E + \chi(2) \, E^2 + \chi(3) \, E^3 + \cdots$$

where $\chi(1)$ represents the linear susceptibility of the medium independent of the field. The linear component gives rise to refraction, absorption, dispersion, and birefringence of the medium. The $\chi(2)$ is the second-order susceptibility, which describes the quadratic relation between the polarization and the electric field. This component gives rise to SHG, sum and difference frequency generation, and parametric generation. The third-order term $\chi(3)$ gives rise to third-harmonic generation, two-photon absorption, stimulated Raman scattering, optical bistability, and phase conjugation.

The relationship between the polarization and the electric field is described by a time-dependant Schrödinger equation in quantum mechanics, with a Hamiltonian that consists of an electric dipole

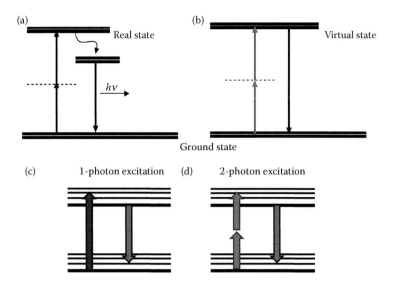

FIGURE 4.1 Energy level diagram for two-photon excited fluorescence and second harmonic generation. (a) Depicts two-photon excited fluorescence. The dotted line is the virtual state; (b) depicts SHG. The emission wavelength is half the excitation wavelength; (c) depicts one photon excitation (1PE); and (d) shows two photon or SHG excitation (2PE).

interaction term $E_\gamma \cdot r$, where E_γ is the electric field vector of the photons and r is the position operator. The Schrödinger equation can be solved using perturbation theory. The first-order solution corresponds to one-photon excitation and the nth-order solution will correspond to nth-order multiphoton excitation. In case of two-photon excitation, the probability that the fluorophore transition from the initial state (i) to the final state (f) through the intermediate state (m) is as follows:

$$P^{(2)} \approx \left| \sum_m \frac{\langle f|E_\gamma \cdot r|m\rangle\langle m|E_\gamma \cdot r|i\rangle}{\varepsilon_\gamma - \varepsilon_m} \right|^2$$

where ε_γ is the photonic energy associated to the electric field E_γ, ε_m is the energy difference between the ground state and the intermediate state m, and the summation is over all intermediate states m. The

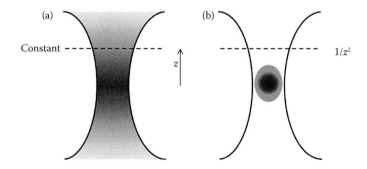

FIGURE 4.2 (a) The volume cross section of one photon excitation (1PE) shows that there is no in-built confocality and all the fluorophores in the beam path is excited unless a pinhole is used. (b) Whereas in two photon excitation (2PE), the excitation volume is restricted and thus reduces photobleaching and photo damage. The emitted SHG signal has certain directionality, while the 2-photon emitted fluorescence is anisotropic.

transition rate, $R^{(2)}$ is the time derivative of the transition probability as $R^{(2)} = d/dt\, P^{(2)} = \sigma^{(2)}I^2$, where $\sigma^{(2)}$ is the cross-section for the two-photon absorption process and I the intensity of the electromagnetic field. The two-photon absorption rate is proportional to the imaginary part of the third-order nonlinear optical susceptibility. In two-photon processes, as there are two dipole terms, the parity of the ground state and the excited state are the same, while in one-photon process, the single dipole term $(f|E_\gamma \cdot r| i)$ will reverse the parity.

The physical principle of SHG can be derived from Maxwell's wave equation for a nonabsorbing, nonconducting dielectric medium (So and Masters, 2008). By taking cross product and expanding the Maxwell's equation with the nonlinear assumptions of no free charge in the medium and no magnetization of the medium, we arrive at the following equation:

$$\nabla^2 E - \frac{n^2}{c^2}\frac{\partial^2 E}{\partial P^{NL} t^2} = \mu_0 \frac{\partial^2 P^{NL}}{\partial t^2}$$

where E is the electric field vector of the electromagnetic field and P^{NL} is the polarization vector that can be expanded as a perturbation series up to nth-order susceptibilities of the medium. μ_0 is the relative permittivity of free space. As we are interested in second-order nonlinear effects, we focus only on the χ^2 susceptibility and take only that polarization vector into account. We also assume that the electromagnetic waves are plane waves propagating only in the z direction. We simplify our analysis by considering a total field E consisting of three infinite uniform plane waves ignoring the effects of double refraction and focusing. The total instantaneous field is therefore of the form $E_i(z,t) = E_i(z)e^{-i\omega_j t} + c.c. = A(z)e^{i(k_j z - \omega_j t)} + c.c$, where $j = 1, 2$, with $\omega_1 = \omega$ and $\omega_2 = 2\omega$, E_j is the complex amplitude of field j, A_j represents the spatially slowly varying field amplitude and $c.c.$ is the complex conjugate. The complex amplitude of the nonlinear polarization is given by $P_2(z,t) = P_2(z)e^{2\pi(k_1 z - w)} + c.c.$ If these expressions are introduced to the above wave equation and utilizing the slowly varying amplitude approximation, we obtain $dA_2/dz = (2i\omega/n_2 c)\,\chi^{(2)}A_1^2 e^{i(2k_1 - k_2)z}$, where k_1 and k_2 are the wave numbers of the fundamental and harmonic wavelengths. Upon direct integration, assuming A_1 to be constant, we find the amplitude of the second-harmonic field after propagation through a distance L to be given by $A_2(L) = (2i\omega\chi^{(2)}A_1^2/n_2 c)(e^{i\sqrt{k}} - 1)/(i\sqrt{k}L)$, where $\Delta k = 2k_1 - k_2$. The intensity of the generated radiation is thus $I_2(L) = \sqrt{\mu_0/\varepsilon_0}\,2\omega^2/n_1^2 n_2 c^2\,[\chi^{(2)}]^2 I_1^2 L^2[\sin c^2(\Delta k L/2)]$.

The second-order nonlinear susceptibility, $\chi^{(2)}$, is nonzero in noncentrosymmetric media, but nonlinear emission dipoles aligned in an antiparallel arrangement producing out-of-phase SHG will result in destructive interference and thus cancellation of signals.

4.1.3 Nonlinear Imaging to Study Biological Problems

Nonlinear imaging has its unique advantages compared to other imaging modalities.

- SHG signals are generated owing to the intrinsic structure and do not need any additional labeling such as dyes or fluorescence proteins.
- The excitation wavelength for SHG/TPEF is usually in the near-infrared range, thus reducing scattering of the light inside the tissue. This increases the penetration depth of few 100 µm (Centonze and White, 1998).
- The excited volume of the tissue is reduced to a few femtoliters. Thus, providing optical sections of a few microns and intrinsic 3D imaging capabilities in complex tissue structures (Campagnola et al., 1999).
- The incident signal is a very short, high-energy pulse in the order of femtoseconds impinging on the sample. This reduces photobleaching and phototoxicity.
- The known excitation and SHG/TPEF emission spectral signatures allow easy separation of signals from collagen and other fluorophores (Brown et al., 2003).

With these distinct advantages, nonlinear imaging has been considered as a viable alternative to conventional confocal microscopy for the imaging of biological samples. SHG can be used to visualize many biological structures that do not have central symmetry (surface materials, chiral materials). Collagen type 1 present in extracellular matrix (ECM) is one such molecule that has noncentrosymmetry and hence generates SHG (Stoller et al., 2002, Zoumi et al., 2002). TPEF can be used to visualize tissue and cellular architecture without extraneous staining just by making use of the autoflorescence generated by certain biological molecules (Zipfel et al., 2003).

The applications of nonlinear imaging in various fields of biology have been growing. We have tried to highlight a few fields of application. In neurology research MPM has been used for imaging electrical activity in the brain cortex (Araya et al., 2006), the 3D blood flow architecture in brain (Nishimura et al., 2006) and detection of plaques in Alzheimer's disease (Bacskai et al., 2004). Application of MPM in immunology has rapidly grown owing to its easy availability as a ready built microscope, which enables the biologist to use it as a tool without the need of in-depth optics knowledge (Phan and Bullen, 2010). The quantitative imaging of immune-cell motility and morphology, such as the *in vivo* interaction of T cells and dendritic cell in lymph nodes (Shakhar et al., 2005), in cranial bone marrow (Cavanagh et al., 2005), and in tumors (Boissonnas et al., 2007, Wang et al., 2009b) have been demonstrated.

MPM in the field of cancer has demonstrated far-reaching advantages over conventional histology as well as confocal microscopy. Live, 3D monitoring of several processes in xenograft models have enlightened many of potential ways to target and treat the disease. MPM imaging of cancer have been demonstrated in melanoma, breast cancer, cervical, and ovarian cancer in animal and human studies (Kirkpatrick et al., 2007, Provenzano et al., 2006, 2008a,b, Zhuo et al., 2009). Especially of mention is the case of melanoma where the skin cancer is directly accessible for imaging (Lin et al., 2005). Rapid progress has been made in this area where diagnosis and staging of skin cancer will soon be made possible with MPM.

Last but not the least, MPM imaging of fibrosis progression in various organs have been successfully demonstrated. The second-harmonic signals generated by collagen molecules in fibrotic tissue are an important indicator of fibrosis progression. Imaging collagen molecules, without the need for staining and sectioning, result in quantitative readout of fibrosis progression rather than a qualitative and semiquantitative assessment of histology ridden with inter- and intraobserver discrepancies (The French METAVIR Cooperative Study Group, 1994, Bedossa et al., 2003, Gronbaek et al., 2002). MPMs quantitative imaging has been demonstrated in liver fibrosis (Gailhouste et al., 2010, Sun et al., 2010), renal fibrosis (Strupler et al., 2008), and skin fibrosis and wound-healing responses (Cicchi et al., 2010).

4.1.4 Contrast Agents for Nonlinear Imaging

The potential of multiphoton imaging of autofluorescence of host tissue and SHG of collagen and other noncentrosymmetric molecules in biological systems is clearly evident. But what if the host tissue lacks autofluorescence or noncentrosymmetric molecules? What if we need to look beyond the autofluorescence of the cell and at specific proteins? Fluorescent and second-harmonic probes help us overcome these hurdles and are being used to answer pertinent biological questions. All fluorescent probes in the UV and visible spectrum developed for 1P fluorescent imaging can be used for 2P imaging as well. But in cases where the autofluorescence of the cells overlap with the UV and visible spectrum probes, near-infrared (NIR) and SHG probes have been developed to specifically visualize only the molecule of interest. In addition to various probes, the fluorescence lifetime of the probes give us additional means to discriminate among them.

In neurology, the fluorescent probe Sulforhodamine B (SRB) has been found to be able to cross the blood–brain barrier and label the astrocytes specifically. The glial–neuronal–vascular communication can be visualized by using SRB to stain the glial cells (astrocytes), fluorescein isothiocyanate (FITC), Dextran to stain the vascular structure and fluorescently transfected neurons. SRB injection to study astrocytes can be useful for studying ischemic injuries in the brain neocortex with the help of deep tissue imaging (Verant et al., 2008). The NIR probes like Heptamethine cyanine dyes, Cypate, a derivative of indocyanine green (Berezin et al., 2007) and 3,3-diethylthiatricarbocyanine iodide (DTTCI) all have 1P excitation from the range of 750–800 nm. These probes can be excited by 2P using the 1550 nm erbium-doped fiber laser. Their application has been preliminarily demonstrated in labeling cancer cells with these probes (Yazdanfar et al., 2010). The advantage of these probes are that their excitation and emission are very well separated from UV and visible probes and autofluorescence from the tissue and a possible improvement in depth of imaging, but the 1550 nm excitation suffer certain limitations such as lower optical resolution, higher heat generation, higher dark noise in the detector in the NIR range, and lower quantum efficiency. The advantage of using gold nanoparticles along with fluorophores in MPM was demonstrated recently (Qu et al., 2008). The emission spectra of the fluorophore and nanoparticles have distinct lifetimes and hence easily distinguishable. With regard to SHG probes, several nanoparticles have been shown to have SHG properties, such as iron iodate ($Fe(IO_3)_3$) (Bonacina et al., 2007), zinc oxide (ZnO) nanocrystals (Kachynski et al., 2008), and barium titanate ($BaTiO_3$) (Hsieh et al., 2010, Pantazis et al., 2010).

4.1.5 Improving the Nonlinear Microscope

There are several advantages of using the SHG microscope to observe noncentro symmetric molecules as described above. Similar to every other microscope, the system can be further improved for added functionality. MPMs resolution takes a beating compared to conventional confocal microscope. But the resolution of MPM can be improved when the technique is combined with stimulated-emission depletion (STED). This supraresolution technique that goes beyond the diffraction limit is very useful in imaging the neural tissue where individual dendritic spines at a depth of a 100 μm within the tissue can be visualized (Ding et al., 2009).

Multifocal multiphoton microscope (MMM) is another improvement over the classical MPM where larger areas of the tissue can be simultaneously scanned using a rotating microlens disk (Bewersdorf et al., 1998) or a microlens array (Bahlmann et al., 2007) or an optical multiplexer–demultiplexer (Chandler et al., 2009) to split the excitation beam into several scanning beamlets. The MMM drastically reduces scanning time for imaging large portions of the tissue.

The need to image live tissue in optically inaccessible regions of the human or animal body has given rise to multiphoton endoscopy. The problem of group velocity dispersion (GVD) and self-phase modulation in optical fibers has to be overcome before achieving a multiphoton endoscope. In 2003, Jung and Schnitzer (2003) developed microendoscopes of 350–1000 μm diameter based on gradient microlenses that overcame the problem of self-phase modulation. A double clad fiber (DCF) can maintain the excitation polarization and in combination with a multiple element lens probe has been shown to be able to visualize rat-tail tendon with an axial resolution of 10 μm and a lateral resolution of 1 μm and upto a penetration depth of 88 μm (Bao et al., 2010).

As with all ultrafast laser set-ups, the SHG microscope also suffers from GVD. As the term suggests, GVD is a dispersion of light on passing through dispersive optical components such as lenses, gratings, and prisms, where the light is stretched temporally. Thus, an ultrafast laser whose temporal profile is a Gaussian of 50 fs will be stretched to a larger time profile. The dispersive components slow down the light with the higher frequency to a slower velocity (Figure 4.3).

Group velocity is defined as the velocity at which the energy of the wave is carried. It is the derivative of the wave number with respect to angular frequency. It is related to phase velocity as $V_g = C (n - \lambda\, dn/d\lambda)^{-1}$, where C is the velocity of light in free space 3×10^8 m/s, n is the refractive index of the traveling medium, and λ is the wavelength of light in vacuum.

GVD can be compensated using commercially available GVD compensators. There are various commercially available dispersion compensators such as the chirped mirrors (Figure 4.4a), pulse compressors (Figure 4.4b), and negative dispersion gratings. These dispersion reversal tools are called pulse compressors or pulse modulators. The basic principle of pulse modulation is to ensure that low-frequency signals travel a longer path compared to the high-frequency signal, such that all frequency components of the

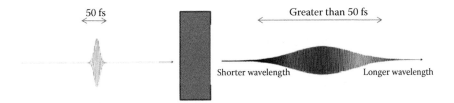

FIGURE 4.3 Group velocity dispersion of a femto-second pulse. A 50 fs laser pulse passing through a dispersive optical component (depicted here as a rectangle) experiences a positive group velocity dispersion, where the longer wavelengths in the laser pulse travels faster than the shorter wavelengths, resulting in a temporal stretch in the pulse.

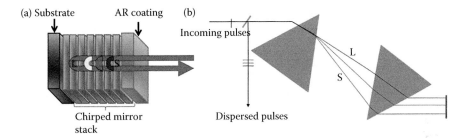

FIGURE 4.4 Reversing group velocity dispersion using pulse modulators such as chirped mirrors and paired prisms. (a) Chirped mirror—a stack of dielectric mirror in which the light of longer wavelength (L) travel deeper before it is reflected while the shorter wavelength (S) light is reflected faster, hence compensating for the negative dispersion. (b) Prism based pulse compressor—a pair of prism spatially arranged in such a way that longer wavelength light (L) travels a longer path compared to the shorter wavelength (S).

resultant wave reaches the sample in a narrow time frame, usually in femtoseconds. To better enunciate this point, a Ti-sa laser with a 700–900 nm wavelength range and pulse width of 70–100 fs, can be distorted to around 370 fs if the material dispersion is around 13,000 fs^2. By placing a pulse compressor or compensator, the pulse width can be readjusted to 100 fs. SHG signals have been pulse-modulated to improve signal-to-noise ratio in qualitatively imaging human skin and mouse kidney (Schelhas and Dantus, 2006, Tang et al., 2006). Pulse modulation (PM) has been used to improve the second-harmonic imaging microscope (SHIM) to quantitatively visualize the collagen changes in tumor interior before and after chemotherapy administration. The tumor response to chemotherapy was only visualized with the PM–SHIM, while the conventional SHIM was not able to detect the collagen changes (Figure 4.5).

Clinicians and medical scientists are always looking for better tools and quantifiable information to assess the severity of a disease as systematically as possible. MPM is an excellent tool for imaging the autofluorescence from liver tissues and collagen from the ECM. The autofluorescence signature and the collagen information obtained are immediately quantifiable. But to ensure the accuracy of the system, we need to ensure that the signal-to-noise ratio and sensitivity of the system is optimal to even pick up the smallest of collagen fibers.

4.2 An Introduction to Liver, Liver Tissue Engineering, and Liver Fibrosis

4.2.1 Structure and Functions of the Liver

Liver is one of the largest and most versatile organ in our body performing myriads of functions including lipid, protein, and carbohydrate metabolism, enzyme and hormone production, vitamin, mineral

Drug treated tumor   Control tumor

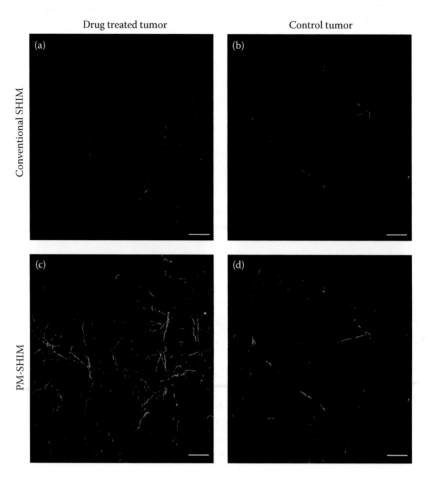

FIGURE 4.5 Changes in collagen distribution in tumors treated with chemotherapy can only be visualized using PM-SHIM. Representative images of tumor samples before and after chemotherapy are shown. The top panels (a, b) show images taken with conventional SHIM and the bottom panels (c, d) show images taken with PM-SHIM. The laser power used to excite the samples with conventional SHIM was 10% higher than with PM-SHIM. The smaller fibers are visualized by the PM-SHIM which are not excited with conventional SHIM. Scale bar: 50μm. (From Raja, A.M. et al. 2010. *Journal of Biomedical Optics,* 15(5), 056016. With permission.)

and water metabolism, bile acid metabolism, glycogen storage, red blood cell decomposition, plasma protein synthesis, and detoxification.

The liver is positioned in the hypochondrium of the abdominal cavity and is held in place by ligaments. A small part of the anterior surface of the liver is in contact with the anterior abdominal wall and the posterior surface is in contact with the diaphragm, esophagus, and inferior vena cava. The human liver is a pyramidal-shaped organ, consisting of the bigger right lobe and the smaller left lobe. The left portion of the right lobe is separated into two smaller caudate and quadrate lobes by the groove of the inferior vena cava. The blood is supplied to the liver from hepatic artery which brings in oxygenated blood from the heart to the liver and the portal vein which brings in blood carrying digested nutrients from the small intestine to the liver. The hepatic artery supplies 25% and portal vein 75% of the liver blood supply. After being processed by the liver, the blood flows out through the central vein in the lobules which merge as hepatic vein and flow out of the liver into the inferior vena cava.

The liver is covered by a peritoneum which protects it from friction caused by other organs. Beyond the peritoneum, the liver is encapsulated by a connective tissue capsule, called the Glisson's capsule that branches and extends throughout its surface. The connective tissue extends from the surface of the liver to continue along the blood vessels, lymphatic vessels, and bile ducts as they traverse the entire liver. The sheets of connective tissue also compartmentalize the liver parenchyma into units called hepatic lobules, the repeating structural units of the liver. The hepatic lobule consists of hexagonal arrangement of single cell thick sheets of hepatocytes or liver cells, radiating outward from a central vein to a portal triad. A portal triad consists of a branch of the portal vein, hepatic artery, lymphatic vessel, and a bile duct. The space between the sheets of hepatocytes is called sinusoid. In the sinusoids, macrophages called Kupffer cells (named after Karl Wilhelm von Kupffer who first observed it in 1876); liver sinusoidal endothelial cells (LSECs) (Wisse, 1970), and natural killer (NK) cells called pit cells (Lotzova, 1993) can be found. The Kupffer cells perform specialized functions such as phagocytosis, antigen processing and presentation, and produce factors such as cytokines, prostanoids, nitric oxide, and reactive oxygen species (Gale et al., 1978, Seki et al., 2000). The LSECs have sieve-like structure called fenestrae that allows selectively certain substances to pass from the blood to the hepatocytes (Fraser et al., 1995). The pit cells are cytotoxic to tumor cells and produce cytokines supporting microbe resistance (Trinchieri, 1989). In the small gap between the sinusoids and the hepatocytes called the space of Disse, there are hepatic stellate cells (also described first by Karl Wilhelm von Kupffer in 1876) (Aterman, 1986) which in their quiescent state, stores vitamin A in lipid droplets.

The structural integrity of the liver is maintained by ECM molecules. The Glisson's capsule around the liver and the sheath around the vessels are mainly composed of a thin basement membrane made of collagen type I, III, V, VI, fibronectin, and elastic fibers. The collagen type I fibers along the vessels continue along the boundaries of the lobules and the structural repetition of fibrous band provides the scaffolding of the liver lobule (Martinezhernandez, 1984). Unlike many organs, where there is a presence of a thick basement membrane upon which the epithelial cells are arranged and where the endothelial and epithelial cells are separated by a dense matrix, within each hepatic lobule the hepatocytes have no basement membrane and the interstitial space of Disse has very sparse ECM consisting mainly of fibronectin and little each of laminin, collagen type I, III, IV, and VI. This distinct feature allows the liver to function as an effective bidirectional exchange unit, where the uptake, processing, and delivery of substances happens very rapidly during digestion and detoxification processes between the plasma and the liver (Martinezhernandez and Amenta, 1993).

The structure of the liver and its function are interconnected. The presence of hepatocytes, hepatic stellate cells, Kupffer cells, and other type of supporting cells, ECM distribution, their systematic arrangement, connected by blood vessels, lymphatic vessels, and bile ducts ensure harmonious functioning of the liver. When this harmony is disturbed by either external factors or internal imbalances, it leads to various forms of liver injuries. When the liver is unable to self correct the imbalances, scarring of the liver can take place which is liver fibrosis, where the ECM homeostasis is altered. The natural regression of fibrosis might be able to restore the liver functions, but many times fibrosis progresses into liver cirrhosis and later liver failure. Thus, far liver transplantation is the only treatment for liver failure. Researchers are exploring various avenues to exploit the potential of liver tissue engineering to emulate the structure of the liver and elicit similar functions from the artificial system. We shall discuss the upcoming liver tissue engineering strategies in the next section.

4.2.2 Liver Tissue Engineering

Tissue engineering as per the definition of Robert Langer and Vacanti is "an interdisciplinary field that applies the principles of engineering and life sciences toward the development of biological substitutes that restore, maintain, or improve tissue function or a whole organ." The elements of tissue engineering includes identifying cell sources, designing and synthesizing the biomaterial scaffold, and assembling the cells and scaffolds together to ensure that the engineered tissue performs the required function (Zheng et al., 2010). The different cell types, matrix proteins, vasculature, and lymphatic system in the liver makes tissue engineering of a functional liver tissue a monumental task.

Several cell sources are being considered including embryonic stem cells guided toward hepatocyte differentiation (Cho et al., 2008), umbilical cord blood cells (Teramoto et al., 2005), liver progenitor cells (Sell, 2001, Tsai et al., 2010), primary hepatocytes manipulated to proliferation (Kang et al., 2004), Oval cells (Wang et al., 2009a), small hepatocytes (Shibata et al., 2006), immortalized hepatocyte cell lines such as OUMS-29, cBAL111, and HC-04 (Ito et al., 2008, Kobayashi et al., 2000, Lim et al., 2007). It is of utmost importance to ensure survival and functionality of these liver cells *in vitro* to be able to construct a functional liver tissue.

Several methods are being explored to construct functional liver tissue using scaffolding materials such as poly lactic acid, poly glycolic acid, Ca-alginate with galactosylated chitosan, chitosan collagen matrix with heparin (Chung et al., 2002, Nam et al., 1999, Wang et al., 2005), polymers, hollow fibers, membranes, and hydrogels (Du et al., 2008, Kaufmann et al., 1997, Rozga et al., 1993). The substrates are modified to provide the cells with matrix cues (Lee et al., 2002) and signaling factor cues (Kang et al., 2004). The potential of culturing the liver cells as hepatic tissue sheets (Ohashi et al., 2007), photopatterning the cells into three-dimensional (3D) structures (Tsang et al., 2007), the use of polymeric linkers to attach the liver cells to one another to form multicellular hepatocyte structures without the need for bulky gels (Mo et al., 2010), use of collagen sandwich culture methods and perfusion-based reactors (Ng et al., 2006, Xia et al., 2009) are exploited to form functional 3D tissue-like units, before they can be transplanted back into the liver. There are still several hurdles to take liver tissue engineering from bench to bedside. The problems with cell sourcing, immunological reactions, biocompatibility of scaffolding materials, vascularization, and functional maintenance remain to be tackled. There have been attempts to introduce these functional liver tissue-like structures as heterotopic transplantation under the kidney capsule or intraperitoneally (Demetriou et al., 1986, Xiangdong et al., 1991) or the cell suspension after *in vitro* culture have been introduced as orthotopic transplantation into the spleen or through the portal vein (Fox et al., 1998, Muraca et al., 2002) with some success.

In times to come, the application of tissue engineering to overcome liver diseases and liver failure will be growing. It is crucial to be able to compare the efficacy of these applications. MPM of implanted engineered constructs in liver tissue samples can provide us insights on the recovery of the liver. The health status of the hepatocytes and the fibrotic level of the liver tissue can be accurately monitored using MPM. Going forward in this chapter, we will focus on monitoring liver fibrosis using MPM. We will discuss the etiology, pathogenesis, and methods to diagnose liver fibrosis to understand how MPM can help us establish a method to systematically study liver fibrosis. This method will enable us to establish quantitative assessment of liver fibrosis progression which will help us study the applicability of tissue engineering approaches to heal liver injuries.

4.2.3 Liver Fibrosis

Liver fibrosis as the name suggests is the deregulation in the formation and degradation of matrix fibers. Liver fibrosis can be caused owing to infection such as hepatitis virus, injury, alcoholism, excess fat in the liver, biliary dysfunction, and autoimmune diseases. Liver fibrosis is an integral part of liver cirrhosis where the liver looses its ability to function normally and a liver transplant might be needed to help the patient survive (Davis et al., 2003) and the patients are a high-risk group for hepatocellular carcinoma. Liver cirrhosis is one of the top 10 causes of death worldwide, and in many developed countries liver disease is now one of the top 5 causes of death in middle-aged group (Bosetti et al., 2007, Griffiths et al., 2005).

Liver fibrosis is a dynamic process in which the balance has tilted to favor excess matrix deposition. This process involves several cell types, cytokines, chemokines, and growth factors. When the liver is injured by any one of the above-stated causes, the liver kick starts a mechanism to heal the liver. Thus, the process of fibrosis can be viewed as a wound-healing response of the liver and this process continues as long as the causal agent of the fibrosis is not removed from the system. In simplified terms, the wound-healing response starts with the release of cytokines that activates the quiescent hepatic stellate cells. The activated stellate cells lose their capacity to store vitamin A. Instead, they assume fibroblast-like phenotype (expressing α–smooth muscle actin) migrate to the site of injury and start producing collagens leading to excessive matrix deposition (Marra, 1999). The fibrosis accumulation further triggers several positive feedback pathways that further amplify fibrosis. The stellate cells can be further activated and migrate owing to membrane receptors signaling (Yang et al., 2003), release of growth factors to stimulate fibrogenesis through activation of cellular matrix metalloproteinases (Schuppan et al., 1998, 2001) and matrix stiffening owing to the enhanced density of ECM helps the stellate cell migration further (Wells, 2008). In advanced stages, liver contains approximately 6 times more ECM overall than normal liver, and there is increasing deposition of collagen types I and IV, undulin, elastin, and laminin. Although collagen types I, III, and IV are all increased, type I increases most and its ratio to types III and IV also increases (Burt et al., 1990, Hahn et al., 1980, Rojkind et al., 1979, Seyer et al., 1977).

The excessive matrix in the liver alters its structure and function. The increased scarring blocks the blood vessel, leading to intrahepatic resistance to blood flow resulting in hepatic insufficiency and portal hypertension (Gines et al., 2004). This increased choking of the blood flow stops nutrients from reaching the hepatocytes leading to their death. The empty spaces created by necrosis fuels further fibrosis. This leads to portal bridging of the fibers, where bands of collagen can be seen connecting the portal triads in the shape of a hexagon.

The key factors that are capable of degrading the excessive matrix are the collagenases or the matrix metalloproteinase (MMP). There are several types of MMPs and they are secreted by the hepatic stellate cells, Kupffer cells, and hepatocytes. Their activity is heightened and lessened during different phases of the fibrogenic cascade based on their activation and deactivation by tissue inhibitors of metalloproteinase (TIMP) or other factors. If the stimulus that induced the fibrogenic response is removed, the activated hepatic stellate cells will undergo apoptosis. The amount of MMPs, TIMPs, and other factors in the liver microenvironment will be modulated resulting in a resolution of fibrosis. Thus, treatment strategies for fibrosis have taken a multipronged approach as shown in Figure 4.6 (Gressner and Weiskirchen, 2006).

4.2.4 Diagnosis of Liver Fibrosis

Currently, liver fibrosis is diagnosed using noninvasive and invasive methods. The noninvasive methods include blood test for serum markers for fibrosis, ultrasound measurements of the liver stiffness, and magnetic resonance elastography measurements of the liver stiffness. The invasive method is to obtain a liver biopsy and to stain the tissue to check for fibrosis. Blood tests for platelet count, aminotransferase serum levels, prothrombin time, and acute-phase proteins serum levels have been proposed (Forns et al., 2002, Imbert-Bismut et al., 2001). But the measurements can only differentiate between presence and absence of fibrosis and no intermediate stages of the disease can be resolved. The ultrasound and elastography is not accurate in obese patients, lacks standardization across all age groups, and highly dependant on operator expertise. Owing to the lack of sensitivity and resolution in noninvasive methods, the liver biopsy still remains as the gold standard for liver fibrosis, even though it has its own disadvantages including patient discomfort and sampling error (Maharaj et al., 1986, Pagliaro et al., 1983, Poniachik et al., 1996, Thampanitchawong and Piratvisuth, 1999).

The liver biopsies are sectioned and stained using different protocols (Hematoxylin and Eosin, Masson's Trichrome, Sirius Red, etc.) to reveal the cellular and matrix architecture of the liver. The pathologists observe the stained slides and use different scales such as Metavir (stage 0–4) (Bedossa and Poynard, 1996) and Ishak score (stage 0–6) to stage the severity of fibrosis. This liver biopsy staging is prone to inter- and intraobserver variability (Bedossa et al., 1994, Gronbaek et al., 2002, Soloway et al., 1971, Theodossi et al., 1980, Westin et al., 1999).

There is a need for quantitative imaging and characterization of biopsy samples for accurate staging of the disease progression and if possible, minimal invasive imaging through an endoscope to obtain cellular level details of fibrosis over a larger area of the liver but at the same time not performs biopsy. The quantitative surface mapping of liver fibrosis will have a remarkable impact on the way liver fibrosis is diagnosed.

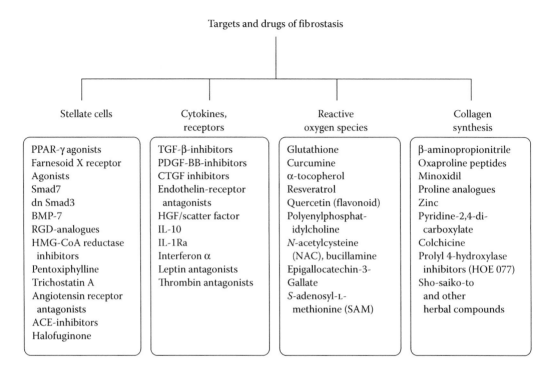

FIGURE 4.6 Compilation of targets and respective drugs used for experimental therapeutic fibrostasis. Abbreviations used are: ACE, angiotensin-converting enzyme; dn, dominant negative; HGF, hepatocyte growth factor; HMG-CoA, 3-hydroxy-3-methylglutaryl coenzyme A; IL-10, interleukin-10; IL-1Ra, interleukin-1 receptor antagonist; PPAR-γ, proliferator-activated receptor; RGD, Arg-Gly-Asp. (Gressner, A. M. and Weiskirchen, R. Modern pathogenetic concepts of liver fibrosis suggest stellate cells and TGF-beta as major players and therapeutic targets. *Journal of Cellular and Molecular Medicine.* 2006. 10, 76–99. Copyright Wiley-VCH Verlag GmbH & Co. KGaA. Reproduced with permission.)

4.3 Two-Photon Microscopy and SHG Imaging in Liver Fibrosis Characterization and Surface Mapping

4.3.1 Nonlinear Imaging for Staging Liver Fibrosis Based on Invasive Liver Biopsies

A perfect marriage between biology and optics takes place when a pertinent biological question finds an apt solution in optics. This is the case for liver fibrosis diagnosis. Nonlinear imaging makes it possible to quantitatively image collagen and liver autofluorescence from NAD(P)H and flavins in hepatocytes, to accurately assess liver fibrosis. This ability to quantify fibrosis progression in the liver, accurately and objectively, is crucial. The improved accuracy will help researchers and pharmaceuticals to develop treatments and intervention strategies and it will help the clinicians to improve their diagnostic accuracy and provide better treatments for the patients.

The noninvasive methodologies used currently provide qualitative readouts of the disease and not precise progression of fibrosis. The current methods have a maximum resolution of four to six stages. Further these results are highly subjected to personal opinion, and hence, inter- and intraobserver discrepancy is as high as 35%. The field is trying to move toward more quantitative assessment methods such as digital morphometric image analysis (FibroXact, FibroQuant, Bioquant Nova Prime), (Gamal et al., 2004, Friedenberg et al., 2005, Masseroli et al., 2000, Matalka et al., 2006, Wright et al., 2003) where the stained collagen in the tissue slices are quantified. There have been some positive results, where collagen modulation in liver fibrosis was observed by quantification, while it was not identified by the conventional histology scoring (Manabe et al., 1993). But even with these methods, the operator faces significant problems with color variations owing to staining, protocols, time-dependent fading, and photobleaching. Thus, for every sample, the operator needs to define the features for thresholding and segmentation, leading to a bias that reduces reproducibility and objectivity of quantification. All this taken together, point to the fact that current techniques and technologies used for liver fibrosis assessment is not able to give the clinicians and researchers the ability to track the disease progression. They are not able to correlate the diagnosis to the functional status of the liver or to physiological changes to the liver such as the onset of portal hypertension or other complications such as bleeding. Thus far there is no FDA-approved liver fibrosis treatment available, as liver fibrosis progression or regression owing to treatment cannot be evaluated with enough resolution or sufficient certainty.

Nonlinear imaging to visualize and quantify liver fibrosis has made it possible to develop a highly objective and reproducible automated image-based liver fibrosis quantification system. The combined TPEF and SHG microscope with systematically optimized parameters for sensitivity, and fully automated quantification algorithms makes it possible to assess liver fibrosis progression and cell necrosis. We have obtained standardized, highly reproducible data to monitor the structural progression of collagen at all stages of liver fibrosis (Sun et al., 2008; Tai et al., 2009) using MPM.

In an attempt to set up nonlinear imaging as a benchmark for liver fibrosis assessment, we have performed extensive animal studies with liver fibrosis animal models such as carbon tetrachloride (CCl$_4$) (Sun et al., 2008) and bile duct ligation (BDL) (Tai et al., 2009) (Figure 4.7). In the CCl$_4$ model, a distinct

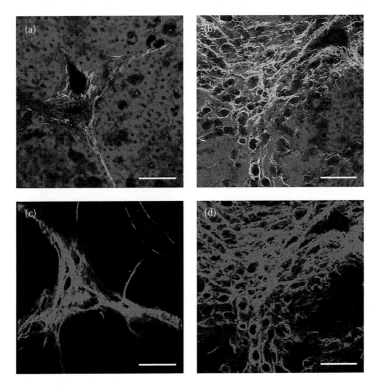

FIGURE 4.7 (**See color insert.**) Images taken from CCl4 and BDL rat models using SHG microscopy. a and b are images obtained from CCl4 and BDL, respectively. c and d are the corresponding 3-D projections of 50-m-thick tissue samples in the SHG channel only. In the CCl4 model, collagen aggregated around vessel walls as liver fibrosis progressed, whereas fine collagen distributed in sinusoidal areas disappeared. In the BDL model, collagen aggregation took place around vessel walls as well as in areas where bile ducts proliferated. Scale bars are 200 μm. (From Tai, D. C. et al. 2009. *Journal of Biomedical Optics*, 14, 044013. With permission.)

collagen remodeling and redistribution phenomenon is observed. In the normal liver, most collagen is present around the vessels, bile ducts, and portal tracts, dubbed as "aggregated collagen" and there are fine fiber collagen distributed all across the liver parenchyma, dubbed as "distributed collage." In the CCl_4 model, during the course of 3 weeks, it was found that the aggregated collagen progressively increased, while the distributed collagen progressively decreased (Figure 4.8). It has to be noted that no significant difference between the total collagen content could be observed in the first 3 days. As a nonstain method, the autofluorescence signature of the hepatocytes can indicate their health status. Also the signature varies between the various cells in the liver, and can be individually identified and

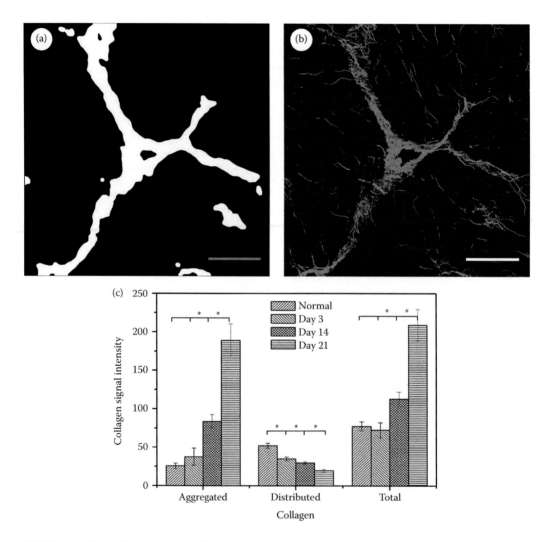

FIGURE 4.8 Identification and quantification of distributed and aggregated collagens: (a) A mask image generated by using low-pass filtering and segmentation from typical collagen images, captured with SHG microscopy. (b) Differentiation results between distributed (small thin fibres) and aggregated (thick corded fibers) collagen. (c) Graph of collagen progression for both collagen types at various time points. It was shown that the distributed collagen fibers decreased significantly from as early as day 3, whereas aggregated collagen only showed slight increase in the first three days. This finding suggested a remodeling between distributed and aggregated collagens during this period. Note that the asterisk in the graph, indicates that the differences reached statistical significance, and the scale bars shown are 200μm. (From Sun, W. et al. 2008. *Journal of Biomedical Optics*, 13, 064010. With permission.)

quantified. MPM images can thus be processed to identify damaged cells in the liver, biliary epithelial cells proliferation, blood vessel changes, and so on. Thus, changes in the liver parenchyma that were never quantified using traditional methods can be tracked now.

To be able to take MPM-based liver fibrosis diagnosis to clinical use, it is imperative to compare it to conventional methods, for establishing realistic quantification algorithms and look for sensible features and properties of liver fibrosis progression. MPM-based liver fibrosis assessment is done by acquiring 3D autofluorescence and collagen images in high spatial resolution and identifying morphological characteristics and spatial distributions of collagen using an adaptive algorithm. MPM imaging is able to extract the morphological information as in from a histological image as well as the quantitative information as in from a protein assay from one single sample. The adaptive algorithm, compared to the existing direct thresholding methods can detect even the small fibers which will be ignored by the human eye, improving the sensitivity of the quantification. Combining the imaging and quantification, a standardized staging system index, Fibro-C-Index, was derived. When Fibro-C-Index was compared with the existing gold standard: morphological staging performed by pathologists (Sun et al., 2008, Tai et al., 2009), it was found that Fibro-C-Index correlated morphologically to all the features that pathologists evaluate (Figure 4.9). It can give the pathologists the power to quantify the different morphological features they are looking for such as number of portal tracts in a biopsy sample, total collagen area, collagen around the vessels, distributed collagen, and so on. It can reduce the time required for diagnosis significantly and removes the staining process and alleviates inter- and intraobserver discrepancies. The use of MPM-based imaging in clinical liver fibrosis diagnosis is being verified by various groups performing trials using human tissue (Figure 4.10) (Gailhouste et al., 2010, Sun et al., 2010).

4.3.2 Nonlinear Surface Mapping of Liver for Noninvasive Liver Fibrosis Assessment

Liver fibrosis assessment will tremendously benefit from the quantitative, stain-free, and 3D nature of MPM imaging. But the sample collection for fibrosis assessment still remains an invasive procedure. A needle with a core diameter of a few millimeters is inserted into the liver and a piece of tissue is pulled out. This procedure can lead to a few complications, prohibiting clinicians to go for serial or repetitive biopsies from the same patients, making it impossible to accurately determine the changes of disease progression or to monitor the treatment effects. The small piece of tissue might not be representative of the entire liver (Forbes et al., 2004, Tuchweber et al., 1996). This motivated the development of surface mapping of the liver to assess liver fibrosis, where a larger surface of the liver can be assessed without the need for invading into the liver parenchyma. Surface scanning of the liver to detect nodules and tissue hardening has been attempted using laparoscopy (Jalan et al., 1995, Poniachik et al., 1996). But as detailed earlier, the liver is surrounded by a thick fibrous cap called Glisson's capsule. This capsule prevents any meaningful visualization of the liver using existing laparoscopic technologies with limited penetration depths. Even during histological examination of the liver biopsy, the capsule portion is not considered to provide any useful information. By exploiting the increased penetration depth of nonlinear imaging, a few hundred microns of the liver can be seen. But it is not known thus far whether the capsular region undergoes significant detectable changes during fibrosis. It thus needs to be ascertained that the capsular and subcapsular regions are a good indicator of fibrosis progression for minimally invasive tools like the endoscope combined with nonlinear imaging capabilities to be a feasible future alternative to invasive liver biopsies.

In a study with BDL animal models it has been demonstrated that the changes in collagen in the capsular and subcapsular regions correlate with the changes in liver interior (He et al., 2010). Liver samples from BDL rats from early to late stages of fibrosis were isolated from the left and right lobe and cross sections perpendicular to the surface of the liver were obtained. MPM images of the sections were acquired. The TPEF from the liver autofluorescence was used to demarcate the edge of the tissue and the collagen band on the liver edge was quantified as the capsule width. The capsule thickness is

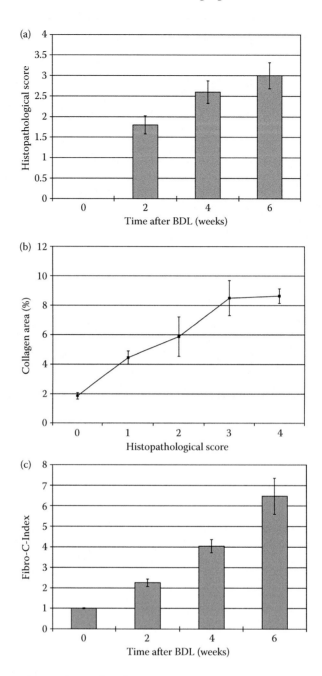

FIGURE 4.9   Fibro-C-Index correlates well with conventional histopathology scoring system. (a) Histopathology scoring of liver tissue at different time points after performing bile duct ligation (BDL). Scoring results are 0, 1, 2, 3, or 4. Values shown are averaged over all the tissue samples at the given time point, showing a liver fibrosis severity increase over 6 weeks. (b) Comparison between scoring results from conventional histopathological scoring systems against areas of collagen detected using SHG microscopy. There is clear overlap of collagen areas between different fibrosis stages. In early stages (1 and 2), overlap occurs by as much as 50%. In later stages (3 and 4), there was no significant difference in the collagen area in the two groups. Within each group, the collagen area varies from 11 to 45%. (c) Fibro-C-Index results obtained at different time points after BDL showed that Fibro-C-Index, a purely quantitative measurement of fibrosis progression, agrees with the conventional scoring results shown in (a). (From Tai, D. C. et al. 2009. *Journal of Biomedical Optics*, 14, 044013. With permission.)

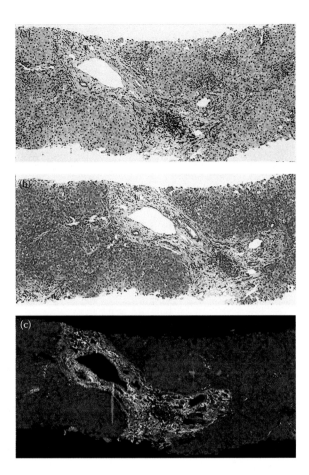

FIGURE 4.10 Comparison of (a) Hematoxylin and Eosin (H&E) and (b) Masson's Trichrome stained liver sample with (c) TPEF/SHG imaged human liver sample shows that TPEF/SHG imaging reveals cellular and stromal information as much as stained samples, without the need for staining. It reduces processing time and improves consistency of images.

a unique parameter that can be used to ascertain the presence of liver fibrosis (Buschmann and Ryoo, 1989). But the capsule thickness by itself lacked the resolution to distinguish between stages of fibrosis (Figure 4.11a). The area beneath the capsule up to a depth of 150 μm was identified as the subcapsular region based on the SHG signal from the collagen distribution (Figure 4.11b). The region beneath the subcapsular region was identified as the liver interior. Morphology-based features such as total collagen, collagen in bile duct areas, bile duct proliferation, and remnant hepatocytes area were quantified in the subcapsular region and the liver interior (Figure 4.11c). These features are related to the parameters that pathologists usually look for in a liver biopsy. Emphasis to bile duct is given as it is a BDL model, but these parameters are applicable to other etiologies of fibrosis as well. The subcapsular region in the anterior side of the liver and posterior side of the liver varied in the parameters quantified. When the anterior and posterior subcapsular region was compared to the liver interior, the parameters correlated well on the anterior side and not on the posterior side. Also, the remnant hepatocytes area had lesser correlation compared to other parameters, but the presence of remnant hepatocytes in the subcapsular region can be an indicator of fibrosis (Figure 4.11d). The differences between the anterior and posterior surfaces demonstrate the heterogeneous progression of liver fibrosis. The endoscopic probing of liver will take place on the anterior surface, thus the correlation has been advantageous for future

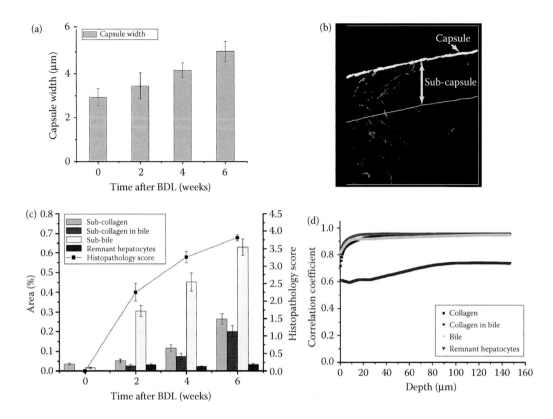

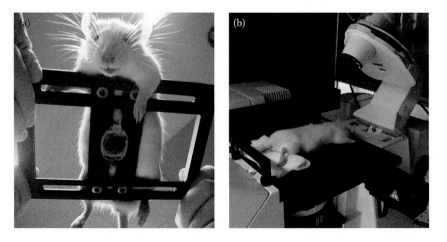

FIGURE 4.11 (a) The liver capsule width was quantified as the fibrosis progressed, but the parameter by itself lacked the resolution to distinguish between the various stages. (b)The area beneath the capsule up to a depth of 150 μm was identified as the subcapsular region based on the SHG signal from the collagen distribution. The region beneath the subcapsular region was identified as the liver interior. (c) Morphology-based features were quantified in the subcapsular region and the liver interior (Figure 4.11c). And their correlation to the histopathology score was calculated, showing that the collagen features correlate well with the histopathology score. (From He, Y et al., 2010. *Journal of Biomedical Optics*, 15, 056007. With permission.)

FIGURE 4.12 Window model for liver surface imaging. Sedated mouse with liver fibrosis with customized glass window on the abdominal surface is placed on the microscope sample holder (a). The holder is now placed on the microscope and images of up to 100 microns from the liver surface can be imaged by reflective SHG/TPEF imaging (b).

MPM application for liver fibrosis. The correlation coefficient converged to the maximum value of 1 at a depth of 20 μm below the capsule. Thus, the penetration depth of MPM imaging is more than sufficient to be able to assess liver fibrosis with certainty. When the parameters in the subcapsular regions were compared with histopathological scoring, the three parameters, except the remnant hepatocytes area compared well to the scores.

To translate the above study to an MPM surface mapping method, window models of liver fibrosis animal models have been developed, where the fibrosis progression is observed periodically through a window in the animal's abdominal area (Figure 4.12). The images obtained from the animals show clear distinction between normal and fibrosed liver (Figure 4.13). The organ characterization and surface mapping of the liver has enabled us to identify zones of interest in the liver and parameters of interest specific to this disease model. Further studies to develop different disease models and to identify specific parameters need to be performed. This will enable us to establish a quantitative liver atlas that provides researchers and clinicians a reference to the structural and functional alterations that the liver undergoes in a particular disease. In future, high-speed, high-resolution, high-fidelity endoscopic tools will exploit the advantages of multiphoton imaging to surface map not only the liver but also other organs to provide accurate diagnosis of disease progression with minimal invasiveness.

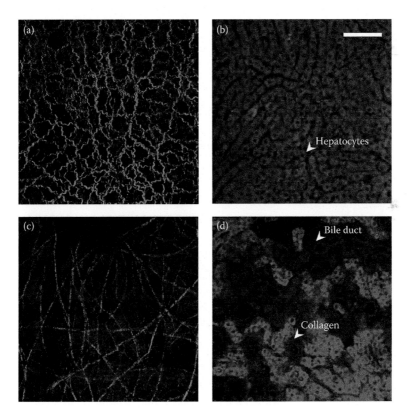

FIGURE 4.13 **(See color insert.)** Liver anterior surface scanning. Reflective SHG and TPEF imaging and 3-D projections of capsule regions shown in both normal (a) and fibrotic liver (c). Irregular and loss alignment of capsule collagen distribution in fibrotic liver is observed compared to normal liver. Sub-capsule region images were obtained 20 μm below the capsule region in normal (b) and fibrotic (d) liver tissues. Features of bile duct and abnormal collagen proliferations were present in fibrotic liver (d) compared to normal liver (b) where only well-organized hepatocytes present. Scale bar is 100 μm. (From He, Y et al., 2010. *Journal of Biomedical Optics*, 15, 056007. With permission.)

4.4 Conclusions

SHG and two-photon imaging has far-reaching applications today in the biomedical world that no one would have thought of 40 years ago. The field is ever expanding with newer applications and scientists improving the technology. There is a need for scientists to not only identify new applications, but also to delve into the biology of the problem and take the application all the way to the clinic. This chapter is devoted to discuss the application of MPM in liver fibrosis assessment. The development of a standardized quantification system for liver fibrosis assessment, both using liver biopsies as well a surface mapping tool in comparison to the conventional histological scoring system in animal models and human samples demonstrates the value of the technology in research and clinical applications. Further studies involving optical engineers, clinicians, and biomedical engineers to tailor the technology for clinical use can take the nonlinear imaging technology to the mainstream.

References

Araya, R., Eisenthal, K. B., and Yuste, R. 2006. Dendritic spines linearize the summation of excitatory potentials. *Proceedings of the National Academy of Sciences of the United States of America,* 103, 18799–804.

Aterman, K. 1986. The parasinusoidal cells of the liver—A historical account. *Histochemical Journal,* 18, 279.

Bacskai, B. J., Skoch, J., Hickey, G. A., Berezovska, O., and Hyman, B. T. 2004. Multiphoton imaging in mouse models of Alzheimer's disease. In: Periasamy, A. (ed.) *Multiphoton Microscopy in the Biomedical Sciences Iv, Proceedings of the SPIE,* 5323, 71–76.

Bahlmann, K., So, P. T. C., Kirber, M. et al. 2007. Multifocal multiphoton microscopy (MMM) at a frame rate beyond 600 Hz. *Optics Express,* 15, 10991–8.

Bao, H., Ryu, S. Y., Lee, B. H., Tao, W., and Gu, M. 2010. Nonlinear endomicroscopy using a double-clad fiber coupler. *Optics Letters,* 35, 995–7.

Baym, G. 1974. *Lecture on Quantum Mechanics,* Menlo Park California, Benjamin Cummins.

Bedossa, P., Bioulacsage, P., Callard, P. et al. 1994. Intraobserver and interobserver variations in liver-biopsy interpretation in patients with chronic hepatitis-C. *Hepatology,* 20, 15–20.

Bedossa, P., Dargere, D., and Paradis, V. 2003. Sampling variability of liver fibrosis in chronic hepatitis C. *Hepatology,* 38, 1449–57.

Bedossa, P. and Poynard, T. 1996. An algorithm for the grading of activity in chronic hepatitis C. The Metavir Cooperative Study Group. *Hepatology,* 24, 289–93.

Berezin, M. Y., Lee, H., Akers, W., and Achilefu, S. 2007. Near infrared dyes as lifetime solvatochromic probes for micropolarity measurements of biological systems. *Biophysical Journal,* 93, 2892–9.

Bewersdorf, J., Pick, R., and Hell, S. W. 1998. Multifocal multiphoton microscopy. *Optics Letters,* 23, 655–7.

Boissonnas, A., Fetler, L., Zeelenberg, I. S., Hugues, S., and Amigorena, S. 2007. *In vivo* imaging of cytotoxic T cell infiltration and elimination of a solid tumor. *The Journal of Experimental Medicine,* 204, 345–56.

Bonacina, L., Mugnier, Y., Courvoisier, F. et al. 2007. Polar Fe(IO3)(3) nanocrystals as local probes for nonlinear microscopy. *Applied Physics B-Lasers and Optics,* 87, 399–403.

Bosetti, C., Levi, F., Lucchini, F., Zatonski, W. A., Negri, E., and La Vecchia, C. 2007. Worldwide mortality from cirrhosis: An update to 2002. *Journal of Hepatology,* 46, 827–39.

Brown, E., Mckee, T., Ditomaso, E. et al. 2003. Dynamic imaging of collagen and its modulation in tumors *in vivo* using second-harmonic generation. *Nature Medicine,* 9, 796–800.

Burt, A. D., Griffiths, M. R., Schuppan, D., Voss, B., and Macsween, R. N. 1990. Ultrastructural localization of extracellular matrix proteins in liver biopsies using ultracryomicrotomy and immuno-gold labelling. *Histopathology,* 16, 53–8.

Buschmann, R. J. and Ryoo, J. W. 1989. Hepatic structural correlates of liver fibrosis: A morphometric analysis. *Experimental and Molecular Pathology,* 50, 114–24.

Callis, P. (ed.) 1997. *The Theory of Two-Photon Induced Fluorescence Anisotropy.* New York, NY: Plenum.

Campagnola, P. J., Wei, M. D., Lewis, A., and Loew, L. M. 1999. High-resolution nonlinear optical imaging of live cells by second harmonic generation. *Biophysical Journal,* 77, 3341–9.

Cavanagh, L. L., Bonasio, R., Mazo, I. B. et al. 2005. Activation of bone marrow-resident memory T cells by circulating, antigen-bearing dendritic cells. *Nature Immunology,* 6, 1029–37.

Centonze, V. E. and White, J. G. 1998. Multiphoton excitation provides optical sections from deeper within scattering specimens than confocal imaging. *Biophysical Journal,* 75, 2015–24.

Chandler, E., Hoover, E., Field, J. et al. 2009. High-resolution mosaic imaging with multifocal, multiphoton photon-counting microscopy. *Applied Optics,* 48, 2067–77.

Cho, C. H., Parashurama, N., Park, E. Y. H. et al. 2008. Homogeneous differentiation of hepatocyte-like cells from embryonic stem cells: Applications for the treatment of liver failure. *Faseb Journal,* 22, 898–909.

Chung, T. W., Yang, J., Akaike, T. et al. 2002. Preparation of alginate/galactosylated chitosan scaffold for hepatocyte attachment. *Biomaterials,* 23, 2827–34.

Cicchi, R., Crisci, A., Cosci, A. et al. 2010. Time- and spectral-resolved two-photon imaging of healthy bladder mucosa and carcinoma in situ. *Optics Express,* 18, 3840–9.

Davis, G. L., Albright, J. E., Cook, S. F., and Rosenberg, D. M. 2003. Projecting future complications of chronic hepatitis C in the United States. *Liver Transplantation,* 9, 331–8.

Demetriou, A. A., Whiting, J. F., Feldman, D. et al. 1986. Replacement of liver-function in rats by transplantation of microcarrier-attached hepatocytes. *Science,* 233, 1190–2.

Denk, W., Strickler, J. H., and Webb, W. W. 1990. Two-photon laser scanning fluorescence microscopy. *Science,* 248, 73–6.

Ding, J. B., Takasaki, K. T., and Sabatini, B. L. 2009. Supraresolution imaging in brain slices using stimulated-emission depletion two-photon laser scanning microscopy. *Neuron,* 63, 429–37.

Du, Y., Han, R. B., Wen, F. et al. 2008. Synthetic sandwich culture of 3D hepatocyte monolayer. *Biomaterials,* 29, 290–301.

Forbes, S. J., Russo, F. P., Rey, V. et al. 2004. A significant proportion of myofibroblasts are of bone marrow origin in human liver fibrosis. *Gastroenterology,* 126, 955–63.

Forns, X., Ampurdanes, S., Llovet, J. M. et al. 2002. Identification of chronic hepatitis C patients without hepatic fibrosis by a simple predictive model. *Hepatology,* 36, 986–92.

Fox, I. J., Chowdhury, J. R., Kaufman, S. S. et al. 1998. Treatment of the Crigler-Najjar syndrome type I with hepatocyte transplantation. *New England Journal of Medicine,* 338, 1422–6.

Franken, P. A., Weinreich, G., Peters, C. W., and Hill, A. E. 1961. Generation of optical harmonics. *Physical Review Letters,* 7, 118.

Fraser, R., Dobbs, B. R., and Rogers, G. W. T. 1995. Lipoproteins and the liver sieve—The role of the fenestrated sinusoidal endothelium in lipoprotein metabolism, atherosclerosis, and cirrhosis. *Hepatology,* 21, 863–74.

Friedenberg, M. A., Miller, L., Chung, C. Y. et al. 2005. Simplified method of hepatic fibrosis quantification: Design of a new morphometric analysis application. *Liver International,* 25, 1156–61.

Göppert-Mayer, M. 1931. Über Elementarake mit zwei Quantensprungen. *Annals of Physics (Leipzig),* 5, 273–94.

Gailhouste, L., Le Grand, Y., Odin, C. et al. 2010. Fibrillar collagen scoring by second harmonic microscopy: A new tool in the assessment of liver fibrosis. *Journal of Hepatology,* 52, 398–406.

Gale, R. P., Sparkes, R. S., and Golde, D. W. 1978. Bone-marrow origin of hepatic macrophages (Kupffer Cells) in humans. *Science,* 201, 937–8.

Gamal, M. D., Mohamed, M. K., Hussien, M. E.-B., Abdel-Motaal, M. F., and Osama, A. S. E.-D. 2004. Digital quantification of fibrosis in liver biopsy sections: Description of a new method by Photoshop software. *Journal of Gastroenterology and Hepatology,* 19, 78–85.

Gines, P., Cardenas, A., Arroyo, V., and Rodes, J. 2004. Management of cirrhosis and ascites. *The New England Journal of Medicine,* 350, 1646–54.

Gressner, A. M. and Weiskirchen, R. 2006. Modern pathogenetic concepts of liver fibrosis suggest stellate cells and TGF-beta as major players and therapeutic targets. *Journal of Cellular and Molecular Medicine,* 10, 76–99.

Griffiths, C., Rooney, C., and Brock, A. 2005. Leading causes of death in England and Wales—How should we group causes? *Health Stat Q,* 6–17.

Gronbaek, K., Christensen, P. B., Hamilton-Dutoit, S. et al. 2002. Interobserver variation in interpretation of serial liver biopsies from patients with chronic hepatitis C. *Journal of Viral Hepatitis,* 9, 443–9.

Hahn, E., Wick, G., Pencev, D., and Timpl, R. 1980. Distribution of basement membrane proteins in normal and fibrotic human liver: Collagen type IV, laminin, and fibronectin. *Gut,* 21, 63–71.

He, Y., Kang, C.H., Xu, S. et al. 2010. Towards surface quantification of liver fibrosis progression. *Journal of Biomedical Optics,* 15, 056007.

He, G. S. and Liu, S. H. 1999. *Physics of Nonlinear Optics.* Singapore: World Scientific Publishing Company.

Hsieh, C. L., Grange, R., Pu, Y., and Psaltis, D. 2010. Bioconjugation of barium titanate nanocrystals with immunoglobulin G antibody for second harmonic radiation imaging probes. *Biomaterials,* 31, 2272–7.

Imbert-Bismut, F., Ratziu, V., Pieroni, L., Charlotte, F., Benhamou, Y., and Poynard, T. 2001. Biochemical markers of liver fibrosis in patients with hepatitis C virus infection: A prospective study. *Lancet,* 357, 1069–75.

Ito, M., Ito, R., Yoshihara, D. et al. 2008. Immortalized hepatocytes using human artificial chromosome. *Cell Transplantation,* 17, 165–71.

Jalan, R., Harrison, D. J., Dillon, J. F., Elton, R. A., Finlayson, N. D. C., and Hayes, P. C. 1995. Laparoscopy and histology in the diagnosis of chronic liver-disease. *Qjm-Monthly Journal of the Association of Physicians,* 88, 559–64.

Jung, J. C. and Schnitzer, M. J. 2003. Multiphoton endoscopy. *Optics Letters,* 28, 902–4.

Kachynski, A. V., Kuzmin, A. N., Nyk, M., Roy, I., and Prasad, P. N. 2008. Zinc oxide nanocrystals for nonresonant nonlinear optical microscopy in biology and medicine. *Journal of Physical Chemistry C,* 112, 10721–4.

Kang, Y. H., Berthiaume, F., Nath, B. D., and Yarmush, M. L. 2004. Growth factors and nonparenchymal cell conditioned media induce mitogenic responses in stable long-term adult rat hepatocyte cultures. *Experimental Cell Research,* 293, 239–47.

Kaufmann, P. M., Heimrath, S., Kim, B. S., and Mooney, D. J. 1997. Highly porous polymer matrices as a three-dimensional culture system for hepatocytes. *Cell Transplantation,* 6, 463–8.

Kirkpatrick, N. D., Brewer, M. A., and Utzinger, U. 2007. Endogenous optical biomarkers of ovarian cancer evaluated with multiphoton microscopy. *Cancer Epidemiology Biomarkers & Prevention,* 16, 2048–57.

Kobayashi, N., Miyazaki, M., Fukaya, K. et al. 2000. Treatment of surgically induced acute liver failure with transplantation of highly differentiated immortalized human hepatocytes. *Cell Transplantation,* 9, 733–5.

Lee, H., Cusick, R. A., Browne, F. et al. 2002. Local delivery of basic fibroblast growth factor increases both angiogenesis and engraftment of hepatocytes in tissue-engineered polymer devices. *Transplantation,* 73, 1589–93.

Lewis, G. N., Lipkin, D., and Magel, T. T. 1941. Reversible photochemical processes in rigid media. A study of the phosphorescent state. *Journal of the American Chemical Society,* 63, 3005–18.

Lim, P. L. K., Tan, W., Latchoumycandane, C. et al. 2007. Molecular and functional characterization of drug-metabolizing enzymes and transporter expression in the novel spontaneously immortalized human hepatocyte line HC-04. *Toxicology in Vitro,* 21, 1390–401.

Lin, S.-J., Wu, R., Jr., Tan, H.-Y. et al. 2005. Evaluating cutaneous photoaging by use of multiphoton fluorescence and second-harmonic generation microscopy. *Optics Letters,* 30, 2275–7.

Lotzova, E. 1993. Definition and functions of natural-killer-cells. *Natural Immunity,* 12, 169–76.

Maharaj, B., Maharaj, R. J., Leary, W. P. et al. 1986. Sampling variability and its influence on the diagnostic yield of percutaneous needle biopsy of the liver. *Lancet,* 1, 523–5.

Maiman, T. H. 1960. Stimulated Optical Radiation IN Ruby. *Nature,* 187, 493–4.

Manabe, N., Chevallier, M., Chossegros, P. et al. 1993. Interferon-alpha-2B therapy reduces liver fibrosis in chronic non-a-hepatitis non-b-hepatitis—a quantitative histological-evaluation. *Hepatology,* 18, 1344–9.

Marra, F. 1999. Hepatic stellate cells and the regulation of liver inflammation. *Journal of Hepatology,* 31, 1120–30.

Martinezhernandez, A. 1984. The hepatic extracellular-matrix.1. electron immunohistochemical studies in normal rat-liver. *Laboratory Investigation,* 51, 57–74.

Martinezhernandez, A. and Amenta, P. S. 1993. The hepatic extracellular-matrix.1. components and distribution in normal liver. *Virchows Archiv a-Pathological Anatomy and Histopathology,* 423, 1–11.

Masseroli, M., Caballero, T., O'valle, F., Del Moral, R. M., Perez-Milena, A., and Del Moral, R. G. 2000. Automatic quantification of liver fibrosis: Design and validation of a new image analysis method: Comparison with semi-quantitative indexes of fibrosis. *Journal of Hepatology,* 32, 453–64.

Matalka, II, AL-Jarrah, O. M., and Manasrah, T. M. 2006. Quantitative assessment of liver fibrosis: A novel automated image analysis method. *Liver International,* 26, 1054–64.

Mo, X. J., Li, Q. S., Lui, L. W. Y. et al. 2010. Rapid construction of mechanically-confined multi-cellular structures using dendrimeric intercellular linker. *Biomaterials,* 31, 7455–67.

Muraca, M., Gerunda, G., Neri, D. et al. 2002. Hepatocyte transplantation as a treatment for glycogen storage disease type IA. *Journal of Hepatology,* 36, 124.

Nam, Y. S., Yoon, J. J., Lee, J. G., and Park, T. G. 1999. Adhesion behaviours of hepatocytes cultured onto biodegradable polymer surface modified by alkali hydrolysis process. *Journal of Biomaterials Science-Polymer Edition,* 10, 1145–58.

Ng, S., Han, R. B., Chang, S. et al. 2006. Improved hepatocyte excretory function by immediate presentation of polarity cues. *Tissue Engineering,* 12, 2181–91.

Nishimura, N., Schaffer, C. B., Friedman, B., Tsai, P. S., Lyden, P. D., and Kleinfeld, D. 2006. Targeted insult to subsurface cortical blood vessels using ultrashort laser pulses: Three models of stroke. *Nature Methods,* 3, 99–108.

Ohashi, K., Yokoyama, T., Yamato, M. et al. 2007. Engineering functional two- and three-dimensional liver systems *in vivo* using hepatic tissue sheets. *Nature Medicine,* 13, 880–5.

Pagliaro, L., Rinaldi, F., Craxi, A. et al. 1983. Percutaneous blind biopsy versus laparoscopy with guided biopsy in diagnosis of cirrhosis. A prospective, randomized trial. *Digestive Diseases and Sciences,* 28, 39–43.

Pantazis, P., Maloney, J., Wu, D., and Fraser, S. E. 2010. Second harmonic generating (SHG) nanoprobes for *in vivo* imaging. *Proceedings of the National Academy of Sciences of the United States of America,* 107, 14535–40.

Phan, T. G. and Bullen, A. 2010. Practical intravital two-photon microscopy for immunological research: Faster, brighter, deeper. *Immunology and Cell Biology,* 88, 438–44.

Poniachik, J., Bernstein, D. E., Reddy, K. R. et al. 1996. The role of laparoscopy in the diagnosis of cirrhosis. *Gastrointestinal Endoscopy,* 43, 568–71.

Provenzano, P. P., Eliceiri, K. W., Campbell, J. M., Inman, D. R., White, J. G., and Keely, P. J. 2006. Collagen reorganization at the tumor-stromal interface facilitates local invasion. *BMC Medicine,* 4, 38.

Provenzano, P. P., Eliceiri, K. W., Yan, L. et al. 2008a. Nonlinear optical imaging of cellular processes in breast cancer. *Microscopy and Microanalysis,* 14, 532–48.

Provenzano, P. P., Rueden, C. T., Trier, S. M. et al. 2008b. Nonlinear optical imaging and spectral-lifetime computational analysis of endogenous and exogenous fluorophores in breast cancer. *Journal of Biomedical Optics,* 13(3), 031220.

Qu, X. C., Wang, J., Zhang, Z. X., Koop, N., Rahmanzadeh, R., and Huttmann, G. 2008. Imaging of cancer cells by multiphoton microscopy using gold nanoparticles and fluorescent dyes. *Journal of Biomedical Optics,* 13(3), 031217.

Raja, A. M., Xu, S., Sun, W., Zhou, J., Tai, D. C., Chen, C. S., Rajapakse, J. C., So, P. T., and Yu, H. 2010. Pulse-modulated second harmonic imaging microscope quantitatively demonstrates marked increase of collagen in tumor after chemotherapy. *Journal of Biomedical Optics,* 15 (5), 056016.

Rojkind, M., Giambrone, M. A., and Biempica, L. 1979. Collagen types in normal and cirrhotic liver. *Gastroenterology,* 76, 710–9.

Rozga, J., Williams, F., Ro, M. S. et al. 1993. Development of a bioartificial liver—properties and function of a hollow-fiber module inoculated with liver-cells. *Hepatology,* 17, 258–65.

Schelhas LT, S. J. and Dantus, M. 2006. Advantages of ultrashort phase-shaped pulses for selective two-photon activation and biomedical imaging. *Nanomedicine,* 2, 177–81.

Schuppan, D., Ruehl, M., Somasundaram, R., and Hahn, E. G. 2001. Matrix as a modulator of hepatic fibrogenesis. *Seminars in Liver Disease,* 21, 351–72.

Schuppan, D., Schmid, M., Somasundaram, R. et al. 1998. Collagens in the liver extracellular matrix bind hepatocyte growth factor. *Gastroenterology,* 114, 139–52.

Seki, S., Habu, Y., Kawamura, T. et al. 2000. The liver as a crucial organ in the first line of host defense: The roles of Kupffer cells, natural killer (NK) cells and Nk1.1 Ag+ T cells in T helper 1 immune responses. *Immunological Reviews,* 174, 35–46.

Sell, S. 2001. Heterogeneity and plasticity of hepatocyte lineage cells. *Hepatology,* 33, 738–50.

Seyer, J. M., Hutcheson, E. T., and Kang, A. H. 1977. Collagen polymorphism in normal and cirrhotic human liver. *Journal of Clinical Investigation,* 59, 241–8.

Shakhar, G., Lindquist, R. L., Skokos, D. et al. 2005. Stable T cell-dendritic cell interactions precede the development of both tolerance and immunity in vivo. *Nature Immunology,* 6, 707–14.

Sheppard, C. J. R. and Kompfner, R. 1978. Resonant scanning optical microscope. *Applied Optics,* 17, 2879–82.

Shibata, C., Mizuguchi, T., Kikkawa, Y. et al. 2006. Liver repopulation and long-term function of rat small hepatocyte transplantation as an alternative cell source for hepatocyte transplantation. *Liver Transplantation,* 12, 78–87.

So, P. T. C. and Masters, B. R. 2008. *Handbook of Biomedical Nonlinear Optical Microscopy.* New York, NY: Oxford University Press.

Soloway, R. D., Baggenstoss, A. H., Schoenfield, L. J., and Summerskill, W. H. 1971. Observer error and sampling variability tested in evaluation of hepatitis and cirrhosis by liver biopsy. *American Journal of Digestive Diseases,* 16, 1082–6.

Stoller, P., Celliers, P. M., Reiser, K. M., and Rubenchik, A. M. 2002. Imaging collagen orientation using polarization-modulated second harmonic generation. In: Periasamy, A. and So, P. T. C., eds. *Conference on Multiphoton Microscopy in the Biomedical Sciences II,* Jan 20–22 2002 San Jose, CA, pp. 157–165.

Strupler, M., Hernest, M., Fligny, C., Martin, J. L., Tharaux, P. L., and Schanne-Klein, M. C. 2008. Second harmonic microscopy to quantify renal interstitial fibrosis and arterial remodeling. *Journal of Biomedical Optics,* 13, 054041.

Sun, W., Chang, S., Tai, D. C. et al. 2008. Nonlinear optical microscopy: Use of second harmonic generation and two-photon microscopy for automated quantitative liver fibrosis studies. *Journal of Biomedical Optics,* 13, 064010.

Sun, T. L., Liu, Y. A., Sung, M. C. et al. 2010. *Ex vivo* imaging and quantification of liver fibrosis using second-harmonic generation microscopy. *Journal of Biomedical Optics,* 15, 036002.

Tai, D. C., Tan, N., Xu, S. et al. 2009. Fibro-C-Index: Comprehensive, morphology-based quantification of liver fibrosis using second harmonic generation and two-photon microscopy. *Journal of Biomedical Optics,* 14, 044013.

Tang, S., Krasieva, T. B., Chen, Z., Tempea, G., and Tromberg, B. J. 2006. Effect of pulse duration on two-photon excited fluorescence and second harmonic generation in nonlinear optical microscopy. *Journal of Biomedical Optics,* 11, 020501.

Teramoto, K., Asahina, K., Kumashiro, Y. et al. 2005. Hepatocyte differentiation from embryonic stem cells and umbilical cord blood cells. *Journal of Hepato-Biliary-Pancreatic Surgery,* 12, 196–202.

Thampanitchawong, P. and Piratvisuth, T. 1999. Liver biopsy: Complications and risk factors. *World Journal of Gastroenterology,* 5, 301–4.

The French METAVIR Cooperative Study Group. 1994. Intraobserver and interobserver variations in liver biopsy interpretation in patients with chronic hepatitis C. *Hepatology,* 20, 15–20.

Theodossi, A., Skene, A. M., Portmann, B. et al. 1980. Observer variation in assessment of liver biopsies including analysis by kappa statistics. *Gastroenterology,* 79, 232–41.

Trinchieri, G. 1989. Biology of natural-killer cells. *Advances in Immunology,* 47, 187–376.

Tsai, H. A., Wu, R. R., Lee, I. C., Chang, H. Y., Shen, C. N., and Chang, Y. C. 2010. Selection, enrichment, and maintenance of self-renewal liver stem/progenitor cells utilizing polypeptide polyelectrolyte multilayer films. *Biomacromolecules,* 11, 994–1001.

Tsang, V. L., Chen, A. A., Cho, L. M. et al. 2007. Fabrication of 3D hepatic tissues by additive photopatterning of cellular hydrogels. *Faseb Journal,* 21, 790–801.

Tuchweber, B., Desmouliere, A., Bochaton-Piallat, M. L., Rubbia-Brandt, L., and Gabbiani, G. 1996. Proliferation and phenotypic modulation of portal fibroblasts in the early stages of cholestatic fibrosis in the rat. *Laboratory Investigation,* 74, 265–78.

Verant, P., Ricard, C., Serduc, R., Vial, J. C., and Van Der Sanden, B. 2008. *In vivo* staining of neocortical astrocytes via the cerebral microcirculation using sulforhodamine B. *Journal of Biomedical Optics,* 13, 064028.

Wang, P., Liu, T. H., Cong, M. et al. 2009a. Expression of extracellular matrix genes in cultured hepatic oval cells: An origin of hepatic stellate cells through transforming growth factor beta? *Liver International,* 29, 575–84.

Wang, Q., Ornstein, M., and Kaufman, H. L. 2009b. Imaging the immune response to monitor tumor immunotherapy. *Expert Review of Vaccines,* 8, 1427–37.

Wang, X. H., Yan, Y. N., Lin, F. et al. 2005. Preparation and characterization of a collagen/chitosan/heparin matrix for an implantable bioartificial liver. *Journal of Biomaterials Science-Polymer Edition,* 16, 1063–80.

Wells, R. G. 2008. The role of matrix stiffness in regulating cell behavior. *Hepatology,* 47, 1394–400.

Westin, J., Lagging, L. M., Wejstal, R., Norkrans, G., and Dhillon, A. P. 1999. Interobserver study of liver histopathology using the Ishak score in patients with chronic hepatitis C virus infection. *Liver,* 19, 183–7.

Wisse, E. 1970. An electron microscopic study of fenestrated endothelial lining of rat liver sinusoids. *Journal of Ultrastructure Research,* 31, 125.

Wright, M., Thursz, M., Pullen, R., Thomas, H., and Goldin, R. 2003. Quantitative versus morphological assessment of liver fibrosis: Semi-quantitative scores are more robust than digital image fibrosis area estimation. *Liver International,* 23, 28–34.

Xia, L., Ng, S., Han, R. B. et al. 2009. Laminar-flow immediate-overlay hepatocyte sandwich perfusion system for drug hepatotoxicity testing. *Biomaterials,* 30, 5927–36.

Xiangdong, W., Arrajab, A., Ahren, B., Andersson, R., and Bengmark, S. 1991. the effect of pancreatic-islets on transplanted hepatocytes in the treatment of acute liver-failure in rats. *Research in Experimental Medicine,* 191, 429–35.

Yang, C., Zeisberg, M., Mosterman, B. et al. 2003. Liver fibrosis: Insights into migration of hepatic stellate cells in response to extracellular matrix and growth factors. *Gastroenterology,* 124, 147–59.

Yazdanfar, S., Joo, C., Zhan, C., Berezin, M. Y., Akers, W. J., and Achilefu, S. 2010. Multiphoton microscopy with near infrared contrast agents. *Journal of Biomedical Optics,* 15.

Zheng, M. H., Ye, C., Braddock, M., and Chen, Y. P. 2010. Liver tissue engineering: promises and prospects of new technology. *Cytotherapy,* 12, 349–60.

Zhuo, S., Chen, J., Luo, T., Jiang, X., Xie, S., and Chen, R. 2009. Two-layered multiphoton microscopic imaging of cervical tissue. *Lasers in Medical Science*, 24(3), 359–63.

Zipfel, W. R., Williams, R. M., Christie, R., Nikitin, A. Y., Hyman, B. T., and Webb, W. W. 2003. Live tissue intrinsic emission microscopy using multiphoton-excited native fluorescence and second harmonic generation. *Proceedings of the National Academy of Sciences of the United States of America,* 100, 7075–80.

Zoumi, A., Yeh, A., and Tromberg, B. J. 2002. Imaging cells and extracellular matrix *in vivo* by using second-harmonic generation and two-photon excited fluorescence. *Proceedings of the National Academy of Sciences of the United States of America,* 99, 11014–9.

5

Atomic Force Microscopy for Cell and Tissue Niches

Wanxin Sun
Bruker Nano-Surface

Since the invention of atomic force microscope (AFM) two decades ago (Binnig et al., 1986), it has been extensively used in almost every branch of science and engineering. This offers an advantage of having the ability to convey nanometer spatial resolution in three dimensions in the absence of vacuum and contrast reagent. Furthermore, it is compatible with chemical and physical environment control in the liquid state. It provides an alternative for researchers to explore cell and tissue dynamics with molecular resolution under a controlled environment, for example, pH, ion strength, temperature, and signaling molecules. This chapter begins with the working principles and suitable applications of different imaging modes, followed by emerging technology, such as quantitative nanomechanical properties measurement of biomaterials, molecular recognition on a substrate or cell membrane, and so on. An uprising trend in characterization techniques is to integrate or tandem different instruments to get information from different aspects. The correlation of different information from a combination of multiple systems provides insight into the underlying mechanism. AFM is not an exception, it has been integrated with advanced fluorescence techniques, including confocal fluorescence microscopy, total internal reflection fluorescence microscopy, and electrophysiological means, for example, patch clamp and ion conductance measurement. The applications and challenges of the integration between AFM and optical techniques will also be discussed in this chapter.

5.1 AFM Working Principles and Imaging Modes Suitable for Cell and Tissue Engineering Applications

5.1.1 AFM Working Principles

In AFM, a sharp probe runs a raster scan across the sample surface with a positioning accuracy in subnanometer level. The probe and sample interaction is maintained at a constant by moving the probe relative to the sample surface up and down. The up/down movement is recorded against XY position to form the topography of the sample surface, as shown in Figure 5.1. The nano positioning in an AFM is

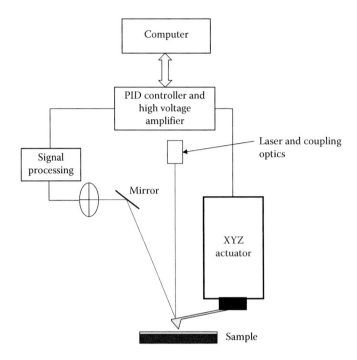

FIGURE 5.1 Schematic diagram of an AFM. (The diagram is not in scale.) The probe runs a raster scan over the sample surface while maintaining a constant probe–sample interaction by moving the probe up and down with the surface topography. The probe–sample interaction is measured by an optical lever. The feedback is implemented by a proportional-integral-derivative (PID) servo loop.

implemented by piezoelectric scanners, where the movement in XYZ is driven by high voltages in the range of hundreds of volts. Intrinsically, piezoelectric material does not respond linearly to the voltage applied. In today's prevailing AFM, XYZ position sensors are used to monitor the actual movement of the scanner and correct the nonlinearity through a close feedback loop. Another approach is to measure and model the nonlinearity first, thereafter using nonlinear voltage to drive the scanner to obtain linear movement. After calibration, both methods are able to produce high-quality good images.

The commonly used AFM probes consist of a sharp tip and a microcantilever, as shown in Figure 5.2. The radius of curve of the today's AFM tip ranges from a few nanometers to 30 nm, depending on the fabrication process and their applications. The microcantilever is 30–40 μm in width and 125–450 μm in length. The thickness ranges from a fraction of μm to a few μm. The cantilever works as a sensor to detect the probe–sample interaction, which is a complicating process, where many types of forces are involved (Israelachvili, 1992). In addition, both the sample and probe deform under the interaction forces, making it difficult to model the interaction from the first principle. In this chapter, we will not dwell in-depth into the origins of interaction forces. We will focus on aspects directly related to AFM instrumentation and applications, for example, the magnitude of the overall interaction forces, and the direct consequences of interactions including energy dissipation and cantilever dynamics changes. The normal force between the probe and the sample is measured by the cantilever bending, which simply follows Hook's law,

$$F = k\Delta z \tag{5.1}$$

where k is the spring constant of the cantilever, Δz is the cantilever bending in nanometers.

With a variety of cantilevers available commercially, AFM can measure forces ranging from a few piconewton (pN) to hundreds of micronewton (μN). By maintaining a constant bending of the

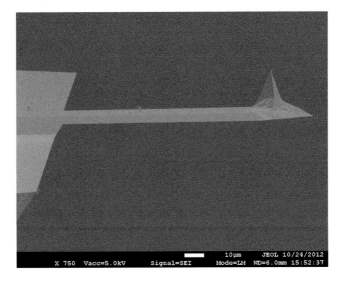

FIGURE 5.2 Scanning electron microscopic image of a silicon probe, which is normally used in the tapping mode. The sharp tip of the probe points up in this image.

cantilever, AFM produces topography under constant force. Alternatively, for a cantilever oscillating at its resonance frequency, the probe–sample interaction can change its dynamics, resulting in amplitude decrease and time lag in oscillation. By maintaining constant oscillation amplitude, the probe can track the surface to form topography of the sample while doing a raster scan. This mode is usually called tapping mode or intermittent contact mode. The time lag is measured by the phase difference between drive signal and cantilever oscillation. Both the amplitude and the phase lag are measured by a lock-in amplifier in today's AFM. Phase lag is caused by the energy dissipation during each oscillation cycle of the cantilever. When the probe contacts the sample surface, the sample surface undergoes both elastic and plastic deformation, causing energy loss. As it retracts from the sample surface, it often needs to overcome the attractive force and adhesive force. The adhesive force is also a reason for energy loss. Therefore, phase lag is related to viscoelastic properties of the sample. Phase lag has been extensively used to differentiate different materials (Babcock and Prater, 1995).

The tiny cantilever bending or oscillation is measured by an optical lever configuration, as shown in Figure 5.3. The laser beam from a laser diode is focused onto the end of the cantilever. The reflected beam from the cantilever surface is redirected to a quadrant photodiode, where the output is proportional to the position of the laser spot on the photodiode. $((A + B) − (C + D))/(A + B + C + D)$ is proportional to the vertical position. $((A + C) − (B + D))/(A + B + C + D)$ is proportional to the lateral position. When the cantilever bends vertically, the direction of the reflected laser beam changes accordingly. Then, the laser spot on the diode shifts vertically, causing the vertical output signal changes. In the same way, lateral position of the laser spot on the diode measures the torsion of the cantilever. It is noted that the optical lever measures the angle of deflection, not the displacement. Therefore, short cantilever is more sensitive to detect height difference on the sample surface.

5.1.2 Contact Mode

In contact mode, the normal force, that is, the vertical deflection of the cantilever, is maintained at a constant during scan. When the probe scans across a protruding feature, the cantilever is pushed up, generating an error in the vertical deflection. To eliminate this error, the control station will lift the probe until the error becomes zero. For recessed features, the probe is lowered to eliminate the deflection error. With an optimized servo loop, the probe tracks the surface with constant force to form the

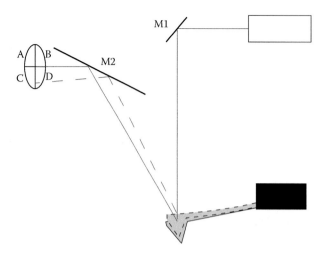

FIGURE 5.3   A schematic diagram of the working principle of an optical lever. A focusing laser beam is delivered to the free end of a cantilever by mirror M1. The reflected laser beam from the cantilever is steered onto a quadrant photodetector, which works as a position-sensitive detector. The adjustable mirror M1 is used to align the laser beam on the cantilever and M2 to align laser to the center of the photodetector. The reflected laser beam from the cantilever changes its direction with cantilever bending up/down. After propagating enough distance, the laser spot on the detector displaces up/down. By measuring lateral displacement of the laser spot on the detector, the cantilever torsion is measured. The quadrant detector usually measures both the vertical and lateral cantilever deflections.

topography of the sample. During the scan, lateral/frictional force always exists between the probe and sample. Lateral force has been successfully used to study self-assembled monolayer (Heaton et al., 2004), where different functional groups show different frictional force, while their heights are almost the same. Lateral force microscopy (LFM) unveils the distribution of functional groups while the topography shows little contrast. As well as providing rich information, lateral force also cause problems, that is, lateral force, if not controlled properly, can cause damage to delicate samples and tip wear. For live cells, the measured Young's modulus is in the magnitude of kilopascal (kPa). Force higher than a few nanonewton (nN) will cause significant deformation, producing lower measured height. On increasing the force further, the probe can poke a hole on the cell membrane. The suitable force for cell imaging is from sub-nanonewton to a few nanonewtons.

To achieve proper force control for live cell imaging, cantilever spring constant is a critical parameter. Softer cantilever deflects more under the same load, which is favorable for force detection sensitivity. However, the thermal noise is also higher in soft cantilever. In practice, cantilevers with a spring constant between 0.01 and 0.06 N/m are normally used to image live cells. This kind of cantilever is made of silicon nitride and in V shape, as shown in Figure 5.4. Silicon nitride is transparent or translucent, depending on the manufacturing process. To increase its reflectivity, a thin layer of Au or Pt is coated on the backside of the cantilever. Cantilevers with spring constant higher than 0.2 N/m are difficult to maintain constant force for live cell imaging in practice. In addition, such cantilevers cause scratches on live cells sometimes. In the other end, cantilevers with a spring constant lower than 0.01 N/m are influenced significantly by environmental temperature change. The heat generated by the AFM laser or microscope illumination often causes significant drift of cantilever deflection, in the range of a few volts in the quadrant photodiode detector. In addition, protein adsorption on the cantilever generates stress on one side, resulting in observable deflection change, which has been used as a sensor for protein detection (Fritz, 2008). The drift in cantilever deflection causes instability in imaging. For example, if the cantilever deflection increases with time, the probe will be lifted from the sample surface gradually. This is because the increase in cantilever deflection means higher repulsive force to the control station although it is caused by cantilever deflection drift rather than the repulsive force. The response of the

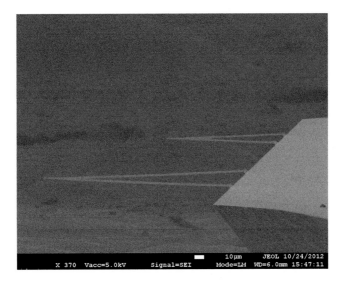

FIGURE 5.4 Scanning electron microscopic image of a typical V-shape silicon nitride probe. The tip points up in this image. This kind of cantilever is normally used for contact mode or tapping mode in liquid.

control station is to lift the cantilever. As a result, the last part of the image does not show any features as the probe does not track the surface. To overcome this issue, higher force must be set to compensate the cantilever drift. The image is not obtained at constant force. The higher force in some part of the image does not fulfill the purpose of lowering force by using soft cantilever. Therefore, too stiff or too soft cantilever is not advisable in live cell imaging.

AFM can image cells under pseudo-physiological conditions with temperature and chemical environment control. In general, AFM cannot measure anything suspended in the liquid. Only after the cells are attached to a support surface, AFM can then image them. Fortunately, a large number of cell types can grow directly on glass cover slip and plastic ware, for example, Petridish. To enhance adhesion, Bunsen burner flame treatment helps in general (Thimonier et al., 1997). For some cell types, it might be difficult to get them adhering to the bare surface of glass and plastics. Adhesives, like polylysine, collagen, laminin, Cell-Tak, and PEG derivatives, can enhance the adhesion (Henderson et al., 1992, Nagayama et al., 1996). After cells are seeded, allow time for the cells to attach to the support surface properly. Before loading into an AFM, the cell preparation should be rinsed with filtered medium or buffer to remove cell debris and unattached cells. Otherwise, it is likely that debris or unattached cells stick to the cantilever during imaging. It usually requires to change the cantilever if this happens during imaging, incurring unnecessary work or causing the experiment to fail. About 70–80% confluence is a good trial for most cell types. If the concentration is too low, it will be difficult to locate a cell for AFM imaging. This is different from confocal microscope, where it is not difficult to find an interesting cell because it is faster and has a large field of view. On the contrary, if the concentration is too high, it causes cells to overlap with each other or become loosely attached. To maintain the viability of cells, today's biological AFM is equipped with temperature control, CO_2 atmosphere control, gas purging, and perfusion apparatus.

As discussed in the previous section, contact force is critically important for live cell imaging. Force adjustment is achieved by changing the setpoint, at which the cantilever deflection is maintained. Force–distance curve is a well-accepted method to determine setpoint. After the cantilever is engaged onto a cell of interest, the XY scan is stopped and the cantilever is ramped up and down at a specific point by the Z scanner. The deflection of cantilever, which is proportional to the force, is recorded against the Z position of the cantilever. Figure 5.5 shows a typical force–distance curve of a live cell. Before the tip contacts the cell surface, the deflection is a constant, shown as a flat baseline in the force–distance

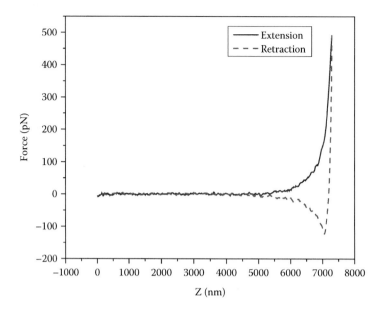

FIGURE 5.5 Typical force–distance curve measured on a live cell. The solid line records the cantilever bending with the probe approaching the cell. The dashed line records the cantilever bending with the probe retracting from the cell. The retraction curve shows an adhesion force of about 150 pN.

curve. After contacting the cell surface, the cantilever bends up as it pushes against the cell. In the mean-time, the cell deforms with the increase in force. After reaching the maximum force set by the user, the cantilever is retracted. It is noticed that the two curves do not overlap with each other. The hysteresis is attributed to the plastic deformation of the cell. It is noticed that the probe is not detached from the cell surface even the deflection has restored to the baseline level. Instead, the cantilever is bent down by the adhesion force between the probe and the cell. The adhesion force is often used to study specific binding when the probe is functionalized with antibodies. It is straightforward to read the repulsive force directly from the extension curve and rupture force of binding pair from the retraction curve. Usually, a repulsive force of 200–500 pN is a good start point to try for contact mode imaging. After the setpoint is chosen, the software is switched back to imaging mode, where the XY scan is enabled and the servo loop controls the probe moving up and down to maintain the setpoint. In today's commercial AFM, the line profiles in both directions (trace and retrace) are displayed in a window. Ideally, the trace and retrace profiles overlap with each other. In the real operation, the servo loop gains and scan speed need to adjust to get the profiles overlapping. Higher servo loop gains make the probe-tracking surface faster. On the other hand, servo loop will oscillate if the gains are too high. As a rule of thumb, gains are increased until slight oscillation is observed. Then the gains are reduced slightly to eliminate the oscillation. Slow scan speed in general makes the probe tracking surface better. However, it takes long time to record an image; 0.5–1 Hz is normally used for cell imaging. During an AFM scan, the deformation on cell membrane makes height less accurate, compared with hard materials. Considering that the cell height is in micrometers, the deformation in nanometers is negligible. AFM has been used to measure cell volume change of neurons during apoptosis. The volume accuracy is better than traditional confocal microscope, where the resolution in Z is in the order of hundreds of nanometers. One of the benefits of cell membrane deformation is that the probe can "feel" the hard cytoskeleton filaments. With a force higher than a few nanonewtons, the deflection error image provides rich information on cytoskeleton, as shown in Figure 5.6. Researchers sometimes increase the force intentionally to get better contrast in the deflection error images.

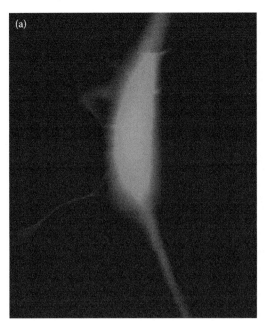
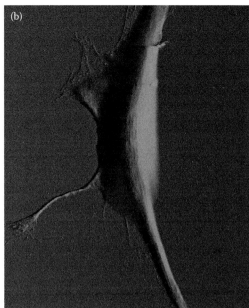

FIGURE 5.6 AFM images of a live fibroblast scanned by contact mode. (a) Topography, (b) deflection image. Details, such as cytoskeleton, are clearly discerned in deflection image. The scan size is 150 μm × 150 μm.

5.1.3 Tapping Mode

Tapping mode is also known as intermittent contact mode or AC mode, where the cantilever is oscillated at its resonance frequency or slightly lower frequency. The oscillation is usually driven by a small piece of piezoelectric material embedded in the probe holder. The resonance frequency is found by sweeping drive frequency and finding the maximum oscillation amplitude. When the probe is brought to the sample surface by the engaging mechanism, the interaction between tip and sample causes decrease in oscillation amplitude. At the bottom-most point of each oscillation cycle, the tip contacts the surface instantaneously. The amplitude decreases further when the probe is brought closer to the surface, and vice versa. Thus, the amplitude is a measure of tip–sample interaction and used to control probe up/down movement. The stronger the interaction, the more the amplitude decreases. The amplitude is maintained a constant in tapping mode. In practical AFM operation, the operation amplitude (setpoint) is chosen at about 80% of the amplitude of free cantilever. In addition to the amplitude, the lag in cantilever oscillation relative to the drive signal provides information about materials, as discussed previously. The lag is measured in phase, not in time, in real AFM instrumentation. The origin of the phase shift is the resonance frequency shift under tip–sample interaction. How much the phase shifts depend on the material viscoelasticity and the operation parameters. Phase image has been used extensively in differentiating different materials. However, there is ambiguity in material identification owing to its intrinsic nature that phase shift is affected by operation parameters as well as material viscoelasticity. The contrast in phase image can be reversed under different operation conditions, as shown in Figure 5.7. Under light tapping, phase contrast is caused by the adhesive force sensed by the probe. With the increase in tapping force, material elasticity contributes more in the phase contrast. Therefore, the phase contrast under light tapping originates from the adhesiveness, which is often affected by relative humidity in ambient environment. In liquid environment, phase contrast under light tapping has been used to recognize molecules (Lv et al., 2010). Under hard tapping, the phase contrast is the combination of adhesive force and elastic modulus of the material under scrutiny. Phase image has been extensively used in copolymer studies (McLean and Sauer, 1997).

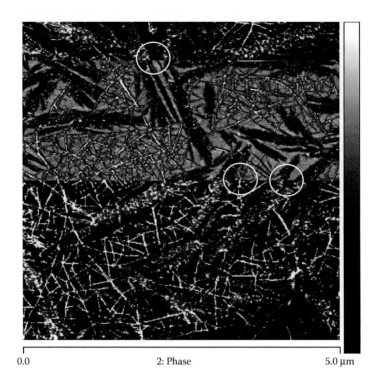

0.0   2: Phase   5.0 μm

FIGURE 5.7 Contrast in phase image changes with setpoint. The phase contrast can be reversed by increase or decrease setpoint. A few regions in the image are marked by circles to indicate the contrast reverse.

Compared with contact mode, the probe contacts sample surface instantaneously in tapping mode. The lateral force is negligible. Tapping mode is suitable for loosely bound samples, which are easily pushed away by probe in contact mode. In general, biomolecules and nanoparticles are bound to a substrate loosely. For example, proteins and DNAs are usually bound to mica by charges. In contact mode, the image is not stable as the molecules move with probe during scan. More than 80% of AFM samples were scanned with tapping mode as this mode causes less damage on samples. Another reason is that silicon probes are in general sharper than silicon nitride probes. Silicon process is not suitable for making soft cantilever because the cantilever is made from a silicon wafer through etching. It is not advisable to make a cantilever with a thickness less than 1 μm from hundreds of micrometers thick wafer. Contact mode requires soft cantilever, making silicon probe not suitable. Silicon nitride cantilever is made through thin film growth technology, it is easy to control the thickness uniformly and form thin and soft cantilevers. That is why contact mode uses silicon nitride cantilever. Compared with silicon process, it is not easy to make sharp silicon nitride probes. In general, silicon probes are sharper than silicon nitride probes. Therefore, tapping mode usually generates sharper images than contact mode. However, tapping mode is slower than contact mode in general. For samples with large feature heights, for example, cells, contact mode is preferred as it can scan faster for a large area. Fundamentally, it is the amplitude in tapping mode that is used to control the movement in the Z direction. Scanning across a high feature, the drop in amplitude takes time in the scale of milliseconds, the cantilever dynamics is the bottleneck in tapping mode. No matter how fast the scanner is, 1–2 Hz is normally used in tapping mode for rough samples. In contrast, the deflection in contact mode changes about 1000 times faster than the amplitude in tapping mode. The bottleneck in contact mode is the scanning mechanism rather than the cantilever itself.

Different cantilevers have different resonance frequencies. For each cantilever mounted into the instrument, its resonance must be found by sweeping frequency, performed by an "autotune" function in a software. The viscoelasticity and density of the medium surrounding the cantilever affects its resonance significantly. In aqueous solution, the resonance frequency drops to about one-third of that in air. For example, the resonance frequency in air of a commonly used silicon nitride cantilever is around 50 kHz, while it is 15 kHz in water. In air, the response curve in "autotune" shows a symmetric Gaussian profile and software can detect its frequency automatically. The same cantilever shows a forest of peaks in aqueous solution. Traditionally, AFM users follow the guideline in the AFM manual to choose a suitable frequency. However, with more and more probes designed for different applications, it may not be easy to find a recommended frequency in manuals or references for new probes. In addition, the cantilever oscillation amplitude affects the imaging stability and resolution. To make the data consistent and comparable among different AFM instruments, it is desirable to keep the oscillation amplitude the same. Therefore, the real amplitude should be measured in nanometers rather than in millivolts because different AFM may have different gains in the position detector. Here, a simple method is described to determine the conversion coefficient from voltage to nanometers, and whether the frequency chosen is suitable for tapping mode imaging in an aqueous solution. Similar to the force–distance curve in contact mode, the cantilever in tapping mode is ramped by the Z scanner while XY position is fixed. The oscillation amplitude is recorded against the Z position, as shown in Figure 5.8. When the probe is far from the sample surface, the amplitude does not change significantly with the Z position. Once the probe touches the surface at the lowest point of each oscillation cycle (shown as mark A in Figure 5.8), the amplitude begins to decrease rapidly as it is pushed against the surface further. Based on the slope between A and B, the coefficient, also called amplitude sensitivity (in nm/V), is obtained. The faster the change in amplitude, the better the cantilever is for imaging. From the forest of peaks, chose one and ramp the amplitude-Z curve. The frequency producing a steep slope in the curve is a suitable one. The amplitude dropping 20% from the corner, shown as mark A in Figure 5.8, is usually a good setpoint.

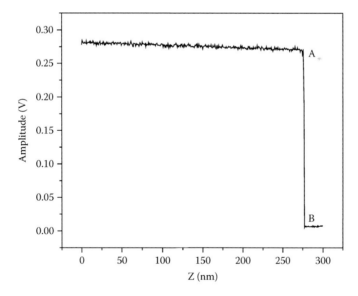

FIGURE 5.8 Cantilever oscillation amplitude changes with the Z position in the ramp mode. When the cantilever is far from the surface, the amplitude does not decrease significantly with the decrease in probe–sample distance. After the tip starts to touch the surface at the bottom-most point in an oscillation cycle (marked as A), the amplitude decreases rapidly with further decrease in probe–sample distance, until reaching B where the probe fully contacts the surface. The slope from A to B is often used to determine the amplitude sensitivity, which is used to convert amplitude in volts to nanometers.

In contact mode, cantilever responses to topographic changes much faster than the tapping mode. As a result, the servo-loop can tolerate higher feedback gains (i.e., proportional gain and integral gain) without causing significant oscillation. Optimization of gains is not critical in contact mode as there is more tolerance in gains. On the contrary, the change of cantilever amplitude in tapping mode is a slow process. It usually causes oscillation in feedback loop if gains are set too high, resulting in noisy image or artifacts. Experimental parameters must be tuned carefully. Among all the parameters, proportional gain, integral gain, setpoint, and scan rate are the top parameters required to take care in tapping mode. The principle of proportional–integral–derivative (PID) controller and tuning procedure can be found in many text books on process control (Bennett, 1993) and articles (Ang et al., 2005). Here, we discuss briefly about the tuning procedure suitable for AFM.

1. Set the scan size to a few micrometers, say 1–2 μm, even the final scan size is tens of micrometers.
2. Increase the integral gain until slight oscillation or noise appears in the trace/retrace profiles. The oscillation or noise can usually be eliminated by an increase in proportional gain gradually. If it cannot be eliminated, integral gain should be decreased until oscillation/noise disappears.
3. Check the trace/retrace profile, if the probe cannot track the falling slope while the rising slope is tracked properly; reduce the setpoint gradually to improve the situation.
4. Reduce scan rate if the probe tracking cannot be improved further by reducing setpoint. It should be noted that reducing scan rate must be taken as the last choice because it increases the time used to scan an image. Slow scan rate gives the servo loop more time to respond to topographic change, making the probe tracking better. However, the system drift is more severe in slow scan rate, resulting in distorted image.
5. Increase the scan size to the required size. If the tracking becomes poor, repeat steps 2 through 4 until the image quality is acceptable.

To improve the resolution, sharper probe and less tapping force are always preferred. Less tapping force is usually achieved by smaller cantilever oscillation amplitude. Therefore, one-third of normal tapping amplitude is used in molecular imaging, for example, DNA, protein, polysaccharide, and other biomolecules. Figure 5.9 shows DNA images obtained in tapping mode.

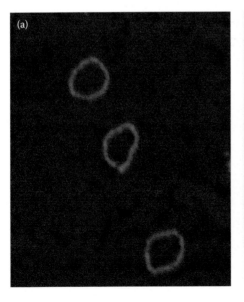 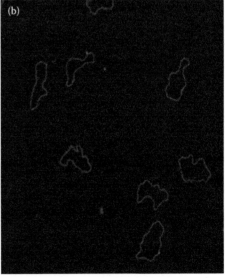

FIGURE 5.9 AFM images of circular DNA scanned by tapping mode. The scan sizes of image (a) and (b) are 450 nm and 2 μm, respectively. DNA molecules are attached to mica by electrostatic force. Both DNA and mica are negatively charged. Through divalent ions, for example, Ni^{2+}, the two negative-charged parts are bound together.

5.1.4 Force Mode

In force mode, the AFM probe is ramped up/down by retracting/extending the Z scanner and the cantilever deflection is recorded against the Z position, forming a force–distance curve. The ramp range is usually from tens of nanometers to a few micrometers. During the extension in a ramp, the probe touches the sample, causing cantilever deflection change as well as sample deformation. While in retraction, the probe is lifted from the surface. In one ramp, material mechanical properties, for example, Young's modulus, adhesion force, energy dissipation can be extracted from the force–distance curve. In force mode, XY scan is stopped and no lateral force is applied on the sample surface. Scratching on sample surface is rarely observed, unlike contact mode, where the lateral force exists always. In today's AFM, ramps can be programmed at user-defined positions or in an array. A series of force–distance curves can be used to construct the topography at a specific force. This method is usually used to image very soft or sticky samples, which are difficult for contact mode and tapping mode.

A typical force–distance curve measured on mica is shown in Figure 5.10. In contrast to the force–distance curve shown in Figure 5.5, the extension and retraction curves in Figure 5.10 overlap with each other. This is because mica is much harder than live cells and the deformation of mica is negligible. Another obvious difference is the adhesion force, as shown in segment 5 in Figure 5.10. Mica is hydrophilic and capillary force is strong when measured in an ambient environment. In case of live cells, nonspecific binding dominates the overall adhesion force. Capillary force is eliminated when both the cantilever and cells are in liquid environment, where no capillary is formed. The magnitude of capillary force is determined by the relative humidity and surface hydrophobicity (Sun et al., 2005). The adhesion force can be used to determine many surface properties, for example, surface hydrophobicity and functional group identification when a functionalized probe is used. If the probe and sample are put into controlled environment, interaction mechanism, for example, the types of interaction force and how the environment changes the interactions, can also be studied.

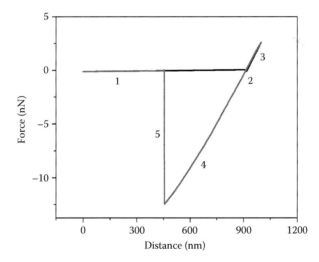

FIGURE 5.10 A typical force–distance curve measured on mica. When the cantilever is far from the surface, there is no detectable interaction force, shown as a horizontal baseline (1) in force–distance curve. After the tip contacts the surface at point (2), the cantilever starts to bend up (3) with further pushing against the surface until reaching the maximum force set by user. The cantilever is then retracted and cantilever deflection restores accordingly. After reaching zero force (2), the cantilever does not go to the baseline. Instead, the cantilever bends down (4) because of adhesion force between the mica surface and probe. After spring force overcomes the adhesion force, the cantilever jumps off the surface and restores to zero force state. Segment (5) is used to determine the adhesion force.

The raw data output from the instrument is the voltage from the position-sensitive detector. To convert the voltage to force, two parameters are needed, that is, deflection sensitivity and spring constant of the cantilever. Deflection sensitivity measures how many nanometers correspond to 1 V in the position sensitive detector output. It is measured by ramping the probe on a hard surface, for example, sapphire. The deformation of such surface is negligible. So the traveling distance of the Z scanner after contact is the same as the bending of the cantilever. In force–distance curve, deflection sensitivity is obtained by fitting the rising slope to a straight line, which is done automatically in today's AFM. The linear range in force–distance curve should be selected for curve fitting to improve the accuracy and repeatability. The snap-in point and the approach/retract turning point are usually excluded from the calculation. Deflection sensitivity is determined by cantilever length and the position sensitive detector. Short cantilever produces better sensitivity and is preferred in measuring displacement in picometer (pm) range. For a given AFM, the same type of cantilever should have the same length and deflection sensitivity. In real operation, the laser spot may not be aligned to the same position on the cantilever every time. It is normal that the deflection sensitivity of the same cantilever varies slightly after realigning the laser. The change in laser spot position affects the effective length of the cantilever. It is a good practice to measure the deflection sensitivity each time after the laser is realigned.

For a cantilever with rectangular cross section, its spring constant can be expressed as

$$k = \frac{Et^3w}{4L^3} \tag{5.2}$$

where E is Young's modulus of the cantilever material, that is, silicon or silicon nitride, t, w, and L are the thickness, width, and length of the cantilever, respectively.

The width and length can be controlled precisely through microfabrication technology, while the thickness bears more deviation owing to its manufacturing process. A 10% thickness error results in about 30% error in spring constant. The nominal spring constant on probe boxes can only be used as an indicative value. Each cantilever must be calibrated to get correct force value. Several methods have been developed to measure cantilever spring constant. The simplest way is to measure the resonance frequency f and check the probe factor b from reference book. The spring constant is calculated by

$$k = b * f^3 \tag{5.3}$$

This method is very easy to use as the resonance frequency can be obtained by "autotune" in tapping mode. The major drawback is its poor accuracy because the dimensions may be slightly different from those in the reference. For example, the cantilever width affects its spring constant, but not its resonance frequency. It will generate error if the real width is different from that used in the reference.

To improve this situation, top view geometry (length and width) is added to into the equation as (Cleveland et al., 1993)

$$k = 2\pi^3 \sqrt{\frac{\rho^3}{E}} L^3 w f^3 \tag{5.4}$$

where ρ and E are density and Young's modulus of cantilever material, L and w are the length and width, respectively, usually measured through a well-calibrated micrograph of scanning electron microscopy or optical microscopy.

A more accurate method was developed in Cleveland et al. (1993) by adding a known mass and measuring the resonance frequency shift, that is,

$$k = (2\pi)^2 \frac{M}{(1/f_i)^2 - (1/f_0)^2} \tag{5.5}$$

where M is the mass added to the cantilever free end, f_0 and f_1 are resonance frequencies before and after the mass is added, respectively.

This method involves gluing a particle with known mass to the end of the cantilever. It is not trivial even with a detailed protocol available (Ducker et al., 1991, Preuss and Butt, 1998). Furthermore, it is even more troublesome to remove the particle after the spring constant is calibrated. Therefore, force measurements are usually done before the calibration. After all experiments are finished, the calibration is carried out and the measurement results are rescaled based on the correct spring constant. This method is complicating and the cantilever cannot be used after calibration. This method is rarely used in biological application in spite of its accuracy.

All the above-discussed methods do not measure cantilever *in situ*. In practice, the laser may not be aligned to the same position as in spring constant measurement. The difference in laser alignment results in difference in cantilever effective length. With the advances in AFM instrumentation, the thermal noise of cantilever has been used to calculate its spring constant. After deflection sensitivity is calibrated, the cantilever is lifted from the sample surface by at least 100 μm and the random motion of cantilever free end is recorded for a period of time. Ten seconds are usually enough to get accurate results. The power spectral density (PSD) is then calculated by Fourier transformation over the noise recorded, as shown in Figure 5.11. The area under the peak is the mean square of free end amplitude. According to Equipartition theorem (Greiner et al., 1995),

$$\left\langle \frac{1}{2} m\omega_0^2 z^2 \right\rangle = \frac{1}{2} kT \qquad (5.6)$$

where
  m is the effective mass of the cantilever,
  ω_0 and z are the angular frequency and noise amplitude of cantilever free end, respectively,
  k and T are Boltzman constant = 1.3805×10^{-23} Joules/K and absolute temperature in Kelvin,
  $\langle\,\rangle$ means averaging over time, which is obtained by integrating PSD over frequency.

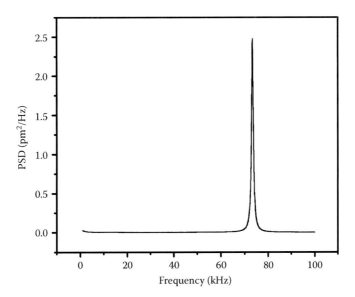

FIGURE 5.11 PSD of thermal noise recorded on a silicon nitride cantilever. Cantilever resonance frequency and Q factor can be obtained by fitting PSD to a harmonic oscillator. By integrating the PSD over the frequency range, the mean oscillation amplitude at the cantilever free end is obtained, which is then used to determine the spring constant.

This method is well accepted because it neither relies on the cantilever geometry, as in methods 1 and 2, nor demands the complicating procedure, as in the mass adding method. After further improvement by Butt and Jaschke (1995) followed by Hutter (2005), the accuracy achieved by the thermal noise method is reliable (Ohler, 2007). In most of today's commercial AFM, this method is automated by software. The frequency range covers from a few kHz to 2 MHz. Virtually, all the commercially available cantilevers can be calibrated with this method. For some ultra-soft cantilever, the resonance frequency in liquid is below 2 kHz. The Z scanner noise and laser pointing noise in some commercial AFM may limit the accuracy in spring constant measurement. In such cases, the spring constant measured in air is a good estimation.

Young's modulus determination through force curve has been extensively studied (Oliver and Pharr, 2004). Hertz, DMT, JKR, and Maugis models are commonly used to measure Young's modulus of materials (Cappella and Dietler, 1999). In all the models, good understanding of the probe shape and size is as important as accurate measurement of force. For sample deformation in a few nanometers, the AFM tip is modeled as a sphere. For polymer samples, a few nanometers deformation produce enough force for accurate measurement in AFM. DMT model with spherical probe is normally used (Israelachvili, 1992).

$$F = \frac{4}{3}E^* \sqrt{R} d^{3/2} + F_{ad} \tag{5.7}$$

where
   F is the force experienced by the probe,
   F_{ad} is the adhesion force between tip and sample,
   R is the radius of the tip,
   d is the sample deformation,
   E^* is the reduced Young's modulus $= E/1 - v^2$, v is the Poisson ratio of the material. Young's modulus
      of the probe material is considered much higher than the sample.

For live cells, the deformation is usually in the range of tens of nanometers, a conical shape is a good estimation of AFM tip. In liquid, adhesion force is negligible. Hertz mode with conical tip shape is normally used for live cell measurement,

$$F = \frac{2}{\pi}E^* (\tan\alpha) d^2 \tag{5.8}$$

where
   F is the force felt by the probe,
   E^* is the reduced Young's modulus of the material,
   d is the sample deformation,
   α is the half-angle of the conical probe.

In practical operation, probes with pyramidal shape are normally used, where the half-angles in the two directions are α and β, respectively. Equation 5.8 should be revised accordingly,

$$F = \frac{4}{\pi\sqrt{\pi}}E^* \sqrt{\tan\alpha \cdot \tan\beta} d^2 \tag{5.9}$$

In AFM measurement, the force is recorded against the scanner Z position rather than the sample deformation. When the cantilever is pushed against the sample, the cantilever bends up while the sample deforms. The sample deformation can be calculated as follows:

$$d = (z - z_c) - (def - def_0)$$ (5.10)

where

z_c is the scanner position at the contact point,

def_0 is the baseline of the cantilever deflection,

z and def are the scanner position and cantilever deflection in the force–distance curve.

Baseline deflection def_0 can be obtained easily from force–distance curve. Contact point z_c for hard materials can be obtained by the intersection of the baseline and linear slope. For soft materials, such as cells, it is not easy to tell accurately where the probe begins to contact the sample, as shown in Figure 5.5. It is a good practice to keep both E^* and z_c as two unknown parameters to be extracted through fitting the force–distance curve to a suitable model. During the fitting procedure, all other quantities are known or measured from the force–distance curve. The reduced Young's modulus extracted from the force curve shown in Figure 5.5 is 50 kPa. In this fitting, Hertz model with conical tip shape is adopted. The half-angles of pyramidal tip are 20° and 17.5°, and spring constant is 0.01 N/m.

Soft cantilever is preferred for force measurement (Shao et al., 1996). Deformation is measured by subtracting cantilever deflection from scanner movement in Z. Soft cantilever usually produce more error in deformation determination. In general, stiff cantilever can produce more accurate Young's modulus results as long as it is sensitive to measure the force that produces 1–2 nm deformation. In real measurement, the error in deflection sensitivity generates more uncertainty in results as it affects the accuracy in both force and deformation. It must be calibrated carefully. In model selection, a rule of thumb is that DMT with spherical tip shape is used for deformation less than the radius of the tip and Hertz with conical tip shape is used for large deformation in tens of nanometers. The Poisson ratio is usually unknown for most of materials. A value of 0.5 is normally used in live cell measurements.

Adhesion force between AFM tip and sample is obtained directly from force–distance curves. There might be a few binding events existing between the AFM tip and substrate. The last force step (jumping off) can be considered single unbinding (Sun et al., 2005). There is a slim chance that two or more unbinding events happen at last force step simultaneously. To rule out multiunbinding event, many force–distance curves are recorded, and the last force steps are extracted and plotted in a histogram. By reading the peak position in the histogram, the most likelihood rupture force of single binding is obtained. Using functionalized probe, adhesion force has been used in measuring hydrophobicity, functional group/molecule identification, antigen–antibody binding force study.

Another application of force–distance curve is to stretch single molecules, including proteins, polysaccharides, and DNAs. By stretching, a protein molecule is unfolded mechanically. This kind of measurements are pursued for a variety of reasons, including some fundamental questions about folding, structure, and how protein sequence contributes to that. Protein structures are traditionally determined by x-ray crystallography. However, it might not be practical to perform this technique on membrane proteins. Force curve is one of the few ways to gain insight into the protein structure. Figure 5.12 is a typical unfolding curve of titin, which is an 8-mer construct of IG27 domain. The saw teeth in the retraction curve are caused by unfolding of the domains. When the stretching force reaches the critical value (marked as A in the figure), one domain is unfolded and force decreases suddenly. After the domain is unfolded (marked as B), the force increases gradually as the unfolded domain extends with further stretching. The second domain starts to unfold upon the critical force reached (marked as C). By fitting the domain extension curve (e.g., from B to C) to the worm-like chain model (Bustamante et al., 1994),

$$F = \frac{kT}{p}\left[\frac{1}{4}\left(1 - \frac{x}{L_c}\right)^{-2} - \frac{1}{4} + \frac{x}{L_c}\right]$$ (5.11)

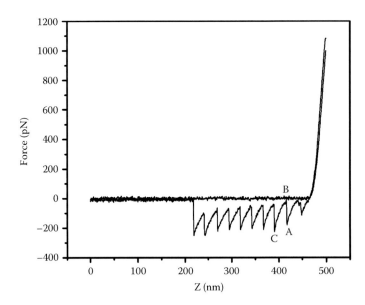

FIGURE 5.12　Typical pulling curve of titin unfolding (8-mer construct of IG27 domain). Each saw tooth corresponds to an unfolding event. The probe is retracted after staying on the surface for a fraction of second to catch a protein molecule. With retraction, the adhesion force pulls the cantilever down until reaching critical force at A, where one domain starts to unfold. The force is released because the molecular length increases suddenly. At point B, the domain is fully unfolded. The unfolded domain extends further with the increase in pulling force. With the continuing pulling, another critical force is reached at point C, where a new domain starts to unfold.

the persistence length p and contour length L_c are obtained. In the fitting, the molecular length x is determined by subtracting cantilever deflection from Z scanner traveling distance, as the deformation calculation described in equation 5.10.

To perform the stretching experiment, one end of the molecule is tethered to a gold substrate by a thiol group. The other end is picked up by a soft probe with spring constant ranging from 0.01 to 0.1 N/m. The probe will pick up many molecules if the concentration is too high. On the other hand, many trial and errors have to be performed to pick up a molecule in case that the concentration is too low. A measure of 50 μg/mL is a good start concentration of titin. Take 25–50 μL of the titin solution and drop onto a fresh gold surface. After 15 min incubation at room temperature, rinse it with 1–2 mL PBS buffer. Then the sample is mounted into an AFM to perform the force measurement. If the chance of picking up is too low, 1 s stay of the probe on the surface will increase the chance significantly.

As well as at a specific point, force measurement can be performed in mapping mode, for example, along a line or in a matrix. After a series of force–distance curves are obtained, information, such as Young's modulus, binding force, deformation, are extracted from the force curves. It is a general routine to map this kind of information to an image to illustrate the spatial distribution. Figure 5.13 is a force mapping over a live cell, (a) is the topography and (b) is Young's modulus distribution over the cell. It is worth noting that force mapping is a slow process. It is usually prohibitive to do a high-density mapping as in normal image mode (512 × 512 pixels). In Figure 5.13, force mapping is done on a 32 × 32 matrix. It takes 1024 s to finish the 1024 force curves.

5.2　New Developments in AFM Technology

AFM was applied to biological materials from its very start (Butt et al., 1990). However, it has less application in biological/biomedical regime than laser scanning confocal microscope and electron microscope in spite of its unique capability of achieving high-spatial resolution even in aqueous solution.

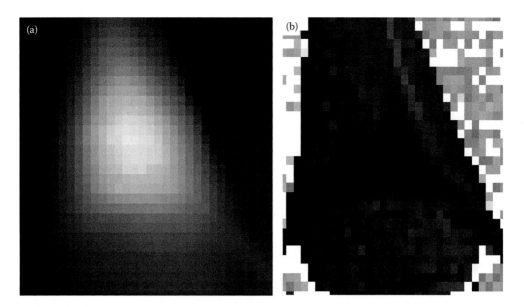

FIGURE 5.13   Force mapping over a live cell. 32 × 32 force–distance curves are ramped with equal spatial interval at a speed of 1 curve/s. (a) is the topography obtained at constant force while (b) is the log Young's modulus mapping. Each modulus is obtained by fitting a force curve to Hertz model with conical probe shape. The scan size is 50 μm × 50 μm. The gray scale of log Young's modulus is from 3(dark) to 5(bright). The edge of the cell is harder than the central part. The bright pixels in (b) are presumably measured on the plastic bottom of the Petridish.

This is because AFM is difficult to use compared with optical microscope and provides less material- or biological-specific information. The recent developments are mainly in easy-to-use and expanding functionalities. AFM is difficult to use because its feedback loop needs to be optimized to maintain constant probe–sample interaction and track the sample surface properly. Force mode can provide biological-specific information by extracting a variety of mechanical properties from force–distance curve, which is typically measured at 1 curve/s. It is not practical to achieve the typical spatial resolution of tapping mode because of the time consumed and poor trigger-force control in traditional force mode. In this section, a new imaging mode based on force–distance curve is discussed.

5.2.1 Peak Force Tapping Enables AFM to Obtain High-Resolution Image as Well as Mechanical Property Mapping at Nanometer Scale

To speed up the force mapping, pulse force mode was introduced in 1997 (Krotil et al., 1999). In this mode, the probe is modulated by fast sinusoidal ramping, unlike the linear ramping in traditional force mode. Pulse force mode improves the efficiency by three orders. However, the poor force control in this mode limits its usage in high-resolution applications, as the probe–sample interaction is in the range of nanonewtons. In force measurement, the cantilever deflection may change with ramping even if there is no probe–sample interaction because,

1. The ramping may not be a perfect axial motion. The nonaxial motion is coupled into the cantilever deflection. This coupling typically includes frequency component of the modulation, and higher harmonics if the piezo response of the Z scanner is nonlinear.
2. When it jumps off from the sample surface in pulling, the cantilever often oscillates at its resonance frequency. The oscillation amplitude is large for soft cantilever. In ambient environment, capillary force increases the total adhesion, making the oscillation more pronounced. The oscillation may not stop at the start of the next ramping.

3. The viscosity of media (liquid or air) can cause deflection even without probe–sample interaction. This effect becomes severe when the cantilever squeezes a thin layer of liquid between the cantilever and sample surface. For example, the gap between the cantilever and sample surface is usually less than 10 μm when the probe is engaged.

With the increase in ramping speed, effects of all these factors become more severe. These parasitic deflections contaminate force–distance curves, making accurate force control difficult, especially at high speed. To overcome this issue, Su et al. implemented a method to characterize and parameterize the parasitic deflections for each instrument (Su and Lombrozo, 2010). After parasitic deflections are removed, a clean force–distance curve can be obtained at the speed of kilohertz by modulating the Z scanner with a sinusoidal wave. Within a modulation cycle, the repulsive force reaches its maximum at the bottom-most point, as in traditional force mode. The peak force (maximum force) is used to control Z scanner to make the probe track the sample surface, as done in force mapping. Compared with force mapping, the peak force tapping mode is more accurate in force measurement (tens of pN can be achieved), and ramping speed is more than three orders faster. It is worth noting that the ramping rate is still far below the cantilever resonance frequency, which is typically from tens of kilohertz to hundreds of kilohertz. The cantilever works in quasi-static mode. This peak force tapping mode has been implemented in commercially available AFM (Pittenger et al., 2010), known as PeakForce QNM.

Peak force tapping mode does not rely on cantilever dynamics, making cantilever resonance frequency searching unnecessary. The operation is also independent of environment. In traditional tapping mode, the cantilever resonance frequency depends on media (liquid or air), temperature and as well as cantilever itself. Therefore, the cantilever must be tuned under the same environment. On contrast, no matter which cantilever, no matter in liquid or air, the operation is the same for peak force tapping mode. As discussed early in this chapter, the cantilever deflection drift owing to environmental influence prohibits the deflection setpoint very close to that of free cantilever, making imaging at piconewtons a challenging job. In peak force tapping mode, cantilever deflection drift is corrected at each ramping cycle, making imaging at piconewtons very stable. Another aspect of easy to use is fully automatic parameter optimization. In peak force tapping mode, the peak force is used in the feedback servo loop and it is linearly related to the Z movement, making it easier to implement automatic parameter tuning with AFM controller. In today's AFM, all the controllers are based on linear algorithm. It is not easy to use linear controller to control nonlinear process, as in tapping mode.

To demonstrate the accurate force control in piconewton range, OmpG membrane (outer membrane protein G) is imaged using peak force tapping mode, shown in Figure 5.14. The OmpG is an intermediate size, monomeric, outer membrane protein from *Escherichia coli* with 14 β-strands. Opening and closing of the membrane pore can be regulated with environment pH. The pore is open at pH = 7, which is clearly shown in Figure 5.14b. Slight force change at 100 pN can cause the membrane structure change. This experiment demonstrates the accurate force control in the range of tens of piconewtons. This kind of image is usually obtained by contact mode with ultra-soft cantilever in almost interaction-free state. Such ultra-soft cantilever is very sensitive to environment change. A slight perturbation will often cause the probe to be lifted or scratch the membrane. With peak force tapping mode, the sample can be imaged continuously for hours.

At each pixel in peak force tapping mode, a clean force–distance curve is obtained after parasitic deflections are removed. With a calibrated deflection sensitivity, force–indentation curve (shown in Figure 5.15) is reconstructed by subtracting deflection from the scanner Z movement, as done in traditional force mode. From this curve, Young's modulus, adhesion force, deformation, energy dissipation in each tapping cycle, and peak force are extracted. These material mechanical properties are mapped into different channels and form images for spatial distribution analysis. With the knowledge of tip geometry the deformation data can be easily converted to indentation hardness (Swadener et al., 2002). Tip geometry can be obtained by either high-resolution electron microscopy or tip deconvolution. In the latter method, a reference sample with sharp features is scanned and the morphological dilation is analyzed to extract the tip geometry (Belikov et al., 2009). It may cause tip wear to scan over such

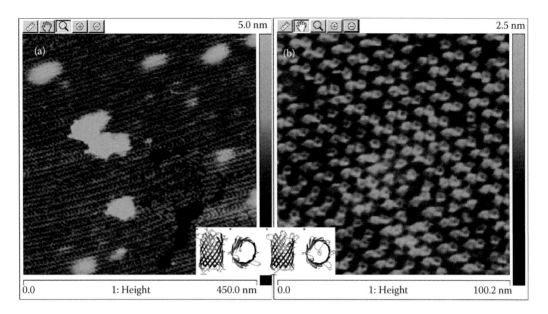

FIGURE 5.14 OmpG membrane imaged with peak force tapping mode in buffer solution at pH = 7. The peak force setpoint is 50 pN. Figure (a) is scanned at 450 nm, showing the stability of the imaging mode. Figure (b) is a 100 nm zoom-in area from A. The open pores in B demonstrate good force control in peak force tapping mode. The inset illustrates the molecular structure of the pores. (Image acquired by C. Bippes and D. Muller, supplied from Bruker image library.)

reference sample as it is rough and very hard. In real practice, deflection sensitivity and spring constant are first calibrated with the same method as in force mode. Tip radius is evaluated by scanning a sample with known Young's modulus. By varying the radius, measured modulus matches the known value. This method is valid though it seems primitive. From the DMT model described by Equation 5.7, the measured Young's modulus E^* is inversely proportional to $\sqrt{R}$ if other parameters are properly

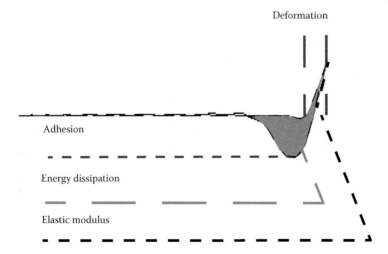

FIGURE 5.15 Mechanical properties extracted from force–deformation curve in peak force tapping mode. The deformation is measured from the zero-force point to the maximum force. Adhesion force is measured from the lowest point to the zero-force baseline. Energy dissipation is the area enclosed by the extension and retraction force curves. Elastic modulus is extracted by fitting the retraction force curve to DMT model.

calibrated. The measured value can be rescaled based on the change in tip radius even after experiments if significant tip wear is observed.

5.2.2 Molecular Recognition by Peak Force Tapping Mode

Recognizing molecules on substrates or cell membrane based on specific binding has been an interest for many researchers since more than a decade ago. Force mapping is the commonly used approach to identify molecules and study the ligand–receptor interaction under different environments (Hinterdorfer et al., 1996, 1998, Baumgartner et al., 2000). As discussed previously, force mapping is in general slow and lack of spatial resolution. To improve the speed, a dynamic recognition microscopy was developed (Raab et al., 1999), where a functionalized probe is oscillated with amplitude (5 nm) smaller than the length of linker molecule (6 nm). When recognition happens, the oscillation amplitude of cantilever is decreased owing to the interaction force. During the operation, the binding is not disassociated by the cantilever oscillation. Disrupture happens only by the lateral pulling force during scan when the distance between the molecule and tip is greater than the linker length. Therefore, the size of recognized molecules is dilated by two times the linker length. With this progress, molecular recognition can be done at the time scale of normal AFM imaging. However, this method cannot provide quantitative binding force. The binding is recognized only by viewing the image.

Another technical challenge is the surface chemistry to functionalize the probe and the substrate. The procedure for probe functionalization includes linkage design and covalently coupling the ligand to the tip surface. The bonding between the ligand and tip surface should be significantly stronger than the ligand–receptor bonds. Otherwise, the functionalized probe may lose its ligand to the substrate, and lose the recognition capability. Poly(ethylene glycol) (PEG) is a well-adopted linker because it is water soluble and nontoxic (Hinterdorfer et al., 2000). PEG has been used in a wide range of applications in surface modification and clinical research. The detailed protocols are described in many articles on force measurement, for example, references (Baumgartner et al., 2000; Hinterdorfer et al., 1996, 1998, 2000; Raab et al., 1999) in this chapter. To illustrate the concept, the procedure is described here briefly. The first step is to clean the probe thoroughly. This is typically implemented by incubating AFM probe in piranha solution (H_2SO_4/H_2O_2, 90/10 (v/v)) for 30 min and rinsing with deionized water afterwards. After being dried by N_2, the probe is subjected to water plasma to generate SiOH groups on the probe surface (Kiss and Golander, 1990). The second step is to bind amines to the tip surface by an esterification protocol described in Swadener et al. (2002) and Belikov et al. (2009). The third step is to conjugate the linker to the amines on the tip surface. The engineered PEG linker has N-hydroxysuccinimidyl (NHS) residue on one end and a functional group to connect the ligand on the other. The NHS group is used to bind the link to the amines. The fourth step is to link the ligand to the functional group on the end of the engineered linker. PDP (2-pyridyldithiopropionyl residue) and NTA (Ni-nitriloacetate) (Conti et al., 2000) are commonly used functional groups.

At the substrate side, different surface-binding strategies must be adjusted to different properties of biological samples. Some receptor proteins strongly adhere to mica surface through hydrophobic or electrostatic forces. In this case, direct adsorption provides sufficiently strong anchoring for recognition experiments. Another option is to use sulfur–gold chemistry. Atomically flat gold surface is prepared as a substrate, as in the titin pulling experiment discussed previously. If silicon or mica is used as substrate for water-soluble receptors, the same surface chemistry as in probe functionalization can be used. For receptors on cells, directly growing the cells on a substrate is typically used. To help the cells attaching to the substrate, some chemistry discussed in the cell imaging section can also be used for force measurement.

To explore the capability of peak force tapping mode in molecular recognition, a model experiment on biotin–avidin binding is carried out. The results are shown in Figure 5.16. Silicon nitride probe with a nominal spring constant of 0.06 N/m is functionalized with biotin. Biotin immobilized on mica is first imaged, the peak force error and adhesion images are shown in Figure 5.16a and b, respectively. Peak force error, similar to the amplitude error in tapping mode, produces fine details about sample topography.

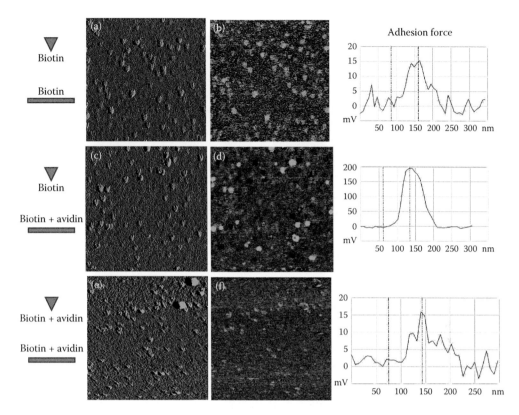

FIGURE 5.16   Avidin recognition by biotin-functionalized probe using peak force tapping mode. (a) and (b) are peak force error and adhesion force images of biotin sample on mica imaged by biotin functionalized probe, respectively. (c) and (d) are peak force error and adhesion force images after avidin is conjugated to the biotin on mica. (e) and (f) are peak force error and adhesion images after the biotin on both mica and probe are blocked with avidin. The line profiles on the right of the images measures the adhesion force of biotin–biotin, biotin–avidin, and avidin–avidin, respectively. (Image courtesy of A. Li, National University of Singapore.)

Individual biotin is clearly discernable in both images. A cross-section analysis on adhesion force image measures the force of 15 mV. After this measurement, the sample is taken out of the AFM and avidin is added. After incubation, the excess avidin is rinsed away. The sample is loaded into AFM and measured again. The peak force error and adhesion images are shown in Figure 5.16c and d, respectively. The adhesion significantly increases to 200 mV. Finally, avidin is added to the probe. After incubation and rinsing, the measurement is conducted. The results are shown in Figure 5.16e and f. The adhesion force reduces to less than 15 mV. It is noticed that force is expressed in millivolts instead of piconewtons. From the previous discussion, force is proportional to the voltage by a scaling factor (deflection sensitivity × spring constant). By calibrating the deflection sensitivity and spring constant, quantitative force measurement is obtained. The biotin–avidin binding is 13 times stronger than biotin–biotin or avidin–avidin bindings. These results show that peak force tapping mode with a functionalized probe is able to differentiate quantitatively the binding forces of different interaction pairs, and it is an ideal tool for molecular recognition.

After the model experiment, CD-36 binding is mapped over red blood cells infected by malaria parasites. After infection, red blood cells develop knobs on the surface, as shown in Figure 5.17b. It is worth noting that the CD-36 binding sites locate solely on the knobs, which agrees with reported results. Another interesting finding is that the adhesion is free of any topographic artifacts, as the debris (marked by white arrow) does not show adhesion force while it is clearly shown in topography. This makes the binding force mapping clean and quantitative.

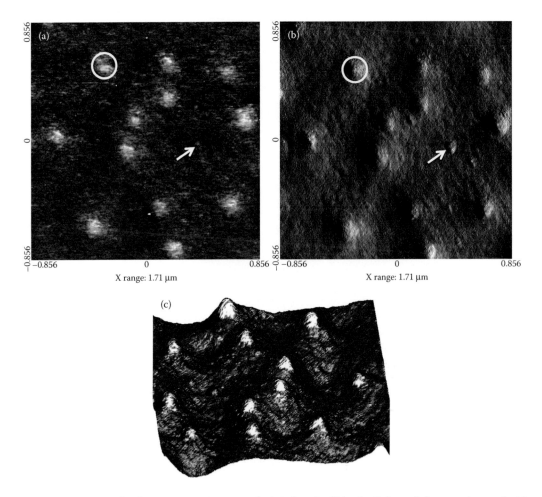

FIGURE 5.17   CD-36 binding site recognition on malaria infected red blood cells by peak force tapping mode. (a) and (b) are adhesion force and peak force error images, respectively. (c) is the overlay of adhesion force over peak force error. Peak force error reveals the detailed topographic information. CD-36 binding sites have one-to-one correspondence with knobs, as shown by the circled area in (a) and (b). The debris marked by an arrow in (b) does not show adhesion force in (a), proving that there is no cross talk between topography and adhesion force. The adhesion force image is topographic artifacts free. (Image courtesy of A. Li, National University of Singapore.)

5.2.3 Integrate AFM with Fluorescence Technology

Advanced fluorescence techniques and a variety of labeling protocols have been extensively used in biological and biomedical applications. Fluorescence techniques in general have high biological specificity and sensitivity, as well as less invasiveness for biological systems under scrutiny. AFM, on the other hand, conveys high spatial resolution than optics in pseudo-physiological environment and mechanical property measurement at nanometer scale. It represents the trend to integrate AFM technology with fluorescence. A hybrid imaging system consisting of AFM and different optical imaging techniques has been reported (Trache and Meininger, 2005). In this report, the transmission optical techniques are sacrificed because the AFM scanner blocks the optical path. Three aspects need to be considered for the integration of AFM with an inverted microscope. (1) *Optical interference*: AFM uses laser beam to detect cantilever deflection or oscillation. The wavelength of this laser should be far away from the fluorescence and excitation wavelengths. Otherwise, fluorescence and AFM will interfere with each other. Biological applications always require liquid environment, which results in many surfaces and interfaces in the optics, causing interference among lasers reflected from different surfaces or interfaces. Laser interference in general has deleterious effect on AFM performance. Therefore, low coherence infrared light source should be chosen in AFM. (2) *Open architecture in AFM design*: In the integration, AFM scanner is mounted onto the inverted microscope. AFM scanner should not block the optical path of microscope or limit the numerical aperture of the whole optical path. Conventionally, AFM is intentionally designed in a compact way to reduce noise and drift. Inverted microscope requires open structure on AFM, requiring AFM designers to compromise AFM performance and compatibility with optical microscope. (3) *Communication between AFM and optical microscope*: The two instruments must be able to exchange control signal to synchronize their data acquisition. In addition, coordinates of AFM and optics must be linked. Otherwise, the colocalization of the two sets of information will be an issue. Fortunately, integrated system is commercially available although there is still some room for improvement.

The applications of the integration between AFM and fluorescence are mainly in three categories.

5.2.3.1 Correlation of AFM Topography with Fluorescence Microscopy

The motivation of this correlation is to use AFM to visualize the detailed structures that are unavailable in fluorescence microscopy, while fluorescence images carry biologically specific information. The two sets of information must come from the same region in order to do the correlation. This is implemented by calibrating the optical imaging system with AFM. A series of optical images are recorded while the AFM scanner moves in a specific pattern. One identical feature in the optical images is then identified either manually or through pattern recognition. Then coordinates in each picture are used to calibrate the XY scales, offsets, angle between AFM, and optical coordinates. In this procedure, AFM is used as reference. This whole process has been automated with user–computer dialog. After the system is calibrated, optical image can be used as a map to navigate the probe to the position of interest. The scanned AFM image is overlaid with optical image automatically, as shown in Figure 5.18. The image is obtained on live endothelial cells at 37°C. The cells are labeled with DiBAC4(3) (bis-oxonol).

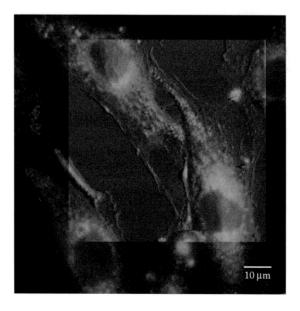

FIGURE 5.18 Overlay of AFM topography with fluorescence image. The green channel is epi-fluorescence of endothelial cells labeled with bis-oxonol. The brownish color is the AFM topography. The fluorescence image is recorded first. The region of interested is then defined on it. AFM navigates its probe to the region and scans accordingly. (Image acquired by C. Callies, H. Oberleithner. Institute for Physiology II, University of Muenster, Germany. Supplied from Bruker image Library.)

5.2.3.2 Force Measurement with the Guidance of Fluorescence Signal

Fluorescence microscopy allows users to identify specific proteins or structures owing to its high biological specificity. From fluorescence microscopy, the position of interest is easily identified. The functionalized probe is easy to navigate to the position accurately if the AFM–fluorescence system has been properly calibrated. This eliminates the necessity to scan the sample before force mapping, preventing the functionalized tip from contamination and saving time as well. In addition, fluorescence signal provides additional information to support that force measurement is really done on specific molecules or structures. Another example is to study cell modulus under anticytoskeletal drug treatment. Fluorescence microscopy monitors the cytoskeleton disruption in real time while AFM measures the cell stiffness. The correlation of the two sets of information can tell which component in the cytoskeleton plays a more significant role in regulating cell stiffness. Figure 5.19 shows that stiffness of live osteosarcoma cells changes with the disruption of tubulin and actin. The results indicate that actin disruption causes more dramatic change in cell stiffness than tubulin.

5.2.3.3 Fluorescence Measurement under Mechanical/Chemical Stimuli by AFM Probe

Mechanical/chemical stimuli usually cause a variety of responses of live cells, including activation of ion channels, dynamics in focal adhesion complex, or cytoskeleton remodeling. In AFM, the probe can move

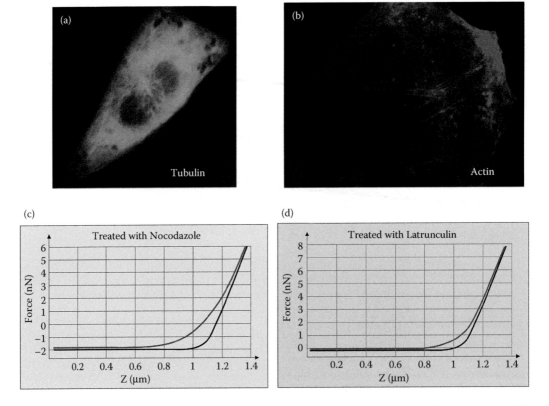

FIGURE 5.19 (See color insert.) Force measurements on live cell after drug treatment. The osteosarcoma cells used in this experiment express green fluorescence tubulin (a) and red fluorescence actin (b). After the cells are treated with Nocodazole, the tubulin fluorescence disappears and force–distance curve is measured, shown in (c). Treated with Latrunculin, the actin fluorescence disappears and force–distance curve is measured, shown in (d). Comparing (c) and (d), the actin disruption causes more significant decrease in cell stiffness. (Image acquired by A. Holloschi, FachHoschule, Mannheim. Supplied from Bruker image Library.)

in XYZ with nanometer positioning accuracy in a user-defined pattern to manipulate nanostructures or do nanolithography on a sample. Vertical and lateral forces can be applied to a predefined position on a sample in a programmed time scale. Under some situations, a sharp probe may damage the sample by poking or scratching, or it is simply too small to sample representative information from the sample. A microsphere is often attached to the free end of a tipless cantilever to increase the contact area. Besides mechanical stimuli, AFM probe is often used to deliver chemicals to a defined spot on the sample. For example, responses to mechanical/chemical stimuli of vascular smooth muscle cells have been studied by using protein-coated bead–cantilever assembly (Sun et al., 2008). With Ca^{2+} fluorescence probe, Ca^{2+} wave propagation under mechanical stimuli of AFM probe has been observed by Oberleithner's group in University of Muenster, Germany. With total internal fluorescene microscopy (TIRF), Reichert's group in Duke University studied the dynamics of cell–substrate contacts under mechanical stimuli of AFM probe. Cytoskeleton remodeling after excessive force stimuli of AFM is observed by Huang's group in Southeast University, China. With the introduction of mechanical/chemical stimuli through the AFM probe, more insight information about the biological process is unveiled.

References

Ang, K. H., Chong, G., and Li, Y. 2005. PID control system analysis, design, and technology. *IEEE Transactions on Control Systems Technology,* 13, 559–576.

Babcock, K. and Prater, C. 1995. Phase imaging: Beyond topography. *Veeco Application Notes,* #11.

Baumgartner, W., Hinterdorfer, P., Ness, W. et al. 2000. Cadherin interaction probed by atomic force microscopy. *Proceedings of the National Academy of Sciences of the United States of America,* 97, 4005–4010.

Belikov, S., Erina, N., Huang, L. et al. 2009. Parametrization of atomic force microscopy tip shape models for quantitative nanomechanical measurements. *Journal of Vacuum Science & Technology B,* 27, 984–992.

Bennett, S. 1993. A History of Control Engineering, 1930–1955. *IEE Control Engineering Series* 47. Peter Peregrinus Ltd, London.

Binnig, G., Quate, C. F., and Gerber, C. 1986. Atomic force microscope. *Physical Review Letters,* 56, 930–933.

Bustamante, C., Marko, J. F., Siggia, E. D., and Smith, S. 1994. Entropic elasticity of lambda-phage DNA. *Science,* 265, 1599–1600.

Butt, H. J. and Jaschke, M. 1995. Calculation of thermal noise in atomic-force microscopy. *Nanotechnology,* 6, 1–7.

Butt, H. J., Wolff, E. K., Gould, S. A., Dixon Northern, B., Peterson, C. M., and Hansma, P. K. 1990. Imaging cells with the atomic force microscope. *Journal of Structural Biology,* 105, 54–61.

Cappella, B. and Dietler, G. 1999. Force–distance curves by atomic force microscopy. *Surface Science Reports,* 34, 1–104.

Cleveland, J. P., Manne, S., Bocek, D., and Hansma, P. K. 1993. A nondestructive method for determining the spring constant of cantilevers for scanning force microscopy. *Review of Scientific Instruments,* 64, 403–405.

Conti, M., Falini, G., and Samori, B. 2000. How strong is the coordination bond between a histidine tag and Ni-nitrilotriacetate? An experiment of mechanochemistry on single molecules. *Angewandte Chemie International Edition,* 39, 215–218.

Ducker, W. A., Senden, T. J., and Pashley, R. M. 1991. Direct measurement of colloidal forces using an atomic force microscope. *Nature,* 353, 239–241.

Fritz, J. 2008. Cantilever biosensors. *The Analyst,* 133, 855–863.

Greiner, W., Neise, L., and Stocker, H. 1995. *Thermodynamics and Statistical Mechanics,* Springer, New York.

Heaton, M., Prater, C., and Kjoller, K. 2004. Lateral and chemical force microscopy mapping surface friction and adhesion. *Veeco Application Notes,* 1–2.

Henderson, E., Haydon, P. G., and Sakaguchi, D. S. 1992. Actin filament dynamics in living glial cells imaged by atomic force microscopy. *Science,* 257, 1944–1946.

Hinterdorfer, P., Baumgartner, W., Gruber, H. J., Schilcher, K., and Schindler, H. 1996. Detection and localization of individual antibody–antigen recognition events by atomic force microscopy. *Proceedings of the National Academy of Sciences of the United States of America,* 93, 3477–3481.

Hinterdorfer, P., Kienberger, F., Raab, A. et al. 2000. Poly (ethylene glycol): An ideal spacer for molecular recognition force microscopy/spectroscopy. *Single Molecules,* 1, 99–103.

Hinterdorfer, P., Schilcher, K., Baumgartner, W., Gruber, H., and Schindler, H. 1998. A mechanistic study of the dissociation of individual antibody–antigen pairs by atomic force microscopy. *Nanobiology-Abingdon,* 4, 177–188.

Hutter, J. L. 2005. Comment on tilt of atomic force microscope cantilevers: Effect on spring constant and adhesion measurements. *Langmuir: The ACS Journal of Surfaces and Colloids,* 21, 2630–2632.

Israelachvili, J. N. 1992. *Intermolecular and Surface Forces.* Academic Press, London.

Kiss, E. and Golander, C. G. 1990. Chemical derivatization of muscovite mica surfaces. *Colloids and Surfaces,* 49, 335–342.

Krotil, H. U., Stifter, T., Waschipky, H., Weishaupt, K., Hild, S., and Marti, O. 1999. Pulsed force mode: A new method for the investigation of surface properties. *Surface and Interface Analysis,* 27, 336–340.

Lv, Z., Wang, J., Chen, G., and Deng, L. 2010. Imaging recognition events between human IgG and rat anti-human IgG by atomic force microscopy. *International Journal of Biological Macromolecules,* 47, 661–667.

Mclean, R. S. and Sauer, B. B. 1997. Tapping-mode AFM studies using phase detection for resolution of nanophases in segmented polyurethanes and other block copolymers. *Macromolecules,* 30, 8314–8317.

Nagayama, S., Morimoto, M., Kawabata, K. et al. 1996. AFM observation of three-dimensional fine structural changes in living neurons. *Bioimages,* 4, 111–116.

Ohler, B. 2007. Cantilever spring constant calibration using laser Doppler vibrometry. *The Review of Scientific Instruments,* 78, 063701.

Oliver, W. C. and Pharr, G. M. 2004. Measurement of hardness and elastic modulus by instrumented indentation: Advances in understanding and refinements to methodology. *Journal of Materials Research,* 19, 3–20.

Pittenger, B., Erina, N., and Su, C. 2010. Quantitative mechanical property mapping at the nanoscale with PeakForce QNM. *Veeco Application Notes,* #128.

Preuss, M. and Butt, H. J. 1998. Direct measurement of particle–bubble interactions in aqueous electrolyte: Dependence on surfactant. *Langmuir: The ACS Journal of Surfaces and Colloids,* 14, 3164–3174.

Raab, A., Han, W., Badt, D. et al. 1999. Antibody recognition imaging by force microscopy. *Nature Biotechnology,* 17, 901–905.

Shao, Z. F., Mou, J., Czajkowsky, D. M., Yang, J., and Yuan, J. Y. 1996. Biological atomic force microscopy: What is achieved and what is needed. *Advances in Physics,* 45, 1–86.

Su, C. and Lombrozo, P. M. 2010. Method and apparatus of high speed property mapping. US Patent.

Sun, W. X., Neuzil, P., Kustandi, T. S., Oh, S., and Samper, V. D. 2005. The nature of the gecko lizard adhesive force. *Biophysical Journal,* 89, L14–L17.

Sun, Z., Martinez-Lemus, L. A., Hill, M. A., and Meininger, G. A. 2008. Extracellular matrix-specific focal adhesions in vascular smooth muscle produce mechanically active adhesion sites. *American Journal of Physiology. Cell Physiology,* 295, C268–C278.

Swadener, J. G., George, E. P., and Pharr, G. M. 2002. The correlation of the indentation size effect measured with indenters of various shapes. *Journal of the Mechanics and Physics of Solids,* 50, 681–694.

Thimonier, J., Montixi, C., Chauvin, J. P., He, H. T., Rocca-Serra, J., and Barbet, J. 1997. Thy-1 immunolabeled thymocyte microdomains studied with the atomic force microscope and the electron microscope. *Biophysical Journal,* 73, 1627–1632.

Trache, A. and Meininger, G. A. 2005. Atomic force-multi-optical imaging integrated microscope for monitoring molecular dynamics in live cells. *Journal of Biomedical Optics,* 10, 064023.

III

In Vivo Applications

6

Magnetic Resonance Imaging to Monitor Implanted Constructs

Nicholas E. Simpson
University of Florida

Athanassios
Sambanis
*Georgia Institute of
Technology*

6.1 Introduction

The field of tissue engineering has opened up novel approaches in the treatment of disease and tissue/organ dysfunction. Oftentimes thought of as futuristic, or science fiction-like, novel tissue engineering approaches from artificial organs to regenerated tissues are poised to come to the forefront of medicine. The reader of this book most likely already possesses a deep appreciation of the significant impact that the field of tissue engineering offers to medicine. What the reader of this book may seek is a deeper understanding of some of the methods and techniques that are currently being used to obtain key information from tissue-engineered substitutes. The reason this topic is of such great interest is because the ability to study tissue constructs while under development *in vitro*, or while functioning *in vivo*, assists researchers and physicians in monitoring function, optimizing design, understanding the interactions of the tissue implant with the host, and possibly also predicting the failure of tissue-engineered substitutes.

In this chapter, we discuss the roles that nuclear magnetic resonance (NMR), and an exploitation of the NMR phenomenon that identifies spatial information, termed magnetic resonance imaging (MRI), play in the monitoring of tissue-engineered devices. To do this instructively, some preliminary effort will be spent on describing the phenomenon of NMR, and how imaging with NMR came to be achieved. This description of NMR will be done in a "classical" rather than a quantum mechanical fashion, so that although the details may not be completely accurate, the reader can still gain a deeper understanding of the advantages and disadvantages of the NMR technique with minimal difficulty, and with no advanced mathematical treatment. Next, we describe some methods that have been developed to enhance the NMR signal and allow users to obtain important information that may otherwise be obscured. The chapter then discusses some approaches to monitor *in vitro* systems, as *in vitro* data can be critical to the basic understanding and development of functional tissue-engineered constructs. The *in vivo* application of MRI to observe implanted constructs is then reviewed, with a mention of several organ/ construct systems of potential interest. The given list of these organ systems is far from complete, and serves merely to demonstrate the power and potential scope of the NMR method. Finally, this chapter concludes with a detailed example of an application, that of monitoring an implantable bioartificial pancreas, as a means to illustrate the rationale for using NMR/MRI, and how this, or a similar approach, can be tailored for the desired application. It is hopeful that through reading the pages of this chapter, the reader will obtain a greater insight into the power of NMR/MRI, as well as gain an understanding of its limitations, so that the techniques described here can be considered for future data gathering. The ultimate goal of this chapter is to provide sufficient information on NMR so that this technique can be considered as a potential tool in the hands of the researcher, physician, and scientist interested in the field of tissue engineering.

6.2 Rudiments of NMR

6.2.1 Definition

For the uninitiated, the term "nuclear magnetic resonance" can bring an uneasiness and fear of the unknown. Often, vague memories of quantum mechanics and difficult mathematical formulae return, and dampen an enthusiasm for the NMR method. However, the rigorous mathematics and physics that can describe the NMR phenomenon do not need to be mastered to appreciate the phenomenon and its utility in spectroscopic and imaging applications. Those "hard-core" aspects can be dealt with later, if necessary. For the audience of this chapter, perhaps persons interested in determining if NMR can be applied to their tissue engineering approach, it is more important to initially describe NMR in uncomplicated terms. To this end, NMR is most easily understood when one considers and defines the three words in this phrase individually.

Nuclear: The word "nuclear" often conveys a negative connotation, and concerns of radioactivity. Yet the word "nuclear" in the term NMR merely refers to the nucleus of the atom, which as the reader will undoubtedly recall, is composed of protons and neutrons. The simplest and most common nucleus used in imaging is that of hydrogen. This simplest nucleus consists of only a proton, and often, one will encounter the term "proton imaging" used interchangeably for MRI. Other nuclei of interest have various combinations of protons and neutrons. Many, but not all, of these combinations result in the nucleus having a nuclear angular momentum, or what is referred to as a nuclear spin (described in more detail in Section 6.2.4). In simple terms, nuclei that have a nuclear spin are observable with NMR; nuclei without spin are not. For example, carbon-12, the most common isotope of carbon (~99% natural abundance), contains a nucleus of six protons and six neutrons. However, it has no nuclear spin, and therefore cannot be observed by NMR. But add another neutron to the nucleus to yield carbon-13 (~1% natural abundance), and this stable nucleus now has a nuclear spin and can be measured by NMR techniques.

Magnetic: Recall that all nuclei must have a positive charge because they contain at least one proton, and each proton has a positive charge (the neutrons have no charge, thus their name). The nuclei that have the nuclear spin can therefore be thought of as a spinning charge. In the macroscopic world, spinning charges generate an electromagnetic field. Although the quantum/atomic realm is not exactly the same as the macroscopic world, the spinning charged nuclei do have a magnetic component. Sometimes, people visualize the spinning nucleus as a tiny bar magnet with the axis of the magnetic field along the axis of rotation. Therefore, spinning nuclei, which have a positive charge, generate a magnetic field, and importantly, can be influenced by magnetic fields. This is the source of the term "magnetic" in NMR.

Resonance: And finally, resonance refers to the particular natural frequency at which this nucleus oscillates or absorbs magnetic energy. In simplistic terms, spinning nuclei each have their own resonant frequency, generally depending on the nuclei, and the magnetic field the nuclei are exposed to, though there are other factors that can influence the resonant frequency, as discussed further in Section 6.2.5.

6.2.2 Requirements

To perform the NMR experiment, the following items are required. The first requirement is a sample that contains the nuclei of interest. In some cases, these nuclei may not be "native" to the sample, but can be introduced, such as nuclei from a tracer. The nuclei of interest must also be in sufficient abundance to be observed readily. This abundance requirement is often at millimolar or higher concentrations. And, of course, these nuclei must be NMR-observable (i.e., have a nuclear spin, as mentioned before). Second, a strong and homogeneous magnetic field is needed, for reasons that are outlined in detail in Sections 6.2.4 and 6.2.5. A third requirement is a radiofrequency (rf) coil. These rf coils come in a variety of shapes and configurations, but their purpose is the same; apply electromagnetic pulses at the resonant frequency of the nuclei of interest, and measure the return of the nuclei to their equilibrium state after these pulses have been removed. The typical NMR/MRI machinery consists of a powerful magnet (today nearly always a superconducting magnet) with magnetic field strengths measured in Tesla (1 T is the equivalent of 10,000 Gauss, and the earth's magnetic field strength is approximately 0.5 Gauss), a console that allows an operator to perform the studies, computer software, powerful gradient systems to manipulate the magnetic field, rf amplifiers, and other electronic equipment that allow for the collection of data as well as the analysis of the signals received from the rf coils. This equipment is highly specialized and quite expensive.

6.2.3 History

In the early twentieth century, fundamental changes in our understanding of the structure and properties of the atom occurred. The resultant standard atomic model, of a positively charged nucleus surrounded by a cloud of electrons, and the mathematical approaches and physics used to describe and calculate atomic behavior, had far-reaching implications that led to the prediction and discovery of the NMR phenomenon. Early work that observed the fine structure of atomic spectra led Wolfgang Pauli, in 1924, to suggest that some nuclei must possess angular momentum, or spin. Because these spinning nuclei are charged, they should produce a magnetic moment, a vector quantity of magnetism with a magnitude and a direction. Importantly, in the presence of a magnetic field, this magnetic moment wobbles (or precesses) at a frequency that depends on the magnetic field strength and local environment. Isidor I. Rabi of Columbia University was a pioneer of the early experiments to attempt to observe the nuclear magnetic phenomenon. In the late 1930s, his successful approach was to send a beam of lithium chloride particles through a homogeneous magnetic field (Rabi et al., 1938a,b). Then, electromagnetic energy was provided by the application of a radio wave. By adjusting the external magnetic field strength, the precessional frequency of the particles in the beam would be altered. When the radio wave was adjusted to this precessional frequency, the nuclei absorbed the radio wave energy, jumped to a higher energy state, and the particle beam was deflected. For these seminal observations using the NMR phenomenon to study nuclear magnetic properties and structure, he was awarded the Nobel Prize in Physics in 1944.

This discovery was the impetus for a number of other studies into the nuclear magnetic resonance phenomenon. The demonstration of NMR in nongaseous materials was first accomplished in 1946, by two independent laboratories on the opposite coasts of the United States. Nearly simultaneously, Felix Bloch of Stanford University and Edwin Purcell of Harvard University demonstrated and described the NMR phenomenon, in water and paraffin wax, respectively (Bloch et al., 1946; Purcell et al., 1946). In 1952, they shared the Nobel Prize in Physics for their pioneering work.

NMR reached deep into the field of chemistry in the early 1950s when the nuclear resonant frequency was determined to be influenced by the local chemical environment. A powerful analytic technique, known as NMR spectroscopy, became an important investigative tool useful in determining chemical composition, environment, and structure. Advances in NMR spectroscopic methods have resulted in increasing power and widespread use, with some of these advances being sufficiently profound to warrant the Nobel Prize. For example, in 1991, Richard Ernst received a Nobel Prize in Chemistry for his seminal work in developing the methodology for high-resolution NMR spectroscopy, including a Fourier analytic approach to MRI (Kumar et al., 1975). And in 2002, Kurt Wuthrich received a Nobel Prize in Chemistry for his pioneering development of multidimensional spectroscopy, whereby the three-dimensional molecular structures of biological macromolecules in solution could be determined (Kline et al., 1988).

6.2.4 Spin

To understand the term "spin," it is important to understand the model of the atom. The atom is composed of a positively charged nucleus surrounded by negatively charged electrons. The components of the nucleus, the protons and neutrons, possess a physical property of angular momentum, or spin. Because these fundamental particles try to form pairs that cancel out their individual angular momentums, not all nuclei have a spin. The net nuclear angular momentum, given the quantum number I, is zero if a nucleus has an even number of both protons and neutrons that are paired (e.g., nuclei with an even atomic mass and number). Fortunately, many elements and their isotopes have nuclei with odd atomic mass or odd atomic number and therefore possess nuclear spin. These nuclei can theoretically be detectable through NMR provided enough of these spins are present in the sample being observed.

This quantum property I can be attributed to the movement of the positive charge over the surface of the nucleus, generating a small magnetic field. Because a spinning charge generates a magnetic field, the spinning nucleus will also generate a magnetic field. When there is no external magnetic field, these nuclei are oriented randomly, so the magnetic fields do not add. But if these nuclei are placed in a magnetic field, they try to align themselves with the field (Figure 6.1). However, they do not all align in the same orientation. In the simple case of protons, there are two alignments possible (i.e., energy states). Under even very strong magnetic fields, these two energy states are nearly identical in population. However, there is a

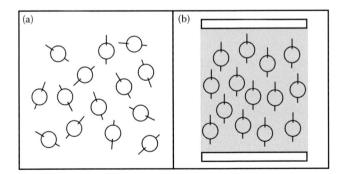

FIGURE 6.1 Representation of nuclei with spin. (a) Without a magnetic field present, the nuclei are randomly oriented, and have no preferred direction. (b) In the presence of a magnetic field (shown in gray), the nuclei align with the field.

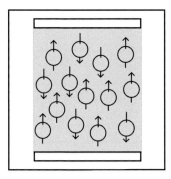

FIGURE 6.2 Representation of the alignment of nuclei in a magnetic field. In this simple case of two energy states, nuclei align either with or against the field. The direction of the magnetic moment is depicted by an arrow. There is a net difference between the two energy states, with a slight preference for the lower-energy state (aligned with the field).

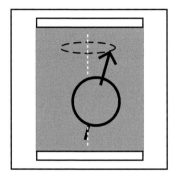

FIGURE 6.3 Representation of an individual magnetic moment. In the presence of a magnetic field (gray), the nuclear magnetic moment wobbles, or precesses, about an axis aligned with the magnetic field. The frequency of this precession is termed the Larmor frequency.

slight preference for the lower energy state that aligns with the field, resulting in a few more spins in this population (Figure 6.2). Therefore, there is a net magnetization with a direction, a vector. This population difference is quite small: at 1 T, there is only about a difference of 10 spins per million. However, this population difference between energy states (the net magnetization moment) increases with an increasing applied magnetic field, thereby increasing the signal observed. This signal increase is a major reason why the NMR machines of today are built with increasingly higher field strengths.

These nuclei not only align with the field, as pictured earlier, their individual magnetic moments also wobble, or precess about an axis at a frequency (termed the precession or Larmor frequency), much like a wobbling top (Figure 6.3). This precession frequency is linearly related to the strength of the external magnetic field, and with the magnets in use today, is a frequency in the megahertz range. When illustrating a number of these precessing moments together (Figure 6.4a), it can be seen that many will cancel others out, leaving a net magnetic moment (Figure 6.4b). It is this magnetization vector that is moved and measured in the NMR experiment (Figure 6.4c), here shown as a consequence of a 90° rf pulse moving the magnetic moment into the xy-plane.

6.2.5 Resonance

Resonance is a term that is probably familiar to the reader, and it was defined earlier in this chapter. Despite this familiarity, a simple example of how resonance energy can be transferred from one system to another may be beneficial, and is provided for those who might be interested. Consider the following

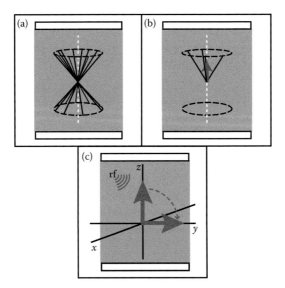

FIGURE 6.4 Representation of the net magnetic moment. (a) In the presence of a magnetic field, the moments of individual nuclei are rotating about a cone. (b) There are more moments in the preferred direction, canceling out those in the opposed state. Moreover, if thought of as a Cartesian space, where the z direction is with the magnetic field, the moments are randomly distributed around the cone, thus canceling out any xy-plane component and leaving a net magnetization (dark arrow). (c) If an rf pulse at the resonant frequency is applied, the net magnetic moment will be tipped, as shown. In this example, a 90° pulse moves the moment into the xy-plane.

macroscopic case of inducing resonance in musical instruments at particular frequencies (or pitches). In this example, a room contains two instruments, a guitar and a trombone. The guitar has six strings of various size, length, and tension to allow each to resonate at particular frequencies. If a matching frequency to one of these strings is played on the other instrument, the trombone, the sound wave generated by the trombone will induce a resonance in the one string, though no finger has plucked this string. This energy transfer through the sound wave only occurs at certain discrete frequencies. Different pitches do not induce the resonance in the string. Matching tuning forks set at a distance from each other are often used to demonstrate this behavior: when one fork is struck, the sound waves will cause the other to resonate. In an analogous manner, an rf wave contains electromagnetic energy that oscillates at a certain frequency. When an appropriate rf wave is applied to a sample that is exposed to a magnetic field, the nuclei having a magnetic resonant frequency that matches the rf pulse will absorb the energy.

6.2.6 NMR Signal

The early NMR machines were called continuous-wave instruments and obtained nuclear resonant information from samples by slowly changing (or "sweeping") the radiofrequency or magnetic field and looking for changes to the magnetization vector. It is analogous to the optical spectroscopy method of sweeping monochromatic radiation through frequencies looking for absorption maxima. In some early NMR machines, the magnetic field was held constant and the rf was changed (much like tuning a radio). In other machines, the rf was held at a steady frequency, and the magnetic field was altered. A major limitation of this "sweeping" technique is the time it takes to complete a single scan: each scan would take minutes to perform. Another limitation of the continuous-wave method is the inherently low signal-to-noise ratio (SNR) obtained through this means.

The continuous-wave method was supplanted in the 1970s by what is called pulsed NMR or Fourier transform NMR. In these machines, the magnetic field is static; thus, the resonance frequencies within

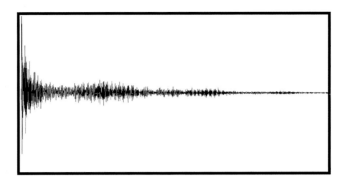

FIGURE 6.5   Free induction decay. Shown is an example of a free induction decay (FID) signal from an NMR experiment. The FID is the *xy*-plane component of the magnetic moments, and the differing frequencies and amplitudes can yield a complex pattern.

the sample are static, and a rapid pulse of rf radiation is applied. This rf pulse has a bandwidth that allows the excitation of all nuclei that resonate within this bandwidth. The moments are moved away from the *z*-axis, and the *xy*-plane component of the magnetization can be measured. As the individual magnetic moments return to equilibrium, the differing frequencies and amplitudes from the moments create a complex decaying wave pattern, called the free induction decay (or FID), an example of which is shown in Figure 6.5. A great advantage of the pulse NMR method is that the process can be rapidly repeated (within seconds, or less), and the signals summed, averaged, and analyzed, thus allowing for large increase in SNR.

6.2.7 Analyzing the Signal

Once the NMR signal is obtained, it must be analyzed. Analysis of the FID is now all done by a computer, but a basic understanding of the general approach can be instructive. Although there are alternative methods used to analyze the FID, such as the statistical Bayesian analysis approach (Bretthorst et al., 1989), the most widely used analytic approach is with the Fourier transform (Ernst and Anderson, 1966). This mathematical operation takes information from the time domain, and converts it to the frequency domain.

This transform method is named after Jean Baptiste Joseph Fourier, a French mathematician and physicist who was born in 1768 and died in 1830. For some time, Fourier held a position as a scientific associate to Napoleon Bonaparte, and his legacy is in his significant contributions using harmonic analysis to solve complex mathematical problems involving heat transfer and vibrations. Until the availability of the electronic computer, application of the Fourier transform (termed FT in the NMR community) to complex data sets such as those generated by the NMR experiment, would be arduous at best, and all but impossible in most cases. However, computers can perform these transforms effortlessly.

In the NMR phenomenon, each discrete magnetic resonance has its own frequency and amplitude, based on the number of spins at that frequency; in essence, each resonance can be considered to be a sinusoidal wave decaying over time. With more resonances combining and interfering, the FID becomes increasingly complex. Fourier believed that time-domain signals (periodic as well as aperiodic) could be considered as a combination of sinusoids. So, complex FID signals can be thought of as a number of interfering sine waves of discrete frequency and amplitude, making them amenable to the FT approach. Consider a simple case of the three notes of differing frequencies, here a major triad, being played simultaneously (Figure 6.6a). In this example, each frequency has the same amplitude. The sum of these tones yields the complex waveform shown in Figure 6.6b. A Fourier transform of this complex wave converts this time domain information (which we would hear as a chord) into the frequency domain, shown in Figure 6.6c. In a similar manner, complex FID signals of an NMR experiment (Figure 6.7a) can be

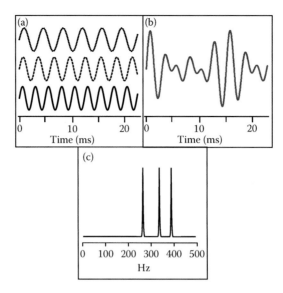

FIGURE 6.6 Simplified description of Fourier analysis. (a) Three sine waves of differing frequency but similar amplitude are shown. They represent the tones of a major chord (such as a C, E, and G), with time moving from left to right. (b) If the three tones are played simultaneously, an interference pattern results from the summing of the peaks and troughs of the individual waves. (c) The Fourier transform converts the time domain information into the frequency domain. Here, analysis of the pattern in (b) results in the three notes being identified by their resonant frequencies (in Hz). As the amplitudes in (a) were the same, the amplitudes in (c) are identical.

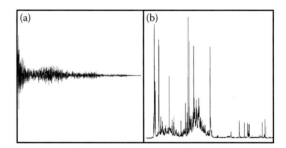

FIGURE 6.7 The FID and FT of NMR spectroscopic data. (a) An NMR proton FID showing the complexity of multiple frequencies and amplitudes of the hydrogen nuclei in the sample. (b) Fourier analysis of this FID yields frequency and amplitude information on the protons within the sample. Certain frequencies arise from the chemical environment, so the type and amount of chemicals can be determined from inspection of the FT data.

separated into the individual frequencies and amplitudes of the contributors through application of the FT (Figure 6.7b).

6.2.8 Dawn of Imaging: Defining Spatial Information

The twentieth century saw the development of many techniques with the ability to see into objects without physically peeling through the layers. X-rays, computed tomography (CT), ultrasound, as well as positron emission tomography (PET) and single photon emission computed tomography (SPECT) methods are among these powerful approaches. Another of the great advances in medical imaging in the last century was the development of MRI. This method, now commonplace in hospitals and research facilities throughout the world, was first proposed and developed in the early 1970s by Paul Lauterbur at the State University of New York, Stony Brook (Lauterbur, 1973), with critical early advances made by Peter Mansfield of Nottingham University, England (Mansfield and Grannell, 1973). These two scientists received the 2003 Nobel Prize in Physiology or Medicine for their discoveries and major contributions to the development of MRI.

The principles behind imaging with NMR are quite different from other imaging techniques. In NMR spectroscopy, a homogenous magnetic field is desired so that the resonances of nuclei in similar environments resonate at the same frequency. With MRI, to create an image, the position of each small picture element, called a pixel, must have its own distinctive resonance or phase so that the nuclear spins it contains can be identified. To achieve this localization, the magnetization of the sample is altered by the use of three orthogonal electromagnet coils, called gradients. These gradient coils impose small linear changes in the magnetic field (thus the name gradient) of the sample (in Cartesian terms, along the x, y, and z directions).

Here is how the most common form of MRI, the 2D Fourier imaging method, works in very simplistic terms. The sample, or subject, is placed in the strong homogenous magnetic field, and this field makes the nuclei of the water protons in the object all resonate at approximately the same frequency, as described in Section 6.2.5. One gradient (the slice selection gradient) is used to select a plane through the sample. It adds a weak linear gradient of magnetism to the field in one direction: this gradient can be along any axis, or even off axis. The gradient creates a band of differing resonant frequencies along its length. Now, an rf pulse applied with a certain bandwidth will induce resonance only in nuclei with frequencies that resonate within the selected bandwidth: those nuclei outside of this slice/plane do not resonate (Figure 6.8). The rf pulse effectively selects the slice from which the image can be formed; in this example, a slice through the z-axis. The slice can be made thinner by using a stronger gradient and/or a narrower range of frequencies (bandwidth). Likewise, the slice can be made thicker by using a weaker gradient and/or a wider range of frequencies. Now that a slice has been selected, the pixels within this plane must be defined. To accomplish this, following the rf pulse, a second weak linear gradient, termed a frequency-encoding gradient, is applied along a direction orthogonal to the slice selection gradient. This second gradient creates bands or columns of spins with distinct resonant frequencies, each column precessing in phase, as shown in Figure 6.9. A third linear gradient, called the phase encoding gradient, is applied for a short period along the last orthogonal direction. When it is applied, the spins along each column speed up by different amounts because of their location in the gradient. When the gradient is turned off, the spins in the column again precess at the same frequency, but now have different phases, depending on position. So in simple terms, each pixel can now be identified by frequency and phase. Application of the Fourier transform identifies these components and can be used to reconstruct the image. A timeline of a simple spin-echo imaging experiment including the timing of the gradients and application of the rf pulse, is shown in Figure 6.10. There are other rf pulses and gradients that can be arranged into an imaging experiment to create 3-D images or add various contrast agents (described

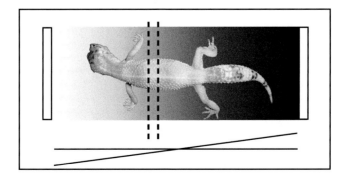

FIGURE 6.8 Depiction of slice selection in NMR imaging. A sample is placed within the magnetic field, causing the spins to precess. A gradient (represented by the crossed lines underneath the magnet) is added along the z-axis. The changing magnetic field is shown as a gradient of gray. A slice of interest is selected by selecting only the frequency range contained in the slice. Other frequencies are too fast or too slow to be affected by future rf pulses.

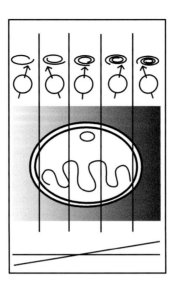

FIGURE 6.9   Representation of another NMR imaging gradient. Once the slice in the sample is selected, another gradient (shown below as crossed lines) orthogonal to the z-axis can be applied. This causes the moments in different columns to precess at different frequencies, shown by the figures on top.

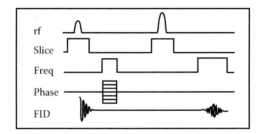

FIGURE 6.10   Simple imaging experiment timeline. A spin-echo sequence used in MRI showing the timing of the rf pulse and gradients. In this sequence, the slice selection gradient is turned on (see Figure 6.8), then the rf pulse applied to affect only the moments within the slice. The FID shows the xy-plane signal. Once the rf and slice gradient are turned off, a frequency encoding gradient (see Figure 6.9) and phase gradient are switched on. After some time, the slice selection gradient is turned on again, and another rf pulse applied (here, twice the power). Lastly, the frequency gradient is turned on, and the moments will refocus to create an echo, shown in the FID. The experiment is repeated n times, with the phase gradient changed with each repetition (thus the phase gradient is shown as a series of steps). A 2-D FT of the entire data set yields the image.

in Section 6.3) to enhance an image and distinguish tissues. For alternative and more complete descriptions of MRI, there are a number of excellent books of varying detail (Abragam, 1961; Farrar and Becker, 1971; Smith and Ranallo, 1989).

6.3 Enhancing the Signal

6.3.1 Contrast

A primary goal of researchers and clinicians alike is to be able to differentiate structure and composition in the tissue construct or bioartificial organ being studied. Although NMR images can theoretically be obtained from any abundant spinning nuclei, typical MRI observes the hydrogen nuclei in the sample, most of which arise from water, with two hydrogen atoms per water molecule. Because of the great amount of water in our bodies, and the high NMR sensitivity of the hydrogen nuclei, images of tissues with exquisite resolution can be obtained. An NMR image can be taken to observe the distribution of this water throughout the sample. This technique is commonly called a proton-density weighted image. This type of image yields anatomical information based on the local concentrations of water.

6.3.1.1 Native Contrast

Proton-density images are extremely useful, and often sufficient for identifying key structures within the sample. Sometimes though, regions that have similar water densities, and thus would appear at the same intensity in a proton-density image, may be differentiated from one another (i.e., given contrast) by exploiting the natural physical differences in these nuclear spins within the sample. This additional tissue contrast is called the native contrast. One characteristic of spins that has been used to enhance the contrast of images is the rate of the return of the spins to their equilibrium (initial) state after they have been moved by an rf pulse. This return to equilibrium is termed relaxation. Because the local environment can have a profound effect on the rate of relaxation, exploiting spatial differences in relaxation of the magnetization is possible. There are two processes that describe this relaxation; they are called the longitudinal and transverse relaxation, or T_1 and T_2, respectively. To better understand these relaxation processes, and how they can be used to enhance an image, a rudimentary description of each of these is provided.

After an rf pulse has moved the net magnetization vector, as explained earlier in Section 6.2.5, this vector will return to its original state. Remember that the magnetization can be thought of as existing in a Cartesian space (with x, y, and z directions). One relaxation process is termed the longitudinal relaxation, and it describes the restoration of the magnetization along the z-axis, as shown in Figure 6.11. This gradual return of the magnetization occurs because the individual spins return from the excited high-energy state back to the lower-energy state. The transfer of energy is to the surrounding areas, or the lattice, and this relaxation is also called spin-lattice relaxation. To measure this, the following pulse sequence can be performed. The magnetization vector is given a 180° pulse, and then a delay time is allowed before the sample receives a 90° pulse (putting the net magnetization vector into the xy-plane). By repeating this process and varying this delay time, a measure of the changing net magnetization vector can be obtained, and a curve of the longitudinal relaxation can be constructed (Figure 6.12). This relaxation curve can be fitted to the following exponential equation:

$$M_z(t) = M_z(0)[1 - 2e^{-t/T_1}] \tag{6.1}$$

The time constant, T_1, gives the process its common name, T_1 relaxation. After the initial 180° flip, at $t = 0$, all magnetization is along the −z-axis. After ~0.69 T_1, the signal goes through the null, and no net magnetization can be measured. After one T_1, ~26% of the magnetization is recovered along the +z-axis; ~90% has returned by three T_1 periods, and nearly 99% by five T_1 periods. If two regions of interest have protons with

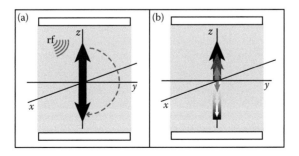

FIGURE 6.11 Demonstration of longitudinal relaxation. (a) In a magnetic field, a net magnetic moment is formed (up arrow). A 180° rf pulse is applied to place this moment along the −z-axis. (b) Over time, this moment makes a return to its equilibrium (initial) state, shown by the series of arrows. To measure the magnitude, a 90° pulse to put the moment into the xy-plane. This experiment is repeated at various times to determine the rate of the longitudinal relaxation.

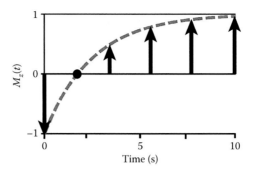

FIGURE 6.12 T_1 relaxation curve. By graphing the net magnetization vector at various times (dark arrows), a curve can be constructed (dashed line). This curve is fitted to an exponential function (see Equation 6.1) that describes the rate of the return of the magnetization along the longitudinal direction.

differing T_1 values, it can be easily understood that contrast between these regions can be attained by giving the sample a pulse of 180°, waiting until one of the T_1 values is nearly nulled (in the example shown, about 3.5 s to null signal from regions with a T_1 curve of "b"), and then giving the net magnetization vectors a 90° pulse (Figure 6.13). Now, spins that are near zero will not give much signal, but the other spins (shown here, those with a T_1 curve of "a") will yield a strong signal. Images that are obtained with pulse sequences that exploit the T_1 differences are called T_1-weighted images, and are commonly encountered in MRI.

A second important relaxation process is the transverse relaxation, and it describes the loss of the net magnetization vector in the xy-plane. This loss occurs due to interactions between the nuclei (due to their spinning, they each produce small local magnetic fields that affect other spins), and is also called spin–spin relaxation. Additionally, local magnetic field inhomogeneities (both internal and external) cause the spins to precess at different speeds and fan out. This "fanning out" of the spins over time makes the net magnetization vector shrink more rapidly than the true relaxation, as shown in Figure 6.14a–c. The rapid dephasing and subsequent exponential decay of the signal yields a transverse relaxation time termed T_2^*, but this decay is not the true transverse relaxation. To measure the true loss of magnetization over time in the xy-plane, the following "spin gymnastics" pulse sequence can be done. Initially, after a 90° pulse (in this example, along the +y-axis), the spins of the net magnetization vector are together, as illustrated in Figure 6.15a. Following some time period t, the spins begin to fan out because of the magnetic inhomogeneities (Figure 6.15b), and the net magnetization in the xy-plane decreases because of the vector sum of the spins, as described previously. To counter

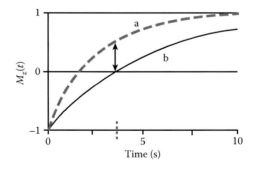

FIGURE 6.13 Using differences in T_1. Two tissues having differing T_1 values (a and b) can be differentiated by choosing a time when there are substantial differences in the magnetizations (double-headed arrow). In this example, by selecting a time when one of the T_1 values is nulled (here ~3.5 s), signal can arise from region "a," while signal from region "b" is essentially zero, resulting in a contrast.

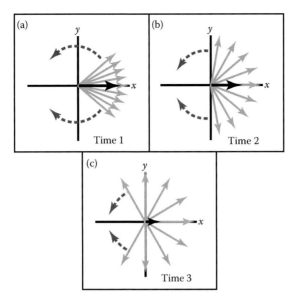

FIGURE 6.14 Magnetization vector in the *xy*-plane. (a) After an rf pulse, the net magnetization vector placed in the *xy*-plane (black arrow) begins to shrink because the individual spins have slightly different frequencies, and begin to fan out. (b) At a later time, more fanning out has occurred, reducing the net magnetization vector. (c) The magnetization vector will shrink toward zero as the spins completely fan out.

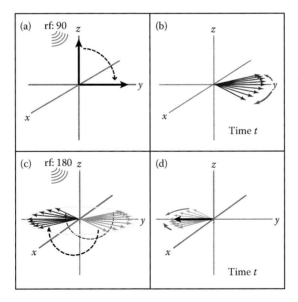

FIGURE 6.15 Demonstration of transverse relaxation. (a) In a magnetic field, a net magnetic moment is formed (up arrow). A 90° rf pulse is applied to place this moment into the *xy*-plane (here, along the *y*-axis). (b) Over time *t*, individual components of this magnetization vector fan out, as described in Figure 6.14. (c) If a 180° rf pulse is applied to these moments, they flip to the opposite axis (−*y*-axis, in this case). (d) These moments continue to move, but now the faster ones on the outside meet up with the slower ones and refocus (black arrow) after another time *t*. The net magnetization vector is smaller than the initial due to transverse relaxation. This experiment is repeated at various times to determine the rate of the transverse relaxation.

this loss, the spins are given a 180° pulse (Figure 6.15c), rotating them to the opposite side of the *xy*-plane. Spins that are faster now catch up with the slower ones. After another time *t*, the spins will all refocus; in this example, along the −*y*-axis (Figure 6.15d). This refocusing of spins is called a spin echo, but the resultant magnetization is smaller due to true transverse relaxation. By repeating this for different times, a curve of this relaxation can be constructed (Figure 6.16) and fitted to another exponential equation:

$$M_{x,y}(t) = M_{x,y}(0)e^{-t/T_2} \tag{6.2}$$

The time constant for this exponential decay, T_2, gives the process its common name, T_2 relaxation. By one T_2, only ~37% of the magnetization remains. This value is around 5% at three T_2 periods, and less than 1% after five T_2 periods. Contrast between regions that have protons with differing T_2 values can be obtained by proper selection of the timing interval (Figure 6.17). Images that are obtained with pulse sequences that exploit the T_2 differences are called T_2-weighted images, and are also commonly encountered in MRI.

Another MRI contrast approach that is becoming more common is diffusion-weighted imaging (LeBihan et al., 1986; Mori and Barker, 1999). This technique uses a series of gradient pulses to exploit

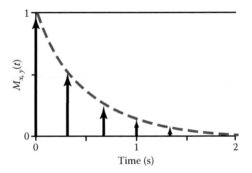

FIGURE 6.16 T_2 relaxation curve. By graphing the net magnetization vector at various times (dark arrows) following the procedure outlined in Figure 6.15, a curve can be constructed (dashed line). This curve is fitted to an exponential function (see Equation 6.2) that describes the rate of the return of the magnetization along the transverse direction.

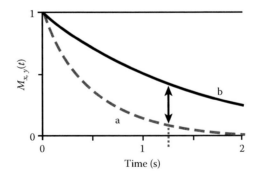

FIGURE 6.17   Using differences in T_2. Two tissues having differing T_2 values (a and b) can be differentiated by choosing a time when there are substantial differences in the magnetizations (double-headed arrow). In this example, by selecting a time when one of the T_2 values is low (here ~1.5 s), much higher signal can arise from region "b" than from region "a," resulting in a contrast.

the differences in the diffusion of water in samples. Regions that have significant barriers will restrict the water motion whereas areas with less restriction allow the water molecules to move greater distances from their origin. In this technique, following the initial rf pulse, additional magnetic gradients are imposed on the sample. Because the phase and precessional frequency are dictated by position, nuclei that diffuse from one region to another will not have the correct phase or frequency to refocus properly, and the resultant signal will be weaker. By varying the time and magnitude of these gradients, and observing the effect on the signal reduction, differentiation of tissues based on their local diffusive character can be achieved. For example, regions of high diffusivity within the sample can be rendered dark (i.e., less signal) because the water molecules that are in an area that is not bounded can diffuse away, taking their contribution to the NMR signal. Similarly, regions of low diffusivity can be made bright (i.e., more signal) because the water molecules that are in an area that is bounded cannot diffuse far, and are not affected by the imposed gradients, thus the region retains its signal. By choosing the magnitude and duration of the additional gradients, flexibility in the extent and type of enhancement of tissues can be accomplished. Furthermore, additional information can be obtained using diffusion approaches. For example, diffusion often is not isotropic, but directional, and the direction of the diffusion can yield important information on the structure of the sample being imaged (Moseley et al., 1990). This imaging approach is termed diffusion tensor imaging. Another NMR technique, termed magnetization transfer (MT), is used to discriminate the structural integrity within a sample (Wolff and Balaban, 1989). The MT method determines this integrity by measuring the rotational freedom of water molecules. Free water molecules (i.e., not bound or hydrating other molecules such as proteins) have fewer restrictions on their molecular motion. These molecules tumble faster, and consequently have a narrow resonance frequency band. Conversely, water molecules bound to other molecules tumble slowly, and have a broader frequency band of resonance. Therefore, rf pulses can be applied at particular frequencies to excite only the restricted water nuclei. When this excitation energy is transferred to free water molecules, the tissue structural integrity can be measured. It should be kept in mind that in MRI, when native contrast methods, which often require extensive repetition and time delays, are added to standard imaging sequences, a trade-off between spatial resolution and total imaging time must be considered.

6.3.1.2 Enhanced Contrast

Contrast can also be achieved through the introduction of a chemical or compound that influences the NMR signal in the area where it is present. These agents are called contrast agents, and they are commonly used to alter the sample's relaxation rates (T_1 and T_2, discussed before) due to the agent's intrinsic paramagnetic or superparamagnetic properties that influence nearby nuclear spins. This NMR approach is often used to spatially discriminate tissues and distinguish tissue characteristics. Most contrast agents include a transition metal; commonly lanthanide series metals such as complexes that contain gadolinium (used extensively in clinical applications), or iron oxide particles, oftentimes specially coated and containing colloidal iron oxide. There are a number of excellent reviews outlining the manufacture, mechanism, and application of NMR contrast agents (Gupta and Gupta, 2005; Gupta et al., 2007; Strijkers et al., 2007; Sun et al., 2008a).

Magnetic nanoparticles/contrast agents have been used in a number of NMR applications: indeed, gadolinium chelates injected directly into the bloodstream are used routinely in many clinical MRI studies. Applications for magnetic particles include clinical detection of cancer, such as lymph node metastases (Harisinghani et al., 2003), liver cancer (Semelka and Helmberger, 2001), and brain tumors (Neuwelt et al., 2004; Sun et al., 2008b). Cardiovascular anomalies such as plaque formation have also been studied using MRI in conjunction with contrast nanoparticles (Kooi et al., 2003; Sosnovik et al., 2007; Wickline et al., 2007). Directly labeling cells with contrast agents and detecting them through NMR methods after implantation is another approach. This has been accomplished in a variety of cell types (Cahill et al., 2004; Dodd et al., 1999; Evgenov et al., 2006; Foster-Gareau et al., 2003; Kriz et al., 2005). Recently, Tseng et al. (2010) created novel gadolinium nanoparticles to successfully track stem cells with MRI. And Moore et al. have used MRI to track labeled immune cells and noninvasively follow them during their infiltration of the pancreas during type 1 diabetes development (Medarova and Moore, 2007; Moore et al., 2002, 2004).

To date, only a few studies have addressed enhanced contrast approaches for implanted constructs (Constantinidis et al., 2009; Terrivitis et al., 2006). However, a novel approach to generate three-dimensional cellular configurations was through guiding labeled cells containing magnetic particles with magnetic fields (Dobson et al., 2006; Ino et al., 2007; Ito et al., 2005). Another recent advance by Karfeld-Sulzer et al. was to include protein polymer contrast agents in their constructs. This material allowed them to use NMR techniques to monitor the structural integrity of the implant over time (Karfeld-Sulzer et al., 2011). Certainly, advancements in approaches to labeling cells and constructs will continue, and NMR studies that take advantage of these advances will provide valuable data concerning the structure and function of implanted constructs.

6.3.2 Coils

Another means to enhance signal is through the proper selection of the rf coil. The coil is an antenna that transmits the radiofrequency into the sample. Recall that there is an electric and magnetic component to radio waves. The coil transmits the magnetic pulse at a frequency of interest into the sample, and receives the signal from the sample, as described earlier. Sensitivity and magnetic field homogeneity are two components of coils that should be optimized for any experiment. Coils come in a variety of shapes, sizes, and designs. It is beyond the scope of this chapter to give a detailed account of NMR coil design, but some information is offered here to introduce the reader to the subject.

There are three frequently encountered rf coil designs. These are the solenoid, which is shaped similar to a spring; the saddle coil (also called a Helmholtz coil); and the surface coil, which is a loop of wire of few turns that is placed onto the sample. Birdcage coils, multiple array coils, and other coils that can transmit power into a sample are also encountered. Each coil type has advantages and disadvantages. For example, the solenoid has the advantage of having great SNRs over a sample volume. However, the magnetic field supplied by a coil must be orthogonal to the magnetic field of the NMR machine: the solenoid may not be ideal for a very large sample within the magnet, and a saddle coil may provide a better solution for the sample under study. The selection of rf coil may depend on the size and orientation of the object to be studied. For isolated tissues, a small solenoid may be best. When performing imaging or spectroscopy on a small implant in a large body, a surface coil may be the better choice. There are even other novel microcoil designs that enhance the signal of implanted cells, such as microcontainers that house small amounts of tissue, and also act to attenuate or boost signal precisely at the site of the cells, allowing for enhanced detection (Gimi et al., 2005, 2007).

Our lab has recently used an implantable coil approach to monitor implanted bioartificial pancreatic constructs (Volland et al., 2010), thus obtaining signal only from the tissue that the implanted coil surrounds. Early *in vivo* studies using implantable coils connected the implanted coil to the outside probe circuitry with wires (Arnder et al., 1996). Although that scheme enhances the sensitivity of the study by obtaining signal right at the source of interest, it can lead to infection at the site of the wire entry. However, it is possible to inductively couple an implanted coil to an external surface coil (Hoult and Tomanek, 2002; Schnall et al., 1986; Silver et al., 2001; Wirth et al., 1993), provided the external coil is of appropriate orientation and proximity, thereby wirelessly transferring the signal from the implanted coil to the external probe circuitry. In our hands, this inductively coupled method results in a twofold sensitivity improvement over that obtainable with a surface coil, allowing for significant information gain from an implanted construct (Volland et al., 2010). This signal gain allows for improved spectroscopy measurements, thereby increasing the potential to study less-sensitive but important nuclei, and the ability to obtain better-quality images of an implanted construct.

6.4 Advantages and Disadvantages of NMR/MRI in Tissue Engineering

There are a number of imaging techniques that can be used to monitor tissue-engineered devices, as outlined in this book. It is not the purpose of this chapter to disparage any of these other useful and powerful techniques, but to give the reader an introduction to the technique of NMR and MRI. However, to better appreciate the advantages and disadvantages of NMR, and determine whether it is a technique that would be appropriate for a given system or application, it is useful to give a brief outline of a few of these alternative methods, and their respective strengths and weaknesses.

6.4.1 Other Techniques

Optical imaging: No longer a method of just a light source and a camera, optical imaging was revolutionized by the incorporation of fluorescent (e.g., green fluorescent protein, aka GFP) (Chalfie et al., 1994; Gee et al., 2002; Hadjantonakis et al., 2003) and luminescent (e.g., luciferase) (Greer III and Szalay, 2002) proteins in cells, the development of novel fluorescent agents (Gee et al., 2002), and new detection techniques (Piston, 1999). Optical imaging techniques are highly sensitive and can generate images with subcellular resolution. These techniques are also excellent for *in vitro* applications. *In vivo* applications are more difficult: they are limited by the short transmission length of light through body tissues (visible light released by fluorescent probes is attenuated nearly 10-fold with each centimeter of tissue depth), and consequently, they are not as well suited to image deep-seated tissues or organs, at least not easily in a noninvasive manner. Additionally, there is tissue autofluorescence that adds to the background signal and greatly reduces SNR. However, significant advances in optical probes, such as quantum dots, and techniques, such as optical coherence tomography (Fujimoto, 2003; Han et al., 2005; Zagaynova et al., 2002) have broadened the field of optical imaging, allowing it to be used more widely in assessing biological function and structure.

Radionuclide imaging: The two chief imaging techniques that detect radionuclides are PET and SPECT. These techniques are commonly used in nuclear medicine to diagnose a variety of pathologies. Though they are highly sensitive, and can detect trace amounts of a radiolabel, both PET and SPECT suffer from low spatial resolution and require radioactive agents to be administered and targeted to the cell or tissue of interest. Even microPET systems designed to provide enhanced resolution in small rodents have a spatial resolution of approximately 1 mm (Cherry et al., 1997; Correia et al., 1999; Tai et al., 2003).

X-irradiation: X-ray and CT are techniques familiar to most. These methods detect ionizing radiation that passes through the subject, and take advantage of the radio-opacity of dense tissues for contrast. They excel at obtaining structural information, particularly in bone, cartilage, and tissue-engineered scaffolds. Because of the radiation required to obtain the information, repeated studies must be planned carefully. They may not be the method of choice for analyzing soft tissues. However, machines have been built to perform combinations of PET, CT, and MRI to take advantage of their complementary strengths (Beyer et al., 2000; Catana et al., 2008; Goetz et al., 2008; Pichler et al., 2008).

Ultrasound: This technique measures reflected high-frequency sound waves to obtain structural information from within tissues. These waves penetrate deeply, and the technique is inexpensive, noninvasive (or minimally invasive), and repeatable. Novel contrast agents (Stride and Saffari, 2004) and microbubble encapsulation approaches (Hope Simpson et al., 1999, 2001) have extended the abilities of ultrasound imaging. Approaches toward monitoring the viscosity, elasticity, and stiffness of implanted devices have been developed (Doyley et al., 2005; Righetti et al., 2005) as a means to measure cellular density, cell rearrangement, and structural integrity. Disadvantages include the need for a skilled sonographer to obtain good images and difficulties getting high-resolution images from deeply embedded tissues due to obstructions. Ultrasonography is continuing to develop methods that may be useful in studying implanted constructs.

6.4.2 Advantages of NMR/MRI

Given the rich history of NMR spectroscopy, it would be imprudent to neglect its power in tissue engineering applications. While other imaging methods have certain strengths, NMR can yield chemical information, in addition to the imaging information. Indeed, these two desirable strengths of NMR can be combined such that chemical information related to a spatial location can be obtained! Arguably the most distinctive advantage of NMR/MRI is the ability to acquire structural and metabolic information from the object under study. Both imaging and spectroscopic information can be obtained through magnetic resonance; it is a versatile technique that is well suited to address many experimental concerns. Because magnetic fields penetrate uniformly throughout the sample, the technique can be used to study surface or deep-seated tissues: there are no issues with depth or interposing obstructions. Another advantage is that it is a noninvasive method that does not require the genetic modification of the cells (e.g., expression of GFP or luciferase) or the introduction of radioactive labels (e.g., PET agents). The method is safe, and can be done repeatedly on the same subject/sample. This benefit allows for the development of long-term studies to observe temporal effects. MRI can also yield images with high resolution.

The spatial resolution attained by MRI is limited by the SNR. Improvements in SNR can be achieved by using stronger magnets. Smaller bore magnets also have stronger gradient coils, enabling high spatial resolutions to be attained. Although current clinical machines have upper magnetic field strengths of 3 T, research instruments are allowed to achieve much higher fields, and therefore greater resolution. For example, a magnet with a 17 T magnetic field strength and gradient strength of 100 Gauss/cm can generate images with an isotropic spatial resolution of ~10 μm within an acceptable image acquisition time. It is this excellent resolution that has sparked the field of NMR microscopy (Blackband et al., 1999) and has allowed the investigation of single cells or neurons (Aguayo et al., 1986; Schoeniger et al., 1994).

Another great advantage of MRI is that one can exploit many different contrast techniques to differentiate tissues, as discussed in Section 6.3. The native NMR contrast techniques such as diffusion, T_1, T_2, and T_2^* allow for images to yield complex and complementary information regarding the structure of the item under study. Other agents added or targeted to tissues can yield additional important information. These contrast approaches do come at a price though, there is a loss of SNR. However, a satisfactory balance between SNR, resolution, and total imaging time can usually be achieved.

Lastly, because the NMR phenomenon occurs for a number of different nuclei, and not just protons, there are many biologically important compounds that can be studied. The acquisition of NMR images from other metabolites is possible (spectroscopic imaging) (Maudsley et al., 1983; Zhou et al., 2008); however, the concentration of the metabolites under study must be in the millimolar or greater range to be of value. There are a number of potentially useful and important metabolites that have such concentrations *in vivo*, including but not limited to choline, *N*-acetyl aspartate, lactate, ATP, and phosphocreatine. The continued advances in magnet and gradient technologies have increased the ability of magnetic resonance approaches to detect these metabolites, and thus have broadened the potential of NMR in the study of physiological events in tissues, organs, and tissue-engineered devices.

6.4.3 Disadvantages of NMR/MRI

NMR is not without its disadvantages, however, and careful consideration should be made to determine if the technique is the appropriate one to apply to the study. A major limitation is the relative insensitivity of the NMR phenomenon. Whereas optical and radionuclide imaging techniques can detect tracer quantities, the NMR signal arises from the magnetic moments of the spinning nuclei. However, the magnetic moments of these nuclei do not all align together; that is, they are not all in the same energy states, or "pointing in the same direction." Rather, the signal is a result of the net magnetization vector (the sum of all the individual magnetic moments' magnitude and direction). Because there is only a slight preference for one energy state (the lower) over another, the resultant magnetic moment is quite weak. For example, protons in a strong homogenous magnetic field of 1 T (approximately 20,000 times the earth's magnetic field), and at room temperature, differ in energy states (or "direction") by only a few per million. It is only because there are so many protons in tissue ($\sim10^{23}$ per g tissue) that NMR imaging is so successful. For nuclei that are not as abundant (nor as sensitive) as protons, the issue of sensitivity is more critical, and concentrations in the millimolar range must be present to be detected.

Another disadvantage is the expense of performing experiments with NMR. Costs to operate and maintain NMR equipment are high, and these costs are often passed on to the users through service fees. The stronger and more "cutting-edge" the magnet, the more expense is to be expected. The equipment is also highly specialized, requires experienced staff to operate, and often not readily available. This limited access may also apply to relaxation agents, as delivery of them to the target of interest may not be possible.

6.5 MRI Monitoring of Tissue Constructs *In Vitro*

6.5.1 Bioreactors and Scaffolds

NMR approaches have been successfully used to study a variety of engineered constructs, including bone, cartilage, bladder, liver, and pancreas. It should be noted that NMR methods may be appropriate for many other tissue types and configurations than those discussed here, provided the construct is composed of materials conducive to NMR techniques. *In vitro* studies utilize NMR-compatible bioreactors for a wide range of research such as fundamental investigations on cellular biochemistry in different culture configurations or tissue constructs; tissue growth, remodeling, and morphogenesis; and monitoring or tracking contrast agents incorporated in cells or biomaterials, or delivered to the cells in a dynamic fashion. In many cases, *in vitro* bioreactor studies are used to determine key sensitivity parameters so that NMR-based monitoring of tissue constructs can be extended to *in vivo* applications.

The low sensitivity of NMR necessitates the presence of high cell densities in the volume from which signal is acquired. Bioreactors that accomplish this are hollow fiber (Gillies et al., 1993; Mancuso and Fernandez, 1990) and fixed-bed bioreactors (Constantinidis and Sambanis, 1995; Papas et al., 1999a,b; Thelwall and Brindle, 1999) supported by perfusion circuits. As dissolved oxygen (DO), but not other nutrients in the medium, is depleted in one pass through the bioreactor, perfusion circuits generally implement a medium recirculation loop with an oxygenator that replenishes the medium DO after each pass through the bioreactor. Simple recirculation results in gradual depletion of the other nutrients in the medium. Hence, for long-term studies, the recirculating medium needs to be replenished with fresh medium either periodically or continuously by a feed/bleed loop. In the latter case, the system may attain a steady state, in the sense that the concentrations of nutrients and metabolites in medium stabilize at certain residual levels. These levels may drift gradually with time because of net cellular growth or death in the bioreactor, which changes the overall metabolic activity in the system.

Although most of the *in vitro* studies have been spectroscopic in nature, several imaging studies have also been performed. Spectroscopic studies involve measuring intra- or extracellular metabolites by ^{1}H, ^{31}P, or ^{13}C NMR to elucidate different aspects of cellular metabolism, whereas imaging studies enable visualization of construct architecture, construct or tissue remodeling, the distribution of tracking agents, or flow patterns of the perfusion medium through the bioreactor.

Hollow fiber units supported by perfusion circuits have been used to study cultures and the metabolism of hybridomas by ^{31}P and ^{13}C NMR spectroscopy (Gillies et al., 1991; Mancuso and Fernandez, 1990; Mancuso et al., 1994, 1998). Similar systems loaded with rat hepatocytes have been implemented to study the transport of hepatobiliary contrast agents by MRI (Planchamp et al., 2004a,b). These culture systems maintained hepatocyte viability, exhibited no adsorption of contrast agents in the absence of hepatocytes, and were able to delineate the dynamics of contrast agents that enter the cells from those that remain extracellular. In other studies with Chinese hamster ovary cells grown on macroporous microcarriers in a fixed-bed bioreactor, diffusion-weighted ^{1}H MRI was used to evaluate the cellular distribution, and diffusion-weighted ^{1}H NMR spectroscopy was implemented to assess the bioreactor volume fraction occupied by the cells. The increase in the latter correlated with the increase in total ATP measured by ^{31}P NMR spectroscopy. The imaging studies revealed that cell growth was maximal at the periphery of the microcarriers, as expected from mass transfer considerations. Engineered bladder tissues grown on collagen scaffolds or previously decellularized matrices have also been studied with MRI methods (Cheng et al., 2005). Using MRI in conjunction with gadolinium-labeled contrast agents determined the positive vascularization effects of vascular endothelial growth factor (VEGF) on the bladder constructs (Cheng et al., 2005). The same group also correlated the VEGF dose to the vascularity by using contrast agents of differing molecular weights (Cheng et al., 2007), and determined that by adding hyaluronic acid to the bladder matrix, cellular development was enhanced, and the strength and hydration of the bladder increased. These changes altered the T_2^* relaxation rate of the tissue, meaning that MRI could be used to monitor and track changes in bladder development (Cheng et al., 2010).

As discussed in Section 6.4.1, NMR techniques are not often the first choice when studying tissues such as bone and cartilage, which are well suited for radioimaging methods. However, NMR techniques have been implemented in studies of both the cells and the matrix deposition they provide. MRI has been successfully implemented to monitor cartilage (Chen et al., 2003; Irrechukwu et al., 2011; Nugent et al., 2010; Potter et al., 1998; Ramaswamy et al., 2009) and bone (Wahburn et al., 2004; Xu et al., 2006) tissue growth and changes in tissue composition in bioreactors. In studies with hollow fiber units inoculated with chondrocytes, neocartilage formation was followed by ^{1}H NMR microimaging, whereas the fixed charge density of the tissue was evaluated by the gadolinium exclusion method (Chen et al., 2003; Potter et al., 1998). Results indicated the ability of these NMR methods to track increases in tissue volume, cellularity, and macromolecular content, and to reveal regional variations in cell density and sulfated glycosaminoglycan content. Isolated chondrocytes cultured in a collagen scaffold have been studied over time to determine NMR endpoints that can be used to compare cultured cartilage with native tissue (Nugent et al., 2010). Ramaswamy et al. (2009) labeled chondrocytes with iron oxide nanoparticles as a contrast agent and monitored them as they were cultured in a hollow fiber bioreactor or in a photopolymerized hydrogel. The labeling allowed for MRI tracking, and did not alter the viability or function of the cells. Similarly, in experiments with hollow fiber bioreactors seeded with primary osteoblasts, magnetic resonance microscopy (MRM) was used to track tissue growth and mineralization around fibers over a period of 9 weeks (Chesnick et al., 2007). It was found that the noninvasive MRM measurements were compatible with the spatial mapping of the tissue by Fourier transform infrared (FTIR) microspectroscopy, thus indicating that MRM is indeed capable of monitoring bone formation *in vitro*. Other recent MRM experiments determined that the NMR signal decreases as the osteoblasts fill scaffolds with matrix (Wahburn et al., 2004; Xu et al., 2006). Also, Buschmann et al. (2011) used MRI techniques to determine the relaxation characteristics of bone tissue growth in constructs seeded with mixed cell populations. As the bony material developed, the capability to perfuse through the construct decreased; this compromised the ability to distribute the contrast agent gadolinium by perfusion into the construct and to alter the relaxation parameters of the NMR signal.

6.5.2 Example Studies: Bioartificial Pancreas

Over the past several years, our laboratories have been studying encapsulated systems of insulin-secreting cells in fixed-bed reactors by ^{1}H MRI and by ^{31}P and ^{1}H NMR spectroscopy, as well as by ^{19}F NMR spectroscopy of perfluorocarbons in the encapsulation matrix. The experiments described here serve to illustrate bioreactor studies of encapsulated insulin-secreting cell systems that can be used as immunoprotected implants for the treatment of insulin-dependent diabetes. NMR imaging and spectroscopy studies in bioreactor configurations were performed to acquire fundamental biochemical information on these systems and to determine critical parameters that would allow noninvasive monitoring of such constructs *in vivo*. The latter is significant, as it permits monitoring of an implant over time in the same animal; it also allows establishing an important link between construct implantation and endpoint animal physiology, in this case regulation of blood glucose level.

Initial studies by Constantinidis and Sambanis (1995) established the feasibility of using a fixed-bed bioreactor supported by a perfusion circuit to study the bioenergetics of alginate-encapsulated, recombinant insulin-secreting murine pituitary AtT-20 cells over more than 60 days in culture. A main objective of that study was to evaluate whether the encapsulated system remained stable over prolonged periods of time, even though the encapsulated cells were capable of robust proliferation. Results showed that the levels of high-energy phosphates measured by ^{31}P NMR, including phosphocreatine (PC) and nucleotide triphosphates (NTPs), remained relatively stable over 40 days, as did also glucose consumption and secretion of insulin-related peptides. ^{1}H MRI was used to confirm the uniform packing of the bed over time and, when processed so as to obtain phase-sensitive images along the bioreactor axis, to visualize flow and ensure the absence of any significant flow channeling across the bed.

To further investigate the stability of encapsulated proliferative cells, Papas et al. (1999a,b) used a similar bioreactor and perfusion circuit to study the bioenergetics of alginate-encapsulated murine insulinoma βTC3 cells in the long term and under changes in glucose and oxygen concentrations. Those changes were implemented to mimic physiologically relevant conditions. Metabolites followed by ^{31}P NMR included PC, NTP, and inorganic phosphate (P_i). In this system, net cell growth occurred during the initial 10 days of culture, as manifested by the approximate doubling of measured metabolic and secretory rates and of the intracellular metabolite levels. These and subsequent (Gross et al., 2007a) studies indicated that in these bioreactor systems, the number of viable, metabolically active cells that can be supported in a certain bead volume is mainly determined by the available oxygen in the perfusion medium. The glucose transients of this and other (Papas et al., 1997) studies allowed us to elucidate fundamental aspects of the fuel hypothesis in βTC3 cells, that is, on the correlation of the rate of insulin secretion and the levels of ATP-related metabolites in cells. In this, the application of NMR enabled the noninvasive and nondestructive assessment of intracellular NTP and P_i. This ensured that these metabolites were not subjected to any inadvertent changes during cell extraction and sample processing procedures, which may occur during conventional assays.

6.6 MRI Monitoring of Implanted Constructs *In Vivo*

6.6.1 Overview of *In Vivo* NMR/MRI Studies

In vivo studies of implanted constructs using MRI are not yet widespread, but they are gaining in number as researchers come to discover the strengths of this technique. Most of the studies using NMR are developmental, and are predominantly *in vitro* efforts to study construct architecture and cell remodeling, as discussed in Section 6.5. Below are some examples, not exhaustive, of *in vivo* NMR studies that have been performed. A major section on the NMR application toward studying the bioartificial pancreas concludes the chapter.

Cardiac: One potential tissue-engineered approach to treat myocardial infarction is to strengthen the weakened heart wall by adding support from a scaffolding biomaterial. Recently, Stuckey et al. (2010) grafted patches into normal and infarcted rat hearts in an *in vivo* MRI study of heart patch scaffold materials. They demonstrated that MRI was useful in visualizing and evaluating the scaffold, and determined that one of the materials fractured due to the stress of cardiac contraction and was not a suitable biomaterial. Through MRI techniques, they could noninvasively locate the patch, observe the degradation of the patch material over time, identify changes in the regrowth of cardiac tissues, and evaluate the impact of the materials on the heart muscle, and importantly, use the technique in the further development of a more suitable heart patch biomaterial.

Bone: Bone loss or traumatic injury can lead to an inability for the body to naturally replace the damaged bone. Because of this, tissue engineering substitutes for bone grafts have been considered. MRI has demonstrated value in monitoring *in vivo* changes in bone formation, and is particularly useful in noninvasive repeated temporal studies (Hartman et al., 2002; Potter et al., 2006). In a rat model, MRI was shown to be able to identify small changes in bone growth (0.5 mm), and could be used to visually display the three-dimensional shape of the new bone formed (Hartman et al., 2002). An athymic mouse model was used to study the development of tissue-engineered phalange constructs compared to normal development (Potter et al., 2006). This study used MRM images to noninvasively visualize and identify changes in size and mineralization, and found that the engineered phalanges were longer and more mineralized than controls.

Cartilage: Osteoarthritis results in the degeneration of cartilage, and impacts the mobility and quality of life of those afflicted. One approach as a therapy for this disabling condition is to create tissue-engineered cartilage constructs to replace lost or damaged cartilage. An *ex vivo* NMR study observed alginate scaffolds containing chondrocytes that had earlier been implanted into patellar defects in a rat model and demonstrated that NMR data (T_2 mapping) allowed discrimination between the extent of repair that occurred due to cartilage formation, termed total, partial, or hypertrophic repairs (Watrin-Pinzano et al., 2004). An *in vivo* study of cartilage construct development used MRI to monitor changes in the cartilage over time, and found that MRI was a sensitive technique suited to monitor compositional changes in the collagen and sulfated glycosoaminoglycan content arising from a treatment by pulsed low-intensity ultrasound (Irrechukwu et al., 2011). These results indicate that NMR is a promising technique toward evaluating cartilage repair processes *in situ*.

6.7 Specific Application: Monitoring Tissue-Engineered Pancreatic Substitutes *In Vitro* and *In Vivo* with NMR/MRI

6.7.1 Pancreas and Diabetes

Regulation of blood glucose levels (glycemic control) is critical to our health and well-being. When sugar levels are too low (hypoglycemia), the brain, which uses glucose as a primary fuel source, can become starved for nutrition, and dizziness, weakness, tremors, heart palpitations, and blurred vision can occur. In severe glucose deprivation, fainting, coma, and death are possible. Elevated blood sugar levels (hyperglycemia) lead to fatigue, excessive thirst, and urination, and can lead to coma. Long-term persistent hyperglycemia causes significant morbidity and shortened life span. In vertebrates, the organ primarily responsible for this critical regulation is the pancreas. The pancreas is a multifunctional organ that has both exocrine and endocrine features. In humans, approximately 98% of the pancreatic tissue functions as an exocrine gland, secreting digestion enzymes directly into the duodenum. These enzymes help break down proteins, carbohydrates, and fats, and include proteases (trypsinogen and chymotrypsinogen), amylase, and lipase. The remaining 1–2% of the pancreas consists of specialized structures known as the islets of Langerhans, distributed throughout the pancreas. These small round collections of cells secrete the endocrine hormones that regulate blood glucose levels. The predominant cell types in islets are alpha cells, which secrete glucagon; delta cells, which secrete somatostatin; and the most numerous and arguably the most important cells in the islet, the insulin-secreting beta cells (β-cells).

Insulin is an essential hormone critical for the regulation of blood glucose levels and the cellular regulation of metabolic processes in the body. β-Cells are stimulated to release insulin by a variety of agonists, in particular, metabolic fuels such as glucose. The presence of glucose or other fuels is sensed by these cells, and insulin is secreted until glucose returns to basal levels. The mechanism of insulin secretion is a complex cascade of events (Malaisse et al., 1979), but the "fuel hypothesis" mechanism, which is broadly accepted as applicable to normal islet β-cells (Liang and Matchinsky, 1994; Malaisse et al., 1979; Meglasson and Matchinsky, 1986) is simplified as follows, and shown graphically in Figure 6.18. The β-cell contains specialized glucose transporters that bring in available glucose, thus sensing a rise in the blood glucose level, for example, after a meal. Metabolism of the glucose increases available ATP and closes ATP-dependent potassium channels. The resultant depolarization opens voltage-sensitive calcium channels, allowing calcium to enter the cell. The influx of calcium stimulates the β-cells to release stored insulin into

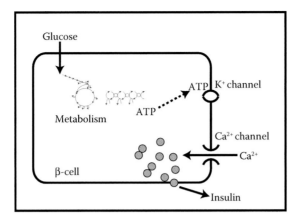

FIGURE 6.18 Representation of the "fuel hypothesis." Glucose enters the beta cell and is metabolized. This produces ATP, which closes an ATP-sensitive K^+ channel. Membrane depolarization follows, causing Ca^{2+} channels to open, allowing calcium to enter the cell. The calcium stimulates the release of insulin from the beta cell.

the bloodstream, and it also induces an increase in insulin biosynthesis. The released hormone reaches other cells in the body, binds to insulin receptors present on their surfaces, and promotes the cellular uptake and metabolism of glucose and amino acids. Insulin also acts by inducing glycogenesis and lipogenesis, and depressing gluconeogenesis. All of these processes result in a lowering of blood glucose levels.

Diabetes mellitus is a term used to define a group of chronic metabolic diseases characterized by hyperglycemia due to the deficiency in insulin secretion, resistance to insulin action, or both (Harris, 1999). Diabetes is a worldwide problem, conservatively predicted to affect over 330 million people by 2025 (King et al., 1998). The prevalence is escalating rapidly, and a recent estimate by the U.S. Centers for Disease Control and Prevention states that nearly 1 in 10 Americans are currently afflicted with diabetes, though many are currently undiagnosed, and perhaps as many as one in three may be afflicted by the year 2050 (Boyle et al., 2010). At present, roughly 8% of diabetics have what is termed type 1 diabetes, formerly known as juvenile or insulin-dependent diabetes. This disease is an autoimmune disorder resulting from the destruction of the β-cells in the islets of Langerhans. It usually presents during the first two decades of life, and accounts for more than 80% of all cases of diabetes in children and adolescents. Type 1 diabetes is an insidious disease, with the classic symptoms of diabetes not usually manifested until after the body's immune system has destroyed 70–90% of the total insulin-producing β-cell mass of the pancreatic islets of Langerhans (Eisenbarth, 1986).

The standard of care for type 1 diabetes consists of insulin delivery either through multiple daily injections or an insulin pump supplemented by injections. Although this treatment strategy affords patients a near normal life, it requires constant vigilance and cannot provide the exquisite physiological regulation offered by native β-cells. Consequently, although great strides have been made in treatment, even insulin-treated type 1 diabetics have disproportionately high morbidity and mortality rates and shortened life spans. Without strict regulation of blood glucose levels, severe short- and long-term tissue damage results, with vascular complications being a major underlying component. Chronic hyperglycemia and/or undetected hyperglycemic excursions lead to microvascular complications (e.g., retinopathy and blindness, nephropathy and renal failure, foot ulcers and amputations, as well as peripheral neuropathy) and macrovascular complications (e.g., cardiovascular, peripheral, vascular, and cerebrovascular disease and stroke). These physiological manifestations are common, even in patients obtaining treatment, though good glucose control slows their progression.

A therapeutic approach toward curing diabetes is to replace the defective or destroyed insulin-secreting β-cells and restore optimum physiological control of blood glucose levels without the need for insulin injections. Methods to accomplish this include pancreatic transplantation (Robertson, 1992;

Sutherland, 1996) and intraportal islet implantations (Burridge et al., 2002; Olberholzer et al., 2003; Ryan et al., 2002; Shapiro et al., 2000, 2001a,b). However, these approaches are not trivial, and while they provide hope to type 1 diabetic patients, widespread clinical use of these approaches requires a steady supply of human pancreatic tissue or islets. Unfortunately, there is a severe shortage of implantable donor pancreata and islets, and at present, neither human nor animal β-cells exhibit efficient proliferative capacity with sustained insulin secretion in culture. Moreover, recipients must receive lifelong immunosuppressive medication following the transplant. Alternatively, a mechanical pancreas approach combining a continuous glucose monitor with a controller and an insulin pump has yet to fulfill its promise due to the difficulty in developing a glucose sensor that is stable and accurate and has a sufficient life span. Hence, there is still a great need for an efficacious treatment that provides physiological blood glucose regulation without immunosuppressive medication, one that is easily administered and readily available.

6.7.2 Tissue-Engineered Pancreatic Substitutes

A treatment that fulfills these requirements is the implantation of a tissue-engineered pancreatic substitute (Lim and Sun, 1980; Sullivan et al., 1991; Sun, 1988; Sun et al., 1980), also called the bioartificial pancreas. This device combines a cellular component with a nonbiologic matrix and takes advantage of the cell's ability to sense glucose levels, secrete insulin appropriately, and maintain tight blood glucose control. Different approaches to the bioartificial pancreas have been pursued. One method entraps cells in thin implantable alginate sheets (Storrs et al., 2001), while another maintains cells within hollow fiber devices (Archer et al., 1980). A number of groups have investigated the use of microencapsulation, in which cells are entrapped into small beads of less than 1 mm diameter, to provide immunoprotection and possible functional support to the cells via the surrounding matrix. The materials used in these microconstructs are varied, but alginate, a polysaccharide obtained from a variety of seaweed that forms a hydrogel when in the presence of divalent cations (Smidsrod, 1974), has been widely studied as a biomaterial for encapsulation of insulin-secreting cells ever since the initial work by Lim and Sun (1980). de Vos et al. (2006, 2010) provide excellent reviews of alginate microcapsules and their ability to immunoprotect islets. Arguably the most practical design, at least for monitoring purposes, is a macroencapsulation device, such as a disk (Lahooti and Sefton, 2000), consisting of insulin-secreting cells encapsulated in materials that are of sufficient mechanical strength (Stabler et al., 2005a,b) and can provide at least partial immunoprotection (Lanza et al., 1992). The site of implantation most often used for such devices in animal models is the peritoneal cavity. There are reports in the literature demonstrating the successful restoration of normoglycemia in diabetic animals for extended periods with encapsulated cells implanted at this site (Soon-Shiong et al., 1992; Weber et al., 1999). The cell source choice for such devices is broad. Using implanted islets as the insulin source has been successful for limited periods of time (Fan et al., 1990; Lanza et al., 1992; Soon-Shiong et al., 1992), although again, there is a scarce supply of donor tissue. A promising alternative to islets is the use of glucose-responsive insulin-secreting transformed β-cell lines (Efrat et al., 1993; Hicks et al., 1991; Newgard, 1994). These have the distinct advantage over native β-cells in that they can be grown to large homogeneous populations, thereby providing an unlimited supply of cells. Cells of non-β origin, genetically engineered to express recombinant insulin in response to metabolic cues or transdifferentiated toward the β-cell phenotype, constitute another promising approach (Cheung et al., 2000; Clark et al., 1997; Hohmeier et al., 1997; Tang and Sambanis, 2003; Thule and Liu, 2000; Wang and Iynedjian, 1997; Zhou et al., 2008).

6.7.3 Goals of NMR Monitoring

Regardless of the source of insulin-secreting cells, the function of an implanted bioartificial pancreas is currently determined by measuring blood glucose levels of the treated host. If the glucose level is normal, the implant is functional; if it is too high, the implant has failed, and the consequences of diabetes will follow. Blood sugar measurements only establish if the implanted construct is working or has failed at the time of the measurement; they cannot predict if the construct will continue to function at its current level or for how long. As discussed earlier, *in vivo* monitoring of implanted constructs is a significant issue in tissue engineering. The ability to obtain imaging and spectroscopic data from constructs can assist us in understanding their function, the way in which they interface with the host, and their optimal design. Another important goal is to predict failure. A hypothesis with regard to the bioartificial pancreas is that NMR monitoring may predict the failure of an implanted construct while it is still functioning and regulating the blood glucose effectively. The reason behind this possibility can be understood when considering the pancreas in a person developing type 1 diabetes. Although the volume of islets in a pancreas is small (2% of the total volume), the number of insulin-secreting cells (β-cells) in a pancreas is more than sufficient to control blood glucose levels. In fact, 70–80% of the β-cells need to be destroyed before blood glucose cannot be well regulated and diabetes becomes evident. This is because of a redundancy in cell number. As β-cells are destroyed by the immune system, the remaining β-cells compensate by secreting more insulin. Only when β-cell numbers dip below this threshold is glucose regulation lost. Similarly, if some redundancy is incorporated into the design of a bioartificial pancreas, the implanted construct will continue to function even if some cells die provided that enough functioning cells remain alive. It is one goal of monitoring to determine the dynamic changes that occur in an implanted construct and to use this information to estimate an expected life span, possibly to suggest corrective actions to prolong the functional life span of the construct, and in any case to allow for its replacement before blood glucose regulation is lost.

6.7.4 Approaches to Monitor

Components that make up the bioartificial pancreas, such as the cells and the encapsulating materials, can also be studied with noninvasive NMR techniques both *in vitro* and *in vivo*. These "preclinical" studies are important for optimizing design and assessing structural features of the construct, and determining effective noninvasive approaches to study the behavior and function of the implanted bioartificial pancreas. *In vitro* NMR studies of the biomaterial used in our cell encapsulation process, alginate, have been performed to meet some of these goals. In a series of studies looking at various alginate compositions and gellation conditions, we implemented NMR microimaging studies to determine some valuable NMR-detectable properties of the resultant alginate microbeads (Constantinidis et al., 2007; Grant et al., 2005; Simpson et al., 2003). One studied the effect that alginate composition and gelling conditions have on the NMR relaxation and diffusion processes, described in Section 6.3.1 (Simpson et al., 2003). It was found that the T_2 relaxation time was proportional to the alginate's guluronic acid content, and that T_2 increased as the concentration of the gelling solution ($CaCl_2$) decreased. Identifying the relationships between NMR methods and the state of the entrapped cells or entrapping biomaterial yields information related to the physical state of the implanted construct and may allow for the development of superior approaches to monitor an implanted and functional tissue-engineered substitutes.

Macroconstructs have an advantage over microconstructs in that they are retrievable, and perhaps could be used as a medical device toward the treatment of type 1 diabetes. A macroconstruct must have the following characteristics. It has to be able to support cell viability and function. To do so, it must allow for the exchange of nutrients (e.g., oxygen and glucose) and waste products to and from the entrapped cells. Protection from immune responses of the body is also a requirement; otherwise, the implanted cellular component may be killed by the cellular and/or humoral component of the immune system of the host. It also has to be biocompatible so that it does not elicit a strong fibrotic response, as a fibrotic capsule may compromise the exchange of nutrients resulting in loss of cell function and cell death. It must be durable, and constructed of materials that do not degrade over time, thereby maintaining its structural integrity. It should also ideally have a size and shape that allows for easy implantation and for efficient NMR monitoring so that structural and metabolic information can be obtained. Our macroconstruct version consists of insulin-secreting cells entrapped in alginate beads, and these beads then entrapped into an alginate core, as depicted in Figure 6.19. Additionally, to add strength and the potential to shield an implantable NMR coil, discussed later, the depicted core is housed within a ring of polydimethylsiloxane, a bioinert and electrically insulating material.

Owing to the inherently low sensitivity of NMR and to the low density of cells in tissue constructs relative to natural tissues, a nucleus much more promising than ^{31}P for *in vivo* spectroscopic studies would be ^{1}H. Using alginate-encapsulated βTC3 cells in the same packed-bed bioreactor and perfusion system, Long et al. (2000) collected water-suppressed ^{1}H NMR spectra in long-term feasibility studies. A total choline (TCho) resonance, composed primarily of phosphorylcholine, was found to correlate with the number of metabolically active cells and hence could be used to track the viable cell number in a volume of interest. To advance the application of this NMR-based method toward *in vivo* studies, Stabler et al. (2005a,b) encapsulated βTC3 insulinomas in disk-shaped agarose constructs, which allowed the collection of localized spectra from the implanted cells to the exclusion of signal from the surrounding tissue of the host. *In vitro* NMR studies with these disk-shaped macroconstructs using a 4.7 T magnet and a surface coil antenna for ^{1}H detection confirmed a strong correlation between TCho levels and the traditional MTT assay, which measures mitochondrial activity, as well as between TCho and the insulin secretion rate. For *in vivo* studies, a cell-free agarose layer had to be added around the cell-containing agarose disk to better separate the implanted cells from the host and ensure the absence of contaminating signal from the host tissue, especially signal from peritoneal fat. Also, as a glucose resonance interfered with the TCho signal, a correction was implemented to account for and remove this interference (Stabler et al., 2005a). The glucose-corrected TCho area obtained noninvasively with constructs implanted in live mice correlated positively and strongly with the MTT assay performed on the same constructs postexplantation. Hence, the TCho resonance allowed for the noninvasive *in vivo* tracking of the number of viable cells in disk constructs over time (Stabler et al., 2005b). To acquire the NMR signal, an external surface coil was used. Spin echo ^{1}H NMR imaging was used to position the volume of interest from which the spectroscopic signal was collected and also to assess the structural integrity of the disk constructs (Stabler et al., 2005b). The resolution was sufficient to clearly delineate the boundary between the inner cell containing the disk and the outer cell-free buffer zone. Imaging also detected the presence of significant fibrosis around the implants, when this occurred.

However, as mentioned earlier, these initial studies were performed using an external surface coil. As discussed in Section 6.3.2, surface coils are not the ideal receiver for NMR studies, though when studying structures *in vivo*, such as implanted constructs, they are a logical choice. Unfortunately, the

FIGURE 6.19 Picture of alginate construct core. Depicted here is the photograph of a core of an implantable bioartificial pancreatic construct. The small insulin-secreting cells are entrapped in alginate microbeads (~500 micron diameter). These microbeads beads are stained dark to allow visualization. The beads are entrapped in the outer alginate, and the entire core is placed within a ring, in which an NMR coil can be placed, as discussed later.

signal-to-noise issues inherent in observing a bioartificial construct with external surface coils limit the utility of NMR in monitoring implants. Additionally, tissues around the implant and between the implant and surface coil can contribute to the problem by making it difficult to obtain specific measures from only the implant. To address this issue, and enhance the NMR signal, we have been implanting an rf coil with the macroconstruct, integrating it into the outer material (Volland et al., 2010). The implantable coil approach is possible, though initially it was achieved by directly connecting the implanted coil to an outside circuit (Arnder et al., 1996), which requires the wiring to come through the skin. Subsequent approaches used an inductively coupled system (Schnall et al., 1986; Wirth et al., 1993), in which the implanted coil is inductively coupled to an external surface coil and circuit (Hoult and Tomanek, 2002; Schnall et al., 1986; Silver et al., 2001; Wirth et al., 1993). This approach has nothing emerging from the implanted coil through the skin, and thus prevents infection. Figure 6.20 depicts a coil integrated with the implanted construct, and inductively coupled to an external tuned surface coil. By implementing an inductively coupled coil system (Volland et al., 2010) at 11 T to monitor a bioartificial construct, signal arises from only the tissue that the coil surrounds, and there is virtually no contaminating signal from interposing or surrounding tissues. The data show that through the use of this system, large gains in SNR (i.e., a twofold sensitivity improvement) can be obtained over that obtained through a surface coil. This improvement will allow for better imaging, and permit spectroscopy of nuclei at smaller concentrations or less inherent sensitivity, thus opening the way for more extensive quantitative analysis of implanted functional bioartificial organs. To enhance our abilities to noninvasively monitor implanted constructs, efforts at developing a miniaturized implantable multifrequency wirelessly tunable coil system are underway. This implanted tunable system uses a "single-resonant" approach, where an array of varactors on a microchip allow the implanted coil system to be remotely and wirelessly switched by a digital controller to a wide range of frequencies of interest (e.g., ^{1}H, ^{13}C, ^{31}P, ^{19}F). Varactors are semiconductor diodes that behave like voltage-dependent capacitors, thus allowing circuit tuning. The tunable system allows the implanted coil to act as a single-resonant frequency coil thus providing optimal signal sensitivity. It is important to emphasize that the implantable coil and multitune system approaches are not restricted to the bioartificial pancreas, but may be applied to other tissue-engineered constructs, enhancing the ability to monitor them through NMR methods.

Microcapsules offer distinct advantages relative to macroconstructs in terms of providing higher surface-to-volume ratio and better nutrient transport, thus sustaining a higher number of viable, functional cells within an implant volume. However, owing to their small size and the small number of cells they contain, localized spectroscopy cannot be easily performed on microcapsules. Furthermore, even if this were possible, data from the few microcapsules that could be obtained while the animal is under anesthesia would not be representative of the implant as a whole.

To maintain our ability to noninvasively monitor microcapsules *in vivo*, we pursued the incorporation of a low concentration of perfluorocarbons (PFC) in the encapsulating hydrogel. The inverse of the

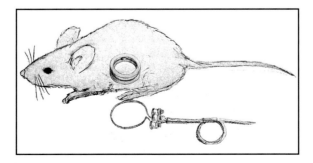

FIGURE 6.20 Drawing of mouse with implant. Depicted is a representation of a coil-containing device implanted into a mouse. Below is a tuned surface coil that inductively couples to the implanted coil to yield NMR information from cells/materials contained within the implant. Sketch by Nicholas E. Simpson.

T_1 relaxation of the natural fluorine isotope, ^{19}F, correlates strongly and linearly with the DO concentration. Furthermore, ^{19}F is the second most sensitive, naturally abundant nucleus after ^1H, and there is no interfering ^{19}F background signal. Hence, with this approach, one may monitor the DO to which cells are exposed and use this parameter to assess cell viability and function, both of which depend strongly on the DO. *In vitro* experiments demonstrated that the incorporation of PFCs in the capsules did not affect the viability and function of cells relative to PFC-free controls (Goh et al., 2010). To validate the use of PFCs in DO monitoring, *in vitro* experiments were performed in a fixed-bed perfusion bioreactor containing encapsulated βTC-tet murine insulinomas (Goh et al., 2011; Gross et al., 2007b). The ability of this approach to monitor *in vivo* the DO in beads containing encapsulated cells, in relation to the DO at the implantation site, was recently demonstrated in experimental mice (Goh et al., 2011).

6.8 Summary and Conclusions

NMR imaging and spectroscopy are actively used in the medical field to diagnose the function of tissues and organs, and they are also finding increasing use in tissue engineering to monitor tissue substitutes *in vitro* and *in vivo*. The major advantages of the NMR method include its noninvasiveness, the ability to monitor deep-seated tissues, and that it is the only modality that can provide both structural and biochemical/functional information on a tissue or organ. Its disadvantages include the relatively low sensitivity, the need for specialized personnel, and the high cost.

In this chapter, we provided an overview of the physics of the NMR phenomenon and discussed specific applications of NMR on monitoring tissue constructs *in vitro* and *in vivo*, with special emphasis on the bioartificial pancreas. *In vitro* studies utilize NMR-compatible bioreactors containing the cells localized at high densities and supported by perfusion circuits; the objectives of these studies range from obtaining fundamental biochemical information, to studying tissue growth and morphogenesis *in vitro*, to evaluating important parameters for advancing the studies to the *in vivo* setting. *In vivo*, NMR has been used to study tissue repair and growth and to monitor the viability and oxygenation of cells in implanted tissue substitutes. Such information is critical in directly assessing the *in vivo* function of a tissue construct and its interaction with the host, and in detecting early processes that may result in construct failure. With the development of improved magnets and coils, the sensitivity of NMR is continuously improving, and this is expected to considerably broaden its application in studying tissue-engineered substitutes.

Acknowledgments

The studies in the authors' laboratories referenced in this chapter have received support by grants from the National Institutes of Health, the Juvenile Diabetes Research Foundation, and from the National Science Foundation through the Georgia Tech/Emory Center for the Engineering of Living Tissues (GTEC). This support is gratefully acknowledged.

References

Abragam, A. 1961. *Principles of Nuclear Magnetism,* Oxford, Oxford University Press.

Aguayo, J. B., Blackband, S. J., Schoeniger, J., Mattingly, M. A., and Hintermann, M. 1986. Nuclear magnetic resonance imaging of a single cell. *Nature,* 322, 190–191.

Archer, J., Kaye, R., and Mutter, G. J. 1980. Control of streptozotocin diabetes in Chinese hamsters by cultured mouse islet cells without immunosuppression: A preliminary report. *J Surg Res,* 28, 77–85.

Arnder, L., Shattuck, M. D., and Black, R. D. 1996. Signal-to-noise ratio comparison between surface coils and implanted coils. *Magn Reson Med,* 35, 727–733.

Beyer, T., Townsend, D. W., Brun, T. et al. 2000. A combined PET/CT scanner for clinical oncology. *J Nucl Med,* 41, 1369–1379.

Blackband, S. J., Buckley, D. L., Bui, J. D., and Phillips, M. I. 1999. NMR microscopy—Beginnings and new directions. *Magn Reson Mater Phys Biol Med,* 9, 112–116.

Bloch, F., Hansen, W. W., and Packard, M. E. 1946. The nuclear induction experiment. *Phys Rev,* 70, 460–474.

Boyle, J. P., Thompson, T. J., Gregg, E. W., Barker, L. E., and Williamson, D. F. 2010. Projection of the year 2050 burden of diabetes in the US adult population: Dynamic modeling of incidence, mortality, and prediabetes prevalence. *Popul Health Metrics,* 8, 29.

Bretthorst, G. L., Kotyk, J. J., and Ackerman, J. J. 1989. 31P NMR Bayesian spectral analysis of rat brain *in vivo. Magn Reson Med,* 9, 282–287.

Burridge, P. W., Shapiro, A. M., Ryan, E. A., and Lakey, J. R. 2002. Future trends in clinical islet transplantation. *Transplant Proc,* 34, 3347–3348.

Buschmann, J., Welti, M., Hemmi, S. et al. 2011. 3D co-cultures of osteoblasts and endothelial cells in DegraPol foam: Histological and high field MRI analysis of pre-engineered capillary networks in bone grafts. *Tissue Engin Part A,* 17, 291–299.

Cahill, K. S., Gaidosh, G., Huard, J., Silver, X., Byrne, B. J., and Walter, G. A. 2004. Noninvasive monitoring and tracking of muscle stem cell transplants. *Transplantation,* 78, 1626–1633.

Catana, C., Procissi, D., Wu, Y. et al. 2008. Simultaneous *in vivo* positron emission tomography and magnetic resonance imaging. *Proc Natl Acad Sci,* 105, 3705–3710.

Chalfie, M., Tu, Y., Euskirchen, G., Ward, W. W., and Prasher, D. C. 1994. Green fluorescent protein as a marker for gene expression. *Science,* 263, 802–805.

Chen, C. T., Fishbein, K. W., Torzilli, P. A., Hilger, A., Spencer, R. G., and Horton, W. E., JR. 2003. Matrix fixed-charge density as determined by magnetic resonance microscopy of bioreactor-derived hyaline cartilage correlates with biochemical and biomechanical properties. *Arthritis Rheum,* 48, 1047–1056.

Cheng, H. L., Chen, J., Babyn, P. S., and Farhat, W. A. 2005. Dynamic Gd-Dtpa enhanced MRI as a surrogate marker of angiogenesis in tissue-engineered bladder constructs: A feasibility study in rabbits. *J Magn Reson Imaging,* 21, 415–423.

Cheng, H. L., Loai, Y., Beaumont, M., and Farhat, W. A. 2010. The acellular matrix (ACM) for bladder tissue engineering: A quantitative magnetic resonance imaging study. *Magn Reson Med,* 64, 341–348.

Cheng, H. L., Wallis, C., Shou, Z., and Farhat, W. A. 2007. Quantifying angiogenesis in VEGF-enhanced tissue-engineered bladder constructs by dynamic contrast-enhanced MRI using contrast agents of different molecular weights. *J Magn Reson Imaging,* 25, 137–145.

Cherry, S. R., Shao, Y., Silverman, R. W. et al. 1997. MicroPET: A high resolution PET scanner for imaging small animals. *IEEE Trans Nucl Sci,* 44, 1161–1166.

Chesnick, I. E., Avallone, F. A., Leapman, R. D., Landis, W. J., Eidelman, N., and Potter, K. 2007. Evaluation of bioreactor-cultivated bone by magnetic resonance microscopy and FTIR microspectroscopy. *Bone,* 40, 904–912.

Cheung, A. T., Dayanandan, B., Lewis, J. T. et al. 2000. Glucose-dependent insulin release from genetically engineered K cells. *Science,* 290, 1959–1962.

Clark, S. A., Quaade, C., Constandy, H. et al. 1997. Novel insulinoma cell lines produced by iterative engineering of GLUT2, glucokinase, and human insulin expression. *Diabetes,* 42, 901–907.

Constantinidis, I., Grant, S. C., Celper, S. et al. 2007. Non-invasive evaluation of alginate/poly-L-lysine/alginate microcapsules by magnetic resonance microscopy. *Biomaterials,* 28, 2438–2445.

Constantinidis, I., Grant, S. C., Simpson, N. E. et al. 2009. Use of magnetic nanoparticles to monitor alginate-encapsulated bTC-tet cells. *Magn Reson Med,* 61, 282–290.

Constantinidis, I. and Sambanis, A. 1995. Towards the development of artificial endocrine tissues: 31P NMR spectroscopic studies of immunoisolated, insulin-secreting AtT-20 cells. *Biotechnol Bioeng,* 47, 431–443.

Correia, J. A., Burnham, C. A., Kaufman, D., and Fischman, A. J. 1999. Development of a small animal PET imaging device with resolution approaching 1 mm. *IEEE Trans Nucl Sci,* 46, 631–635.

de Vos, P., Faas, M. M., Strand, B., and Calafiore, R. 2006. Alginate-based microcapsules for immunoisolation of pancreatic islets. *Biomaterials,* 27, 5603–5617.

de Vos, P., Spasojevic, M., and Faas, M. M. 2010. Treatment of diabetes with encapsulated islets. *Adv Exp Med Biol,* 670, 38–53.

Dobson, J., Cartmell, S. H., Keramane, A., and El Haj, A. J. 2006. Principles and design of a novel magnetic force mechanical conditioning bioreactor for tissue engineering, stem cell conditioning, and dynamic *in vitro* screening. *IEEE Trans Nanobiosci,* 5, 173–177.

Dodd, S. J., Williams, M., Suhan, J. P., Williams, D. S., Koretsky, A. P., and Ho, C. 1999. Detection of single mammalian cells by high-resolution magnetic resonance imaging. *Biophys J,* 76, 103–109.

Doyley, M. M., Srinivasa, S., Prendergrass, S. A., Wu, Z., and Ophir, J. 2005. Comparative evaluation of strain-based and model-based modulus elastography. *Ultrasound Med Biol,* 31, 787–802.

Efrat, S., Leister, M., Surana, M., Tal, M., Fusco-Demane, D., and Fleischer, N. 1993. Murine insulinoma cell line with normal glucose-regulated insulin secretion. *Diabetes,* 42, 901–907.

Eisenbarth, G. S. 1986. Type 1 diabetes: A chronic autoimmune disease. *N Engl J Med,* 314, 1360–1368.

Ernst, R. R. and Anderson, W. A. 1966. Application of Fourier transform spectroscopy to magnetic resonance. *Rev Sci Instr,* 37, 93–102.

Evgenov, N. V., Medarova, Z., Dai, G., Bonner-Weir, S., and Moore, A. 2006. *In vivo* imaging of islet transplantation. *Nat Med,* 12, 144–148.

Fan, M. Y., Lum, Z. P., Fu, X. W., Levesque, L., Tai, I. T., and Sun, A. M. 1990. Reversal of diabetes in BB rats by transplantation of encapsulated pancreatic islets. *Diabetes,* 39, 519–522.

Farrar, T. C. and Becker, E. D. 1971. *Pulse and Fourier Transform NMR: Introduction to Theory and Methods,* New York, Academic Press.

Foster-Gareau, P., Heyn, C., Alejski, A., and Rutt, B. K. 2003. Imaging single cells with a 1.5T clinical MRI scanner. *Magn Reson Med,* 49, 968–971.

Fujimoto, J. G. 2003. Optical coherence tomography for ultrahigh resolution *in vivo* imaging. *Nat Biotechnol,* 21, 1361–1367.

Gee, K. R., Zhou, Z.-L., Qian, W.-J., and Kennedy, R. 2002. Detection and imaging of zinc secretion from pancreatic b-cells using a new fluorescent zinc indicator. *J Am Chem Soc Comm,* 124, 776–778.

Gillies, R. J., Galons, J. P., and Al, E. 1993. Design and application of NMR-compatible bioreactor circuits for extended perfusion of high-density mammalian cell cultures. *NMR Biomed,* 6, 95–104.

Gillies, R. J., Scherer, P. G., and Al, E. 1991. Iteration of hybridoma growth and productivity in hollow fiber bioreactors using 31P NMR. *Magn Reson Med,* 18, 181–192.

Gimi, B., Artemov, D., Leong, T. G. et al. 2007. Cell viability and non-invasive MRI tracking of 3D cell encapsulating self-assembled microcontainers. *Cell Transplant,* 16, 403–408.

Gimi, B., Leong, T., Gu, Z. et al. 2005. Self-assembled three dimensional radio frequency (RF) shielded containers for cell encapsulation. *Biomed Microdev,* 7, 341–345.

Goetz, C., Breton, E., Choquet, P., Israel-Jost, V., and Constantinesco, A. 2008. Spect low-field MRI system for small-animal imaging. *J Nucl Med,* 49, 88–93.

Goh, F., Gross, J. D., Simpson, N. E., and Sambanis, A. 2010. Limited beneficial effects of perfluorocarbon emulsions on encapsulated cells in culture: Experimental and modeling studies. *J Biotechnol* 150, 232–239.

Goh, F., Long, R. C., Simpson, N. E., and Sambanis, A. 2011. Dual perfluorocarbon method to noninvasively monitor dissolved oxygen concentration in tissue constructs *in vitro* and *in vivo. Biotechnol Prog,* 27, 1115–1125.

Grant, S. C., Celper, S., Gauffin-Holmberg, I., Simpson, N. E., Blackband, S. J., and Constantinidis, I. 2005. Alginate assessment by NMR microscopy. *J Mater Sci,* 16, 511–514.

Greer III, L. F. and Szalay, A. A. 2002. Imaging of light emission from the expression of luciferases in living cells and organisms: A review. *Luminescence,* 17, 43–74.

Gross, J. D., Constantinidis, I., and Sambanis, A. 2007a. Modeling of encapsulated cell systems. *J Theor Biol,* 244, 500–510.

Gross, J. D., Long, R. C., and Al, E. 2007b. Monitoring of dissolved oxygen and cellular bioenergetics within a pancreatic substitute. *Biotechnol Bioeng,* 98, 261–270.

Gupta, A. K. and Gupta, M. 2005. Synthesis and surface engineering of iron oxide nanoparticles for bio-medical applications. *Biomaterials,* 26, 3995–4021.

Gupta, A. K., Naragelkar, R. R., Vaida, V. D., and Gupta, M. 2007. Recent advances on surface engineering of magnetic iron oxide nanoparticles and their biomedical applications. *Nanomedicine,* 2, 23–39.

Hadjantonakis, A.-K., Dickinson, M. E., Fraser, S. E., and Papaioannou, V. E. 2003. Technicolour trans-genics: Imaging tools for functional genomics in the mouse. *Nat Genet,* 4, 613–625.

Han, S., El-Abbadi, N. H., and Al, E. 2005. Evaluation of tracheal imaging by optical coherence tomogra-phy. *Respiration,* 72, 537–541.

Harisinghani, M. G., Barentsz, J., Hahn, P. F. et al. 2003. Noninvasive detection of clinically occult lymph-node metastases in prostate cancer. *New Engl J Med,* 348, 2491–2499.

Harris, M. I. 1999. *Newly Revised Classification and Diagnostic Criteria for Diabetes Mellitus,* Philadelphia, PA, Current Medicine.

Hartman, E. H. M., Pikkemmat, J. A., Vehof, J. W. M., Heerschap, A., Jansen, J. A., and Spauwen, P. H. M. 2002. *In vivo* magnetic resonance imaging explorative study of ectopic bone formation in the rat. *Tissue Eng,* 8, 1029–1036.

Hicks, B. A., Stein, R., Efrat, S., Grant, S., Hanahan, D., and Demetriou, A. A. 1991. Transplantation of b cells from transgenic mice into nude athymic diabetic rats restores glucose regulation. *Diabetes Res Clin Prac,* 14, 157–164.

Hohmeier, H. E., Beltran Del Rio, H., Clark, S. A., Henkel-Rieger, R., Normington, K., and Newgard, C. B. 1997. Regulation of insulin secretion from novel engineered insulinoma cell lines. *Diabetes,* 46, 968–977.

Hope Simpson, D., Burns, P. N., and Averkion, M. A. 2001. Techniques for perfusion imaging with micro-bubble contrast agents. *IEEE Trans Ultrason Ferroelectr Freq Control,* 48, 1483–1494.

Hope Simpson, D., Chin, C. T., and Burns, P. N. 1999. Pulse inversion Doppler: A new method for detect-ing nonlinear echoes from microbubble contrast agents. *IEEE Trans Ultras Ferroelectr Freq Control,* 46, 372–382.

Hoult, D. I. and Tomanek, B. 2002. Use of mutually inductive coupling in probe design. *Concepts Magn Reson,* 15, 262–285.

Ino, K., Ito, A., and Honda, H. 2007. Cell patterning using magnetite nanoparticles and magnetic force. *Biotechnol Bioeng,* 97, 1309–1317.

Irrechukwu, O. N., Lin, P.-C., Fritton, K., Doty, S. B., Pleshko, N., and Spencer, R. G. 2011. Magnetic reso-nance studies of macromolecular content in engineered cartilage treated with pulsed low-intensity ultrasound. *Tissue Engin Part A,* 17, 407–415.

Ito, A., Shinkai, M., Honda, H., and Kobayashi, T. 2005. Medical application of functionalized magnetic nanoparticles. *J Biosci Bioeng,* 100, 1–11.

Karfeld-Sulzer, L. S., Waters, E. A., Kohlmeir, E. K. et al. 2011. Protein polymer MRI contrast agents: Longitudinal analysis of biomaterials *in vivo. Magn Reson Med* 65, 2220–2228.

King, H., Aubert, R. E., and Herman, W. H. 1998. Global burden of diabetes. 1995–2025. Prevalence, numerical estimates and projections. *Diabetes Care,* 21, 1414–1431.

Kline, A. D., Braun, W., and Wuthrich, K. 1988. Determination of the complete three-dimensional struc-ture of the a-amylase inhibitor tenamistat in aqueous solution by nuclear magnetic resonance and distance geometry. *J Molec Biol,* 204, 675–724.

Kooi, M. E., Cappendijk, V. C., Cleutjens, K. et al. 2003. Accumulation of ultrasmall superparamagnetic particles of iron oxide in human atherosclerotic plaques can be detected by *in vivo* magnetic reso-nance imaging. *Circulation,* 107, 2453–2458.

Kriz, J., Jirak, D., Girman, P. et al. 2005. Magnetic resonance imaging of pancreatic islets in tolerance and rejection. *Transplantation,* 80, 1596–1603.

Kumar, A., Welti, A. D., and Ernst, R. R. 1975. NMR Fourier zeugmatography. *J Magn Reson,* 18, 69–83.

Lahooti, S. and Sefton, M. V. 2000. Agarose enhances the viability of intraperitoneally implanted microen-capsulated L929 fibroblasts. *Cell Transplant,* 9, 785–796.

Lanza, R. P., Sullivan, S. J., and Chick, W. L. 1992. Islet transplantation with immunoisolation. *Diabetes,* 41, 1503–1510.

Lauterbur, P. C. 1973. Image formation by induced local interactions: Examples employing nuclear magnetic resonance. *Nature,* 242, 190–191.

Lebihan, D., Breton, E., Lallemand, D., Grenier, P., Cabanis, E., and Laval-Jeantet, M. 1986. MR imaging of intravoxel incoherent motions: Application to diffusion and perfusion in neurologic disorders. *Radiology,* 161, 401–407.

Liang, Y. and Matchinsky, F. M. 1994. Mechanisms of action of nonglucose insulin secretagogues. *Ann Rev Nutr,* 14, 59–81.

Lim, F. and Sun, A. M. 1980. Microencapsulated islets as bioartificial endocrine pancreas. *Science,* 210, 908–910.

Long, R. C., Papas, K. K., and Al, E. 2000. *In vitro* monitoring of total choline levels in a bioartificial pancreas: (1)H NMR spectroscopic studies of the effects of oxygen level. *J Magn Reson,* 146, 49–57.

Malaisse, W. J., Sener, A., Herchuelz, A., and Hutton, J. C. 1979. Insulin release: The fuel hypothesis. *Metabolism,* 28, 373–385.

Mancuso, A. and Fernandez, E. J. 1990. A nuclear magnetic resonance technique for determining hybridoma cell concentration in hollow fiber bioreactors. *Biotechnology,* 8, 1282–1285.

Mancuso, A., Sharfstein, S. T., and Al, E. 1994. Examination of primary metabolic pathways in a murine hybridoma with carbon-13 nuclear magnetic resonance spectroscopy. *Biotechnol Bioeng,* 44, 563–585.

Mancuso, A., Sharfstein, S. T., and Al, E. 1998. Effect of extracellular glutamine concentration on primary and secondary metabolism of a murine hybridoma: An *in vivo* 13C nuclear magnetic resonance study. *Biotechnol Bioeng,* 57, 172–186.

Mansfield, P. and Grannell, P. K. 1973. NMR "diffraction" in solids. *J Phys C,* 6, L422–L426.

Maudsley, A. A., Hilal, S. K., Perman, W. H., and Simon, H. 1983. Spatially resolved high resolution spectroscopy by four dimensions. *J Magn Reson,* 51, 147–152.

Medarova, Z. and Moore, A. 2007. *Cellular and Molecular Imaging of the Diabetic Pancreas,* Boca Raton, FL, Taylor & Francis.

Meglasson, M. D. and Matchinsky, F. M. 1986. Pancreatic islet glucose metabolism and regulation of insulin secretion. *Diabetes Metab Rev,* 2, 163–214.

Moore, A., Grimm, J., Han, B., and Santamaria, P. 2004. Tracking the recruitment of diabetogenic Cd8+ T-cells to the pancreas in real time. *Diabetes,* 53, 1459–1466.

Moore, A., Sun, P. Z., Cory, D., Hogemann, D., Weissleder, R., and Lipes, M. A. 2002. MRI of insulitis in autoimmune diabetes. *Magn Reson Med,* 47, 751–758.

Mori, S. and Barker, P. B. 1999. Diffusion magnetic resonance imaging: Its principle and applications. *Anatom Rec,* 257, 102–109.

Moseley, M., Cohen, Y., Kucharczyk, J. et al. 1990. Diffusion-weighted MR imaging of anisotropic water diffusion in cat central nervous system. *Radiology,* 176, 439–445.

Neuwelt, E. A., Varallyay, P., Bago, A. G., Muldoon, L. L., Nesbit, G., and Nixon, R. 2004. Imaging of iron oxide nanoparticles by MR and light microscopy in patients with malignant brain tumours. *Neuropathol Appl Neurobiol,* 30, 228–232.

Newgard, C. B. 1994. Cellular engineering and gene therapy strategies for insulin replacement in diabetes. *Diabetes,* 43, 341–350.

Nugent, A. E., Reiter, D. A., Fishbein, K. W. et al. 2010. Characterization of *ex-vivo*-generated bovine and human cartilage by immunohistochemical, biochemical, and magnetic resonance imaging analysis. *Tissue Eng Part A,* 16, 2183–2196.

Olberholzer, J., Shapiro, A. M., Lakey, J. R. et al. 2003. Current status of islet cell transplantation. *Adv Surg,* 37, 253–282.

Papas, K. K., Long, R. C., Constantinidis, I., and Sambanis, A. 1997. Role of ATP and Pi in the mechanism of insulin secretion in the mouse insulinoma betaTC3 cell line. *Biochem J,* 326, 807–814.

Papas, K. K., Long, R. C., Constantinidis, I., and Sambanis, A. 1999a. Development of a bioartificial pancreas: I. Long-term propagation and basal and induced secretion from entrapped betaTC3 cell cultures. *Biotechnol Bioeng,* 66, 219–230.

Papas, K. K., Long, R. C., Constantinidis, I., and Sambanis, A. 1999b. Development of a bioartificial pancreas: II. Effects of oxygen on long-term entrapped betaTC3 cell cultures. *Biotechnol Bioeng,* 66, 231–237.

Pichler, B. J., Judenhofer, M. S., and Pfannenberg, C. 2008. Multimodal imaging approaches: PET/CT and PET/MRI. *Handb Exp Pharmacol,* 185, 109–132.

Piston, D. W. 1999. Imaging living cells by two-photon excitation microscopy. *Trends Cell Biol,* 9, 66–69.

Planchamp, C., Gex-Fabry, M., Dornier, C. et al. 2004a. Gd-BOPTA transport into rat hepatocytes: Pharmacokinetic analysis of dynamic magnetic resonance images using a hollow-fiber bioreactor. *Invest Radiol,* 39, 506–515.

Planchamp, C., Ivancevic, M. K., Pastor, C. M. et al. 2004b. Hollow fiber bioreactor: New development for the study of contrast agent transport into hepatocytes by magnetic resonance imaging. *Biotechnol Bioeng,* 85, 656–665.

Potter, K., Butler, J. J., Adams, C., Fishbein, K. W., Mcfarland, E. W., and Spencer, R. G. 1998. Cartilage formation in a hollow fiber bioreactor studied by proton magnetic resonance microscopy. *Matrix Biol,* 17, 513–523.

Potter, K., Sweet, D. E., Anderson, P. et al. 2006. Non-destructive studies of tissue-engineered phalanges by magnetic resonance microscopy and X-ray microtomography. *Bone,* 38, 350–358.

Purcell, E. M., Torrey, H. C., and Pound, R. V. 1946. Resonance absorption by nuclear magnetic moments in a solid. *Phys Rev,* 69, 37–38.

Rabi, I. I., Millman, S., Kusch, P., and Zacharias, J. R. 1938a. The magnetic moments of $_3Li^6$, $_3Li^7$ and $_9F^{19}$. *Phys Rev,* 53, 495.

Rabi, I. I., Zacharias, J. R., Millman, S., and Kusch, P. 1938b. A new method of measuring nuclear magnetic moment. *Phys Rev,* 53, 318.

Ramaswamy, S., Greco, J. B., Uluer, M. C. et al. 2009. Magnetic resonance imaging of chondrocytes labeled with superparamagnetic iron oxide nanoparticles in tissue-engineered cartilage. *Tissue Eng Part A,* 15, 3899–3910.

Righetti, R., Ophir, J., and Krouskop, T. A. 2005. A method for generating permeability elastograms and Poisson's ratio time-constant elastograms. *Ultrasound Med Biol,* 31, 803–816.

Robertson, R. P. 1992. Pancreatic and islet transplantation for diabetes–cures or curiosities? *N Engl J Med,* 327, 1861–1868.

Ryan, E. A., Lakey, J. R., Paty, B. W. et al. 2002. Successful islet transplantation: Continued insulin reserve provides long-term glycemic control. *Diabetes,* 51, 2148–2157.

Schnall, M. D., Barlow, C., Subramanian, V. H., and Leigh, J. S. J. 1986. Wireless implanted magnetic resonance probes for *in vivo* NMR. *J Magn Reson,* 68, 161–167.

Schoeniger, J. S., Aiken, N., Hsu, E., and Blackband, S. J. 1994. Relaxation-time and diffusion NMR microscopy of single neurons. *J Magn Reson B,* 103, 261–273.

Semelka, R. C. and Helmberger, T. K. 2001. Contrast agents for MR imaging of the liver. *Radiology,* 218, 27–38.

Shapiro, A. M., Lakey, J. R., Ryan, E. A. et al. 2000. Islet transplantation in seven patients with type 1 diabetes mellitus using a glucocorticoid-free immunosuppressive regimen. *N Engl J Med,* 343, 230–238.

Shapiro, A. M., Ryan, E. A., and Lakey, J. R. 2001a. Clinical islet transplant–state of the art. *Transplant Proc,* 33, 3502–3503.

Shapiro, A. M., Ryan, E. A., and Lakey, J. R. 2001b. Pancreatic islet transplantation in the treatment of diabetes mellitus. *Best Pract Res Clin Endocrinol Metabol,* 15, 241–264.

Silver, X., Ni, W. X., Mercer, E. V. et al. 2001. *In vivo* 1H magnetic resonance imaging and spectroscopy of the rat spinal cord using an inductively-coupled chronically implanted RF coil. *Magn Reson Med,* 46, 1216–1222.

Simpson, N. E., Grant, S. C., Blackband, S. J., and Constantinidis, I. 2003. NMR properties of alginate microbeads. *Biomaterials,* 24, 4941–4948.

Smidsrod, O. 1974. Molecular basis for some physical properties of alginates in gel state. *J Chem Soc Faraday Trans,* 57, 263–274.

Smith, H.-J. and Ranallo, F. N. 1989. *A Non-Mathematical Approach to Basic MRI,* Madison, WI, Medical Physics Publishing.

Soon-Shiong, P., Feldman, E., Nelson, R. et al. 1992. Successful reversal of spontaneous diabetes in dogs by intraperitoneal microencapsulated islets. *Transplantation,* 54, 769–774.

Sosnovik, D. E., Nahrendorf, M., and Weissleder, R. 2007. Molecular magnetic resonance imaging in cardiovascular medicine. *Circulation,* 115, 2076–2086.

Stabler, C. L., Long, R. C., Constantinidis, I., and Sambanis, A. 2005a. *In vivo* noninvasive monitoring of a tissue engineered construct using 1H NMR spectroscopy. *Cell Transplant,* 14, 139–149.

Stabler, C. L., Long, R. C., Sambanis, A., and Constantinidis, I. 2005b. Noninvasive measurement of viable cell number in tissue-engineered constructs *in vitro,* using 1H nuclear magnetic resonance spectroscopy. *Tissue Eng,* 11, 404–414.

Storrs, R., Dorian, R., King, S. R., Lakey, J. R., and Rilo, H. 2001. Preclinical development of the islet sheet. *Ann NY Acad Sci,* 944, 252–266.

Stride, E. and Saffari, N. 2004. Theoretical and experimental investigation of the behaviour of ultrasound contrast agent particles in whole blood. *Ultrasound Med Biol,* 30, 1495–1509.

Strijkers, G. J., Mulder, W. J. M., Van Tilborg, G. A. F., and Nicolay, K. 2007. MRI contrast agents: Current status and future perspectives. *Anticancer Agents Med Chem,* 7, 291–305.

Stuckey, D. J., Ishii, H., Chen, Q.-Z. et al. 2010. Magnetic resonance imaging evaluation of remodeling by cardiac elastomeric tissue scaffold biomaterials in a rat model of myocardial infarction. *Tissue Engin Part A,* 16, 3395–3402.

Sullivan, S. J., Maki, T., Borland, K. M. et al. 1991. Biohybrid artificial pancreas: Long-term implantation studies in diabetic, panreatectomized dogs. *Science,* 252, 718–721.

Sun, A. M. 1988. Microencapsulation of pancreatic islet cells: A bioartificial endocrine pancreas. *Meth Enzymol,* 137, 575–580.

Sun, C., Lee, J. S. H., and Zhang, M. 2008a. Magnetic nanoparticles in MR imaging and drug delivery. *Adv Drug Deliv Rev,* 60, 1252–1265.

Sun, A. M., Parisius, W., Macmorine, H., Sefton, M. V., and Stone, R. 1980. An artificial pancreas containing cultured islets of Langerhans. *Artif Organs,* 4, 275–278.

Sun, C., Veiseh, O., Gunn, J. et al. 2008b. *In vivo* MRI detection of gliomas by chlorotoxin-conjugated superparamagnetic nanoprobes. *Small,* 4, 372–379.

Sutherland, D. E. 1996. Pancreas and islet cell transplantation: Now and then. *Transplant Proc,* 28, 2131–2133.

Tai, Y. C., Chatziioannou, A. F., Yang, Y. et al. 2003. MicroPET II: Design, development and initial performance of an improved microPET scanner for small-animal imaging. *Phys Med Biol,* 48, 1519–1537.

Tang, S. C. and Sambanis, A. 2003. Development of genetically engineered human intestinal cells for regulated insulin secretion using raav-mediated gene transfer. *Biochem Biophys Res Commun,* 303, 645–652.

Terrivitis, J. V., Bulte, J. W., Sarvananthan, S. et al. 2006. Magnetic resonance imaging of ferumoxide-labeled mesenchymal stem cells seeded on collagen scaffolds–relevance to tissue engineering. *Tissue Eng,* 12, 2765–2775.

Thelwall, P. E. and Brindle, K. M. 1999. Analysis of CHO-K1 cell growth in a fixed bed bioreactor using magnetic resonance spectroscopy and imaging. *Cytotechnology,* 30, 121–132.

Thule, P. M. and Liu, J. M. 2000. Regulated hepatic insulin gene therapy of STZ-diabetic rats. *Gene Ther,* 290, 1959–1962.

Tseng, C. L., Shih, I. L., Stobinski, L., and Lin, F. H. 2010. Gadolinium hexanedione nanoparticles for stem cell labeling and tracking via magnetic resonance imaging. *Biomaterials,* 31, 5427–5435.

Volland, N. A., Mareci, T. H., Constantinidis, I., and Simpson, N. E. 2010. Development of an inductively-coupled MR system for imaging and spectroscopic analysis of an implantable bioartificial construct at 11.1T. *Magn Reson Med,* 63, 998–1006.

Wahburn, N. R., Weir, M., Anderson, P., and Potter, K. 2004. Bone formation in polymeric scaffolds evaluated by proton magnetic resonance microscopy and x-ray microtomography. *J Biomed Mater Res A,* 69, 738–747.

Wang, H. and Iynedjian, P. B. 1997. Modulation of glucose responsiveness of insulinoma b-cells by graded overexpression of glucokinase. *Proc Natl Acad Sci,* 94, 4372–4377.

Watrin-Pinzano, A., Ruaud, J. P., Cheli, Y. et al. 2004. Evaluation of cartilage repair tissue after biomaterial implantation in rat patella by using T2 mapping. *Magma,* 17, 219–228.

Weber, C. J., Kapp, J., Hagler, M. K., Safley, S., Chrtssochoos, J. T., and Chaikof, E. 1999. *Long-Term Survival of Poly-L-Lysine Alginate Microencapsulated Islet Xenografts in Spontaneous Diabetic NOD Mice,* Boston, MA, Birkhauser.

Wickline, S. A., Neubauer, A. M., Winter, P. M., Caruthers, S. D., and Lanza, G. M. 2007. Molecular imaging and therapy of atherosclerosis with targeted nanoparticles. *J Magn Reson Imag,* 25, 667–680.

Wirth, E. D. I., Mareci, T. H., Beck, B. L., Fitzsimmons, J. R., and Reier, P. J. 1993. A comparison of an inductively coupled implanted coil with optimized surface coils for *in vivo* NMR imaging of the spinal cord. *Magn Reson Med,* 30, 626–633.

Wolff, S. D. and Balaban, R. S. 1989. Magnetization transfer contrast (MTC) and tissue water proton relaxation *in vivo. Magn Reson Med,* 10, 135–144.

Xu, H., Othman, S. F., Hong, L., Peptan, I. A., and Magin, R. L. 2006. Magnetic resonance microscopy for monitoring osteogenesis in tissue-engineered construct *in vitro. Phys Med Biol,* 51, 719–732.

Zagaynova, E. V., Streltsova, O. S., and Al, E. 2002. *In vivo* optical coherence tomography feasibility for bladder disease. *J Urol,* 167, 1492–1496.

Zhou, Q., Brown, J., Kanarek, A., Rajagopal, J., and Melton, D. 2008. *In vivo* reprogramming of adult pancreatic exocrine cells to beta-cells. *Nature,* 455, 627–632.

7

Application of Imaging Technologies to Stem-Cell Tracking *In Vivo*

Sheng-Xiang Xie
*Singapore Bioimaging
Consortium (SBIC)*

Kishore Kumar
Bhakoo
*Singapore Bioimaging
Consortium (SBIC)*

7.1 Introduction

Stem-cell research is undergoing a critical transition from being a discipline of the basic sciences to being recognized as a potential component of medical practice. Cell transplants to replace cells lost owing to injury or degenerative diseases, for which there are currently no cures, are being pursued in a wide range of experimental models.

Moreover, stem-cell therapy for degenerative disease is now a clinical reality. However, a key question in cell-based therapy is to assess the migration and retention of these transplanted at the site of injury. Moreover, there is a need to develop noninvasive technologies to assess their *in vivo* efficacy. Thus, Singapore Bioimaging Consortium (SBIC) is developing cell-tracking methodologies that can be used in long-term preclinical studies as well as translating these technologies to a clinical environment.

In this short review, an attempt will be made to summarize the latest advances in the field of molecular imaging of stem-cell transplantation and describe various cell labeling and imaging techniques.

Once implanted, it is clear that the migrational dynamics of cells will determine the extent of tissue regeneration at the site of implantation and surrounding tissue. Methods for monitoring implanted cells noninvasively *in vivo* will greatly facilitate the clinical realization and optimization of the opportunities for cell-based therapies. Owing to the seamless integration into the host parenchyma, and migration over long distances, cell grafts cannot be detected based on their mass morphology alone. To monitor cell migration and positional fate after transplantation, current models use either reporter genes or chimeric animals. These methods are cumbersome, involving sacrifice of the animal and removal of tissue for histological procedures, and cannot be translated to human studies (Lagasse et al., 2000). However, this approach lacks the temporal analysis of the donor cells. Hence, its practical uses are limited. The monitoring of stem-cell grafts, noninvasively, is an important aspect of the ongoing efficiency and safety assessment of cell-based therapies. Molecular imaging is potentially well-suited for such an application. However, for transplanted cells to be visualized and tracked by imaging technologies, they need to be tagged so that they are "visible." Moreover, imaging and biosensor technologies are moving from diagnostic toward therapeutic and interventional roles.

7.2 Stem-Cell Therapy

Stem cells have two defining properties: (1) they can produce more stem cells and they can generate specialized cell types such as nerve, blood, or liver cells; (2) they can proliferate indefinitely in culture, while retaining the potential to differentiate into virtually any cell type when placed in the right environment. Thus, in principle, stem cells can generate large quantities of any desired cell for transplantation into patients.

Until recently, the interest in stem-cell biology was as an academic curiosity. However, there is now substantial interest in stem cells as a clinical challenge. Most research effort has been devoted on how to obtain sufficient quantities of pure stem cells, how to differentiate these into appropriate cell types for therapy, and how to circumvent the immune response. However, if cell-based therapies are to find clinical applicability, there are a number of major challenges that have received little consideration, namely: (1) mapping of the spatial distribution, (2) the rate of migration *in situ*, and (3) monitoring the survival of cell grafts.

Therefore, development of methods for monitoring stem-cell grafts noninvasively, with sufficiently high sensitivity and specificity to identify and map the fate of transplanted cells, became an important aspect of application and safety assessment of stem-cell therapy. Imaging methodologies are potentially well-suited for such applications as these technologies allow *in vivo* interrogation of opaque tissues or structures inside the body.

7.3 Ideal Molecular Imaging Technology for Transplanted Stem Cells

The ideal imaging modality should provide integrated information relating to the entire process of cell engraftment, survival, and functional outcome. The *in vivo* monitoring of stem cell after grafting is essential to our understanding the first steps in cellular replacement process, that is, cellular migratory potential toward a lesion and subsequent incorporation into the damaged tissue, especially when islet cells are administrated systemically.

The imaging technologies of magnetic resonance imaging (MRI), radionuclide, and optical imaging are emerging as key modalities for *in vivo* molecular imaging because of their ability to detect molecular events. However, in order to exploit sensitivity, specificity, temporal resolution, and spatial resolution offered by these modalities they require an interaction with respective contrast agents that exerts an "effect size" enough to be visualized by complementary imaging hardware. Effective use of the tools of molecular imaging requires knowledge of the basis of detection of the imaging modality, mechanism of contrast agent interactions, and the biological environment (Bengel et al., 2005). Obviously, the use of a contrast agent for *in vivo* studies needs to be biocompatible, safe, and nontoxic. Moreover, it is necessary to avoid any kind of genetic modification of the stem cell that could perturb its genetic program. Sensitivity and specificity of imaging stem cells is another important aspect that influences its suitability for clinical applications. Indeed, the possibility of quantifying the exact cell number at any anatomical localization is particularly important for the study of the resident time of implanted stem cells. Finally, the ideal imaging should permit longitudinal tracking of implanted cells for months to years allowing long-term follow-up of tissue function and donor survival. So far, no imaging technologies satisfy all these criteria, although some do come close.

7.4 Radionuclide Imaging

Nuclear medicine is the main form of molecular and cellular imaging in current clinical use. Because of their exquisite (10^{-11}–10^{-12} mol/L) sensitivity, single-photon emission CT (SPECT) and positron emission tomography (PET) imaging modalities are able to detect tracer quantity of radioisotopes for studying biological processes in living subjects. Technological developments of both PET and SPECT have led to the implementation of specialized systems for small animal imaging with much greater spatial resolution (1–2 mm) (Massoud and Gambhir, 2003, Weber and Ivanovic, 1999, Acton and Kung, 2003), which has dramatically advanced the field of cell tracking in animal models *in vivo*.

A number of strategies for cell labeling have been described with the use of radioactive atoms for SPECT imaging. They include the direct loading with a radiometal (Gao et al., 2001, Barbash et al., 2003, Chin et al., 2003), the enzymatic conversion and retention of a radioactive substrate (Gambhir et al., 2000), and receptor-mediated binding (Gambhir et al., 2000, Simonova et al., 2003). All these strategies of cell labeling have inherent disadvantages limiting the use of radionuclide technique for cell tracking. Enzymatic conversion and retention of a radioactive substrate requires genetic manipulation of cells *ex vivo* and the administration of a substrate intravenously for each imaging session. Receptor-mediated binding requires stable expression of a receptor not found elsewhere in the body and intravenous injection of a radioactively labeled receptor ligand. Direct labeling of cells with radionuclides appears to be the only choice in radionuclide technique for cell tracking. However, the major limitations are the exposure to ionizing radiation and the short half-life time of radioisotopes leading to loss of the imaging signal within a few days. A more advanced technique for cell imaging with radionuclides over the course of months has been described with PET. It consists of the stable integration of a mutant herpes simplex type-1 thymidine kinase (TK) into islets and periodic intravenous injection of TK substrate 9-(4-[18F]-fluoro-3-hydroxy-methylbutyl) guanine (^{18}F-HBG) (Lu et al., 2006). Unfortunately, the need for genetic manipulation of islets and exposure to radiation of the recipient limits the use of this technique in clinical practice. Other application of targeted PET include noninvasive imaging of β-cell mass (Simpson et al., 2006, Souza et al., 2006). So far, PET and SPECT techniques, already used in diagnostic clinical practice for specific diseases, represent an important method to evaluate the functional effects of stem-cell-based therapy.

7.5 Optical Imaging

This imaging modality measures the intensity of the emitted light with a detector system (generally CCD cameras in a black box). Two distinct contrast mechanisms can be used for molecular and genomic imaging studies, one involving fluorescence and the other bioluminescence. Both can be used for cell tracking, but several methodological limitations preclude the use of these techniques in a clinical setting. In fluorescence imaging an external organic light source (Becker et al., 2001), or organic/inorganic hybrids (quantum dots) as exogenous contrast agents (Frangioni, 2003), is used to excite fluorescent molecules inside the subject. Green fluorescent protein (GFP) or small molecule polymethines are the most commonly used optical reporters. Conversely in bioluminescence there is no need for external light stimulation. In this case *in vivo* imaging involves introducing reporter genes that encode for enzymes (known as luciferases) that can catalyze a light-producing chemical reaction using exogenous substrates (Sato et al., 2004). Imaging of pancreatic islet transplantation was first reported for luciferase-transduced islet grafts monitored by bioluminescence optical imaging (Lu et al., 2004, Fowler et al., 2005). This method requires genetic modification of islets to stably express nonhuman genes, and the injection of strong immunogenic nonhuman substances such as luciferin and coelenterazine. The method also has limited tissue penetration. However, in spite of the opaque nature of tissues, optical imaging is an increasingly important technique for molecular imaging in small animals (Tang et al., 2003, Wang et al., 2003). It is also being developed for clinical use, in particular using near-infrared (NIR) wavelengths (~650–900 nm), which can theoretically travel 5–15 cm through tissue (Harisinghani et al., 2003). A notable theoretical advantage of optical techniques is the fact that multiple probes with different spectral characteristics can be used for multichannel imaging, similar to *in vivo* karyotyping (Weissleder and Ntziachristos, 2003).

7.6 Noninvasive Tracking of Stem Cell Using MRI

Stem-cell transplantation is becoming an effective therapy for a number of degenerative diseases. Currently, there are no methods available to visualize stem-cell grafts noninvasively. Such methods would provide spatial and temporal information regarding location, quantity, and viability of transplanted cells. A number of studies have shown the feasibility of longitudinal noninvasive monitoring of transplants in mice using MRI. This approach could potentially be translated into clinical practice for evaluating stem-cell survival and for monitoring therapeutic intervention during tissue rejection.

MRI is well-suited for stem-cell tracking because it can provide both whole-body and detailed information of host organs with near microscopic anatomical resolution and excellent soft-tissue contrast. MRI provides good anatomical information (resolution <100 µm). Conventionally, MRI relies on differences in proton density and the local magnetic environment of the hydrogen atoms to detect differences between and within tissues. Images can be enhanced using contrast agents. MR only indirectly detects most contrast agents; they are "seen" as a result of their effect on water molecules. These characteristics have led to this imaging modality to become the method of choice for following anatomical changes in soft tissue. Current developments of contrast agents are based on the use of gadolinium-analogs (Gd^{3+}-DTPA) or iron-oxide particles as contrastophores. They are designed to alter either T_1 or T_2 relaxation time and, in turn, providing either hyper- or hypointense imaging, respectively. Despite the inherent reduced sensitivity of MR compared to techniques using radionuclides (SPECT/PET), MR is more sensitive to cell detection owing to the higher concentration of achievable intracellular contrastophores.

There are numerous ways of increasing the signal-to-noise ratio in micro-MRI when imaging small animals. These include working at relatively high magnetic fields (4.7–14 T), using hardware and software customized to the small size of animals of interest, and the relative flexibility of much longer acquisition times during imaging. MRI can provide detailed morphological and functional information and, therefore, seems ideally suited to integrate efficacy assessments with the capability of cell tracking. Yet studies show that the lowest detectable number of cells is 10^5 with the use of conventional MRI scanners without any sequence modification. This threshold of detection can be lowered using high-field magnets (11.7 T) such that single cells containing a single iron particle can be detected and tracked as well as other approaches appeared in recent literature (Heyn et al., 2005, Zacharakis et al., 2005).

MRI is now a rapidly evolving molecular and cellular imaging strategy. The era of modern molecular imaging began with the first successful demonstration of a functionalized Gd-complex to visualize gene expression in 2000 (Shapiro et al., 2006). Since then successful linking of receptor molecules to contrast agents for a broad variety of applications has been published.

For stem cells to be visualized and tracked by MRI, they need to be tagged so that they are "MR visible." At present there are two types of MR contrast agent used clinically. These are gadolinium-analogs (e.g., Gd^{3+}-DTPA) or iron-oxide nanoparticles. However, these reagents were designed as blood-pool contrast agents and are impermeable to cells. Thus, several approaches have been deployed to enhance cell labeling to allow *in vivo* cell tracking by conjugating MRI contrast agents to a range of ancillary molecules to enhance their uptake. With the growing array of cell-labeling techniques, cells tagged with various monocrystalline MR probes have been evaluated both *in vitro* and *in vivo* (Josephson et al., 1999, Bhorade et al., 2000, Lewin et al., 2000, Louie et al., 2000).

Thus, the development of methods for monitoring β-cell grafts noninvasively, with sufficiently high sensitivity and specificity to identify and map the fate of transplanted cells, is an important aspect of application and safety assessment of islet therapy.

More recently, there has been growing interest in using perfluorocarbon compounds and ^{19}F-MRI as an alternative cell-tracking technique. Its advantages include direct detection, more quantifiable results, and decreased toxicity—perfluorocarbons were originally developed as blood substitutes. Unfortunately, its poor sensitivity requires further evaluation before it is assessed for widespread use (refer to Srinivas et al. for a comprehensive review (Chapon et al., 2009)); hence, the review will focus on the traditional proton MRI techniques.

7.7 Intracellular MRI Contrast Agents

Recent work in the design of MRI contrast agents has opened the possibility of combining the spatial resolution available in MRI for anatomic imaging with the ability to "tag" cells, and thus enable non-invasive detection and study of cell migration from the site of implantation. *In vivo* monitoring of stem cells after grafting is essential for understanding their migrational dynamics, which is an important aspect in determining the overall therapeutic index in cell therapies. Despite recent advances in both the synthesis of paramagnetic molecules and the basic cell biology of β-cells, methods for achieving effective cell labeling using molecular MR-tags are still in their infancy.

MRI contrast agents are used extensively in the clinic to improve sensitivity in order to facilitate diagnosis of an underlying pathology. MRI contrast agents either alter the T_1 and/or T_2 relaxation time, making the local tissue hyper- or hypointense, respectively. The most extensively used contrast agents can be classed as either paramagnetic or superparamagnetic.

Examples of paramagnetic agents are Fe^{3+}, Mn^{2+}, and Gd^{3+}. The effect of the magnetic moment in solution results in a dipolar magnetic interaction between the paramagnetic ion and the neighboring water molecules. Fluctuations in this magnetic interaction cause the decrease in T_1/T_2 relaxation time (Srinivas et al., 2010). Paramagnetic compounds produce, predominantly, T_1 effect, giving a hyperintense region.

The other class of contrast agents is superparamagnetic reagents. These consist of an iron-oxide core, typically 4–10 nm in diameter, where several thousand iron atoms are present. A biocompatible polymer surrounds the core to provide steric and/or electrostatic stabilization.

There are two types of superparamagnetic contrast agents, superparamagnetic iron-oxide (SPIO) and ultrasmall superparamagnetic iron-oxide (USPIO). The difference between the two is that the SPIOs consist of several magnetic cores surrounded by a polymer matrix whereas USPIOs are individual cores surrounded by a polymer. Superparamagnetic contrast agents provide predominantly a T_2 effect, but smaller particles have shown to act as a T_1 agent (Merbach and Toth, 2001). A new class of USPIO has been produced known as cross-linked iron-oxide (CLIO), whereby the dextran coat of the USPIO is cross-linked in the presence of epichlorohydrin, and then subjected to amination in the presence of ammonia to produce amine-terminated nanoparticles suitable for conjugation (Merbach and Toth, 2001). Thus, owing to their biocompatibility and strong effect on T_2 relaxation time, iron-oxide nanoparticles are the MR contrast agent of choice for cell labeling.

However, none of the contrast agents used in the clinic was designed for cellular internalization. To cross the cell membrane, contrast agents must either be used in conjunction with a transfection agent such as poly-L-lysine (Wunderbaldinger et al., 2002) or protamine sulfate (Bulte et al., 2004), or conjugated to a biological entity, such as a peptide transduction domain. For example, incubation with non-derivatized dextran-coated iron-oxide particles (Yeh et al., 1993, Weissleder et al., 1997, Schoepf et al., 1998, Dodd et al., 1999, Sipe et al., 1999, Arbab et al., 2004), incubation with liposome-encapsulated iron-oxide particles (Yeh et al., 1995), and lectin-mediated uptake (Bulte et al., 1993). In general, uptake is low, and further improvements in magnetic nanoparticle uptake are needed. Several approaches have been described to optimize the internalization process (Bulte et al., 1996, Daldrup-Link et al., 2003), including the link between nanoparticles and the highly cationic HIV-tat peptide (Bulte et al., 2004) or the use of an antitransferrin receptor monoclonal antibody covalently linked to nanoparticles (MION-46L) (Josephson et al., 1999). Unmodified SPIOs and USPIOs have been also successfully used at high concentration without transfection agents to label cells (fluid phase-mediated endocytosis) (Bulte et al., 1999, Daldrup-Link et al., 2003). However, the need for transfecting agent to improve cell labeling is dependent on the cell type.

Uptake of nanoparticles is time-dependent, but primary cell viability decreases with increasing incubation time. An incubation of 2–4 h seems to be an optimal compromise for labeling hematopoietic stem cells (Bulte et al., 1999). Iron-oxide-labeled cells showed a gradual decline of intracellular iron particles, owing to cell division and exocytosis or release of iron from nonviable cells. Moreover, the gradual loss of signal from labeled cells may also be attributed to the incorporation of iron into metabolic pathways.

Some superparamagnetic contrast agents have been approved by the US Food and Drug Administration (FDA) for noncell applications. The same FDA-approved iron-oxide particles have been shown to affect neither hematopoietic nor mesenchymal stem cell (MSC) function or differentiation capacity (Dunning et al., 2004). The problem with the use of these substances for cell tracking is the dilution of contrast with cell division, making it difficult to quantify owing to the susceptibility artifact, the potential transfer to host cells, such as macrophages after stem-cell death. Furthermore, a significant clinical problem common to all MR methods is the contraindication of scanning when implantable devices, such as pacemakers or defibrillators, or prosthesis are present. However, owing to its safety profile and noninvasive property MR represents, so far, the major imaging method of tracking stem cells *in vivo* (Daldrup-Link et al., 2003, Arbab et al., 2005).

7.8 MRI Tracking of Stem Cells in the Heart

Myocardial infarction is by nature an irreversible injury. The extent of the infarction depends on the duration and severity of the perfusion defect (Bulte et al., 2004). Beyond contraction and fibrosis of myocardial scar, progressive ventricular remodeling of nonischemic myocardium can further reduce cardiac function in the weeks to months after initial event (Reimer et al., 1977).

Many of the therapies available to clinicians today can significantly improve the prognosis of patients following an acute myocardial infarction (Pfeffer, 1995). However, no pharmacological or interventional procedure used clinically has shown efficacy in replacing myocardial scar with functioning contractile tissue. Cellular agents such as fetal cardiomyocytes (Ryan et al., 1999), MSCs (Etzion et al., 2001), endothelial progenitor cells (Shake et al., 2002), skeletal myoblasts (Kawamoto et al., 2001), embryonic stem (ES) cells (Jain et al., 2001), or bone marrow stromal cells (Min et al., 2002) have already shown some efficacy to engraft in the infarct, differentiate toward a cardiomyocyte phenotype by expressing cardiac-specific proteins, and preserve left ventricular function and inhibit myocardial fibrosis. Besides, *in vitro* exposure of stem cells, specifically MSCs, to specific signal molecules prior to transplantation into infarcted myocardium allows the differentiation into cardiomyocytes (Stuckey et al., 2006), and may facilitate a successful engraftment (Makino et al., 1999).

However, verifying the status of transplanted stem cells in animal models has been performed with histological analysis. Clinical data on stem-cell transplantation is in its infancy and is very limited. But preliminary clinical data have showed that stem-cell transplantation for the treatment of ischemic heart failure is feasible and promising (Tomita et al., 1999). MRI was used in some clinical studies to assess the improvement of contractile function after cell transplantation (Hamano et al., 2001), but without any possibility to visualize the transplanted stem cells. Therefore, the ability to label stem cells with MRI-visible contrast agents (Britten et al., 2003) should enable serial tracking with high spatial resolution and quantification, noninvasively by MRI, of transplanted stem cells. The programmable nature of the imaging planes allows reproducible and volumetric coverage of the heart. Moreover, this technology scales well with subject size ranging from mouse to human.

Visualization of magnetically labeled endothelial progenitor cells transplanted intra-myocardially for therapeutic neovascularization in infarcted rats has been demonstrated with *ex vivo* MRI at 7.5 T on T2-weighted images (Bulte et al., 2001).

Garot et al. (Weber et al., 2004), have demonstrated the feasibility of *in vivo* MRI tracking of skeletal muscle-derived myogenic precursor cells (MPC) preloaded with iron-oxide nanoparticles (Endorem®) injected into healthy and infarcted porcine myocardium. Iron loaded cell in the infarcted region were all detected by T2-weighted spin-echo MRI at 1.5 T. In addition, MRI guided the catheter for the injection of the labeled cells, into the ischemic myocardium, using Gd-DTPA delayed-enhancement of the site. Moreover, postmortem analysis demonstrated the presence of iron-loaded MPC at the center and periphery of the infarcted tissue as predicted by MRI.

MSCs derived from bone marrow can be detected and tracked by MRI for up to 3 weeks (Garot et al., 2003, Hill et al., 2003, Kraitchman et al., 2003). Allogeneic ferumoxides (Dick et al., 2003) or iron fluorescent particle (Hill et al., 2003, Kraitchman et al., 2003) were given by intramyocardial injection in a pig model of myocardial infarction. A minimum quantity of MSCs per injection was required to be MR-detectable on T2* (Dick et al., 2003) or T2-weighted images (Hill et al., 2003) as hypointense lesions. Indeed, Kraitchman et al. have shown that using a limited number of MSC injection per animal, only a part (~70%) of the injections performed in each animal can be visualized. Nonetheless, the implantation of MSCs in rats (Kraitchman et al., 2003) and pigs (Jaquet et al., 2005) was shown to reduce myocardial scar size, although the exact mechanism is unclear.

More recently, Chapon et al. have used a preclinical model of myocardial infarction, to track iron-oxide (IO) labeled bone marrow-derived stem cells (BMSCs) by MRI and determined their functional recovery using 2-deoxy-2-[18F]-fluoro-D-glucose-PET (FDG-PET) (Amado et al., 2005). The animals were imaged posttransplantation, from 2 days to 6 weeks postcell implantation, using concurrent MRI at 9.4 T and FDG-PET studies. Implanted IO-BMSCs were visible in the heart by MRI for the duration of the study. Histological analysis confirmed that the implanted IO-BMSCs were present for up to 6 weeks postimplantation. At 1 week post-IO-BMSC transplantation, PET studies demonstrated an increase in FDG uptake in infarcted regions implanted with live IO-BMSC compared to controls. However, there was a significant loss of the early efficacy at later time points; thus, recapitulating the early advantage seen in clinical trials using stem cells to repair ischemic lesions in the heart.

These promising experiments demonstrate the need for future studies to delineate the fate of injected stem cells by incorporating noninvasive tagging methods to monitor myocardial function following cells engraftment in the myocardial infarct area. Consequently, MRI may lead to a better understanding of the myocardial pathophysiology as well as assessment of proper implantation and the effects of stem-cell therapy by allowing a multimodal approach to evaluating anatomy, function, perfusion, and regional contractile parameters in a single noninvasive examination.

7.9 MRI Tracking of Stem Cells in the CNS

In neurodegenerative diseases where cell loss is the predominate feature of the pathology and, for which there are currently no cures, cellular replacement therapy using stem cells provide a beneficial alternative. The efficacy of cell-replacement therapy was first demonstrated using engraftment of human fetal ventral mesencephalic tissue into the brain of patient with Parkinson's disease (Chapon et al., 2009). Functional recovery and L-DOPA withdrawal followed by an increase in released dopamine demonstrated the functional integration of the grafted tissue. Although this was not a true stem-cell transplant, it nevertheless indicated that the adult brain provides local environmental cues to undifferentiated cells to produce neuronal cell types capable of providing functional recovery.

Since this pioneering experiment, many different populations of stem cells have shown to differentiate into neural phenotypes. The most obvious choice of stem-cell population would be those already derived from the neural phenotype. These include neural stem cells, found in the adult subventricular zone (SVZ) and glial-restricted precursors, found in the embryonic spinal cord. Neural stem cells have been shown to differentiate into dopaminergic neurons (Piccini et al., 1999), astrocytes and oligoden-drocytes (Ourednik et al., 2002), and spinal cord motor neurons (Yamamoto et al., 2001). Surprisingly, these cells are also able to transdifferentiate into other nonneural cell types such as skeletal muscle (Shibuya et al., 2002). Glial-restricted progenitors, on the other hand, are restricted to the glial lineage and produce oligodendrocytes and type-1 and type-2 astrocytes (Galli et al., 2000).

ES cells are the most pluripotent of all the stem cell populations, giving rise to many cell types in the body; thus have the greatest regenerative capacity. ES cells differentiate into a variety of neural phe-notypes, including dopaminergic neurons (Noble and Mayer-Proschel, 2001), serotoninergic neurons (Kawasaki et al., 2000), neuronal precursors (Lee et al., 2000), oligodendrocytes (Reubinoff et al., 2001), and astrocytes (Liu et al., 2000). There are several problems, aside from the ethics, that makes ES cell therapy difficult, including the risk of inappropriate cellular differentiation and the potential for tumor formation.

Mesenchymal cells derived from bone marrow have also been shown to differentiate into neural phenotypes (Stavridis and Smith, 2003). Additionally, rat MSCs differentiate into a mixture of neu-ral phenotypes, including astrocytes, oligodendrocytes, and neurons. Upon further differentiation, GABAergic, dopaminergic, and serotoninergic neurons may also develop (Weimann et al., 2003).

Neural progenitor cells have innate migratory properties. For example, neural progenitor cells iso-lated from the SVZ of adult or neonatal rats, when implanted into the different regions of the neonatal brain, migrate and differentiate within regions such as the *olfactory bulb*, cortex, and *striatum*. In con-trast, when grafted into the adult brain, the SVZ cells migrated only to the olfactory bulb, but not to the cortex or *striatum* (Jiang et al., 2002).

MRI was used to longitudinally track the migration of SVZ cells after implantation into the healthy rat striatum using prelabeled cells with BrdU and lipophilic dye-coated ferromagnetic particles (Herrera et al., 1999). Furthermore, MRI revealed that the area grafted with live cells appeared to expand, whereas the area implanted with dead cells decreased in size. Immunohistochemical analysis showed that the SVZ cells differentiated into neurons (MAP-2^+ NeuN$^+$) and migrated within the striatum after being cultured with bFGF. These studies revealed only localized migration. Migration over greater distances was demonstrated using noninvasive MRI studies in a stroke-lesioned animal model. It was hypoth-esized and later confirmed that stroke damage functions as a "chemoattractant" to neural stem cells (Zhang et al., 2003). Neural stem cells derived from the Maudsley Hippocampal Clone 36 (MHP36) cell line, were labeled with the bimodal contrast agent, GRID. This enables detection both by MRI and fluorescent histology. Following a middle cerebral artery occlusion (a rat model of stroke), the neural stem cells were grafted unilaterally in the hemisphere contralateral to the lesion. Using a combination of GRID-labeled cells and MRI, it was demonstrated that following 14 days posttransplantation; most of the cells had migrated to the ipsilateral hemisphere along the *corpus callosum* and populated the sur-rounding lesion area. Moreover, upon fluorescent immunohistochemistry, these cells were found to be GFAP$^+$ astrocytes and NeuN$^+$ neuronal precursor.

Hoehn et al. (Modo et al., 2004) used a similar stroke model, to demonstrate the migratory properties of implanted ES cells into the brain, by MRI. The cells were prelabeled with USPIO encapsulated with a lipofection reagent to enhance cellular uptake. The ES cells also expressed GFP as a reporter gene for immunohistochemical procedures. The labeled cells were implanted into two regions of the unaffected hemisphere: the border between the cortex and the *corpus callosum*, and the striatum. The labeled cells migrated toward the lesion in the opposite hemisphere. Strong hypointense "lines" were seen by MRI in the *corpus callosum*. These were later confirmed as migrating blankets of cells traveling toward the isch-emic lesion. GFP-expressing cells were also seen in the ischemic penumbra, and in contrast to the study by Modo et al. (Hoehn et al., 2002), the majority were NeuN$^+$ suggesting that the cells had differentiated into neurons. Astrocytes and oligodendrocytes were also seen populating the surrounding area.

MRI studies have also been used to track glial progenitors, labeled with contrast agent. However, in these studies MRI was used to scan postmortem tissues, following cellular transplantation. Oligodendrocyte progenitor cells (OPCs) from the CG4 cell line, have greater migratory and myelinating capacity than mature oligodendrocytes. The cells were labeled with monocrystalline iron-oxide nanoparticles targeted to the transferrin receptor to aid internalization of the particles (Modo et al., 2004). The progenitor cells were grafted into the spinal cord of a myelin-deficient rat. Migration was seen by MRI, especially in the dorsal column. Moreover, iron-oxide-labeled cells, fixed with paraformaldehyde and implanted in the same way, did not migrate at all. MRI contrast was seen only at the site of injection. This also suggests that the iron oxide remains localized and is not taken up by other host cells. This is of great importance if iron-oxide labeling and tracking of cells is to be used clinically. The MR images were verified by histological analysis and the lesion was found to include astrocytes, microglia, and myelin. Importantly, the Prussian blue staining correlated with that for myelin, whereas it did not overlap with the GFAP[+] astrocytes or microglia present. Obviously reactive gliosis and an immune reaction had occurred, but the inflammatory cells had not taken up iron oxide; the labeled OPCs were able to infiltrate the inflamed area and produce myelin.

Jendelova et al. (Bulte et al., 1999) used MRI to study the differential response of MSCs and ES cells in rodent models of stroke (photochemical lesion) and spinal cord compression (balloon inflation). Prior to implantation, both MSCs and ES cells were labeled with Endorem and additionally colabeled with either BrdU or GFP, respectively.

Following the induction of the lesions, either ES cells or MSCs were grafted contralateraly to the ischemic lesion. In another set of animals, either ES cells or MSCs were administered intravenously into rats with an ischemic lesion. The animals with spinal cord compression lesion were infused intravenously with MSCs. ES cells given to rats with ischemic lesions, regardless of whether given intravenously or intracerebral implantation, migrated to the lesion site within 2 weeks, as observed by MRI and subsequently confirmed with GFP immunohistochemistry. Additionally, at the site of implantation, hyperproliferation was seen in 10% of the animals. This suggests that tumor formation had taken place, and was detected using MRI as a very large hypointense area, much larger than the other cellular transplants (Jendelova et al., 2004).

Implanted MSCs also migrated to the lesion area but gathered in the necrotic tissue surrounding the lesion. Few cells entered the actual lesion and of those very few differentiated into neurons, as seen 4 weeks after implantation. Additionally, MSCs injected intravenously also migrated to the lesion site and were visible for 7 weeks postimplantation. Similarly, MSCs injected intravenously also migrated to the lesioned spinal cord, which was also confirmed with Prussian blue staining.

More recently, Jackson et al. have reported the visualization of IO-BMSCs implanted into the *striatum* of hemi-parkinsonian rats by MRI. Functional efficacy of the donor cells was monitored *in vivo* using the PET radioligand [^{11}C]-raclopride. The cells were visible for 28 days by *in vivo* MRI. BMSCs provided functional recovery demonstrated by a decreased binding of [^{11}C]-raclopride (Jendelova et al., 2004). This study demonstrates the potential of using multimodal imaging, where it is not only possible to track BMSCs but also establish their effects in a preclinical model of Parkinson's disease.

Recently, there has been concern over the effect of contrast agents on the therapeutic properties of stem cells. Guzman et al. studied the behavior of human central nervous system stem cells grown as neurospheres (hCNS-SCns) in the rat brain. It was found that SPIO-labeled hCNS-SCns were as likely as nonlabeled stem cells to migrate, survive, and differentiate following transplantation into the rat brain. Interestingly, the transplanted hCNS-SCns displayed targeted migration in mature rat brains only in the event of an injury (Jackson et al., 2009). In Modo et al.'s study, however, MHP36 cells labeled with the bimodal contrast agent GRID was found to have insignificant effect on rats with a middle cerebral artery occlusion, unlike the same cells labeled with the red fluorescent dye PKH26 (Guzman et al., 2007).

7.10 MRI of Transplanted Islets

There is presently no cure for diabetes. The pharmacological management of type-2 diabetes resulting from insulin resistance has been a success story of modern medicine. Thus, the development of cell-based therapies is not urgently indicated for these patients. However, although the incidence of early-onset type-1 diabetes mellitus, which results from loss of insulin-secreting b-cells, is relatively low, the transition from adult late-onset type-2 diabetes to type-1 diabetes is growing at an alarming rate; hence, there will be a pressing need for the transplantation of insulin-producing tissues or cells.

The encouraging results obtained from numerous recent experimental studies have allowed this field to move toward clinical translation. The growing number of early phase human studies aims to demonstrate the feasibility and potential efficacy of cell/tissue therapy in a clinical environment. At present, the transplantation of pancreatic islet transplantation has shown much promise in the treatment of type-1 diabetes (Shapiro et al., 2000, Modo et al., 2009). However, following the fate of transplanted islet cells remains a major challenge.

Imaging pancreatic islets is an important area in diabetes research, with most studies reporting on *in vitro* or *ex vivo* attempts to image pancreatic β-cells using various imaging techniques. Moreover, the noninvasive visualization and estimation of β-cell function *in vivo* is an emerging research area that has been made possible by recent advances in the field of molecular and cellular imaging. Determining β-cell mass and function *in vivo* is necessary for the assessment of either tissue or cell-based therapies for diabetes that target β-cell neogenesis, regeneration, and functional restoration.

MRI remains the most widely used imaging technology to appraise the outcome of islet transplantation noninvasively. It has been shown not only to be a useful modality to track islet cell mass, but also to demonstrate loss of intrahepatically transplanted human islets. The islets were prelabeled with Feridex, a commercial iron-oxide-based contrast agent (Shapiro et al., 2003) and tracked for up to 14 days. Later studies found that diabetic animals with transplanted islets showed a significantly higher rate of islet death than their healthy counterparts on *in vivo* MR images (Evgenov et al., 2006b). Other studies relevant to MRI of transplanted pancreatic islets provide further support for the feasibility of the imaging method in the intrahepatic transplantation model, using paramagnetic beads (Evgenov et al., 2006b), the T_1 agent, GdHPDO3A (Koblas et al., 2005), and similar SPIOs at the 1.5 T clinical field strength in a rodent model of islet transplantation (Biancone et al., 2007).

Manganese (Mn^{2+})-enhanced MRI represents an emerging modality that has been used to visualize cellular function in heart and brain *in vivo* (Tai et al., 2006); however, its use in assessing β-cell function is limited to studies *in vitro* (Silva et al., 2004). Mn^{2+} is a longitudinal relaxation time (T_1)-shortening MRI contrast agent that enters cells such as cardiomyocytes and pancreatic β-cells through voltage-gated Ca^{2+} channels (Gimi et al., 2006). Mn^{2+}-enhanced MRI has been used successfully for estimating β-cell function *in vivo*, where glucose-stimulated Mn^{2+} enhancement was seen within the pancreas of normal and streptozotocin (STZ)-induced diabetic C57BL/6J mice.

7.11 Multimodality

The challenge for understanding the very complex phenomena behind the β-cell therapy is multimodality. In fact, no unique technique is capable of addressing all the critical issues and their complementary role has to be fully exploited. So a multidisciplinary approach is needed to validate the therapy.

Most available molecular imaging techniques are applicable not only in animals but also in humans, but some problems arise. Optical techniques are not applicable to humans (and large animals); so an intelligent combination of it with MRI and radionuclide (SPECT or PET) is needed to validate the human technologies and protocols to close existing gaps between basic science and clinical trials of cardiac cell therapy. The goal of imaging is to monitor the β-cell trafficking, homing and fate, viability, and differentiation providing also quantifiable information. However, a recent report demonstrated the noninvasive detection and tracking of pancreatic islet grafts by dual-modality fluorescence/MRI. In this study, isolated human pancreatic islets were labeled with SPIO magnetic nanoparticles modified with the NIR fluorescent Cy5.5 dye, and transplanted under the kidney capsule in immunocompromised mice. It was possible to monitor pancreatic islets up to 188 days after transplantation under the kidney capsule (Gimi et al., 2006). As discussed previously the studies by Chapon et al. (Evgenov et al., 2006a) and Jackson et al. (Chapon et al., 2009) have demonstrated the feasibility of combining different imaging modalities providing synergistic information, on the repair of a lesion. Imaging-based cell-tracking methods can potentially evaluate the short-term distribution of infused cells or their long-term survival and differentiation status. Therefore, these methods would play an indispensable role in detailed preclinical studies to optimize the cell type, delivery methods, and strategies for enhancing cell survival.

7.12 Conclusions

Moreover, imaging and biosensor technologies are moving from diagnostic toward guided therapeutic and interventional roles. Molecular and cellular imaging has enormous potential for transplantation. In the first place, this will largely be as a research tool in both animal and clinical studies—providing early response markers for trials and allowing the development of new therapies. Eventually, these applications may translate into more routine clinical applications, allowing monitoring of individual patients. While this review can provide some idea of the potential, it cannot show the difficulties of developing this technology for a particular application. As well as the input of clinicians and biologists, there is a need for chemists to help develop the contrast agents, physicists for refining the imaging process itself and mathematicians to improve the data analysis and handling. None of these elements is trivial! However, in the next decade some of the approaches outlined in this review may become essential tools for research and management of transplants.

Safety and ethical issues imposed on a human study would limit the clinical translation of those methods that require extensive manipulation of cells; however, most direct labeling methods have already been applied in patients or would be suitable for the clinic.

These promising experiments demonstrate the need for future studies to delineate the fate of injected stem cells by incorporating noninvasive tagging methods to monitor organ function following cells engraftment. Consequently, MRI may lead to a better understanding of the tissue pathophysiology as well as the assessing of a proper implantation and the effects of stem-cell therapy by allowing a multimodal approach to evaluating anatomy and function in a single noninvasive examination.

In conclusion, the use of stem-cell therapy to treat degenerative and traumatic injuries is a realistic possibility in the near future. However, the need for noninvasive imaging techniques is a prerequisite in order to monitor these transplants to determine clinical efficacy. Examination by MRI ensures that the transplants are not only administered to the relevant site, but it also allows the monitoring of inappropriate cellular migration, and furthermore identifies damage to surrounding tissues.

Methods for monitoring implanted stem cells noninvasively *in vivo* will greatly facilitate the clinical realization and optimization of the opportunities for cell-based therapies. In writing this short review, we have concentrated on the application of cell tracking in a few examples. There are, however, numerous examples where similar methodologies of cell tracking can aid in clinical diagnosis or can be used to trace other cells types, such as those from a tumor or following an inflammatory response (Taupitz et al., 2003, Jackson et al., 2009).

Acknowledgment

This work was funded by A*Star, Singapore.

References

Acton, P. D. and Kung, H. F. 2003. Small animal imaging with high resolution single photon emission tomography. *Nucl Med Biol,* 30, 889–95.

Amado, L. C., Saliaris, A. P., Schuleri, K. H. et al. 2005. Cardiac repair with intramyocardial injection of allogeneic mesenchymal stem cells after myocardial infarction. *Proc Natl Acad Sci USA,* 102, 11474–9.

Arbab, A. S., Yocum, G. T., Kalish, H. et al. 2004. Efficient magnetic cell labeling with protamine sulfate complexed to ferumoxides for cellular MRI. *Blood,* 104, 1217–23.

Arbab, A. S., Yocum, G. T., Rad, A. M. et al. 2005. Labeling of cells with ferumoxides-protamine sulfate complexes does not inhibit function or differentiation capacity of hematopoietic or mesenchymal stem cells. *NMR Biomed,* 18, 553–5.

Barbash, I., Chouraqui, P., Baron, J. et al. 2003. Systemic delivery of bone marrow-derived mesenchymal stem cells to the infarcted myocardium: Feasibility, cell migration, and body distribution. *Circulation,* 108, 863–8.

Becker, A., Hessenius, C., Licha, K. et al. 2001. Receptor-targeted optical imaging of tumors with near-infrared fluorescent ligands. *Nat Biotechnol,* 19, 327–31.

Bengel, F. M., Schachinger, V., and Dimmeler, S. 2005. Cell-based therapies and imaging in cardiology. *Eur J Nucl Med Mol Imaging,* 32(Suppl 2), S404–16.

Bhorade, R., Weissleder, R., Nakakoshi, T., Moore, A., and Tung, C. H. 2000. Macrocyclic chelators with paramagnetic cations are internalized into mammalian cells via a HIV-tat derived membrane translocation peptide. *Bioconjug Chem,* 11, 301–5.

Biancone, L., Crich, S. G., Cantaluppi, V. et al. 2007. Magnetic resonance imaging of gadolinium-labeled pancreatic islets for experimental transplantation. *NMR Biomed,* 20, 40–8.

Britten, M. B., Abolmaali, N. D., Assmus, B. et al. 2003. Infarct remodeling after intracoronary progenitor cell treatment in patients with acute myocardial infarction (TOPCARE-AMI): Mechanistic insights from serial contrast-enhanced magnetic resonance imaging. *Circulation,* 108, 2212–8.

Bulte, J. W., Arbab, A. S., Douglas, T., and Frank, J. A. 2004. Preparation of magnetically labeled cells for cell tracking by magnetic resonance imaging. *Methods Enzymol,* 386, 275–99.

Bulte, J. W., Douglas, T., Witwer, B. et al. 2001. Magnetodendrimers allow endosomal magnetic labeling and *in vivo* tracking of stem cells. *Nat Biotechnol,* 19, 1141–7.

Bulte, J. W., Laughlin, P. G., Jordan, E. K., Tran, V. A., Vymazal, J., and Frank, J. A. 1996. Tagging of T cells with superparamagnetic iron oxide: Uptake kinetics and relaxometry. *Acad Radiol,* 3(Suppl 2), S301–3.

Bulte, J. W., Ma, L. D., Magin, R. L. et al. 1993. Selective MR imaging of labeled human peripheral blood mononuclear cells by liposome mediated incorporation of dextran-magnetite particles. *Magn Reson Med,* 29, 32–7.

Bulte, J. W., Zhang, S., Van Gelderen, P. et al. 1999. Neurotransplantation of magnetically labeled oligodendrocyte progenitors: Magnetic resonance tracking of cell migration and myelination. *Proc Natl Acad Sci USA,* 96, 15256–61.

Chapon, C., Jackson, J. S., Aboagye, E. O., Herlihy, A. H., Jones, W. A., and Bhakoo, K. K. 2009. An *in vivo* multimodal imaging study using MRI and PET of stem cell transplantation after myocardial infarction in rats. *Mol Imaging Biol,* 11, 31–8.

Chin, B., Nakamoto, Y., Bulte, J., Pittenger, M., Wahl, R., and Kraitchman, D. 2003. 111In oxine labelled mesenchymal stem cell SPECT after intravenous administration in myocardial infarction. *Nucl Med Commun,* 24, 1149–54.

Daldrup-Link, H. E., Rudelius, M., Oostendorp, R. A. et al. 2003. Targeting of hematopoietic progenitor cells with MR contrast agents. *Radiology,* 228, 760–7.

Dick, A. J., Guttman, M. A., Raman, V. K. et al. 2003. Magnetic resonance fluoroscopy allows targeted delivery of mesenchymal stem cells to infarct borders in Swine. *Circulation,* 108, 2899–904.

Dodd, S. J., Williams, M., Suhan, J. P., Williams, D. S., Koretsky, A. P., and Ho, C. 1999. Detection of single mammalian cells by high-resolution magnetic resonance imaging. *Biophys J,* 76, 103–9.

Dunning, M. D., Lakatos, A., Loizou, L. et al. 2004. Supeparamagnetic iron oxide-labeled Schwann cells and olfactory ensheating cells can be traced *in vivo* by magnetic resonance imaging and retain functional properties after transplantation into the CNS. *J Neurosci,* 24, 9799–810.

Etzion, S., Battler, A., Barbash, I. M. et al. 2001. Influence of embryonic cardiomyocyte transplantation on the progression of heart failure in a rat model of extensive myocardial infarction. *J Mol Cell Cardiol,* 33, 1321–30.

Evgenov, N. V., Medarova, Z., Dai, G., Bonner-Weir, S., and Moore, A. 2006a. *In vivo* imaging of islet transplantation. *Nat Med,* 12, 144–8.

Evgenov, N. V., Medarova, Z., Pratt, J. et al. 2006b. *In vivo* imaging of immune rejection in transplanted pancreatic islets. *Diabetes,* 55, 2419–28.

Fowler, M., Virostko, J., Chen, Z. et al. 2005. Assessment of pancreatic islet mass after islet transplantation using *in vivo* bioluminescence imaging. *Transplantation,* 79, 768–76.

Frangioni, J. V. 2003. *In vivo* near-infrared fluorescence imaging. *Curr Opin Chem Biol,* 7, 626–34.

Galli, R., Borello, U., Gritti, A. et al. 2000. Skeletal myogenic potential of human and mouse neural stem cells. *Nat Neurosci,* 3, 986–91.

Gambhir, S., Herschman, H., Cherry, S. et al. 2000. Imaging transgene expression with radionuclide imaging technologies. *Neoplasia,* 2, 118–38.

Gao, J., Dennis, J., Muzic, R., Lundberg, M., and Caplan, A. 2001. The dynamic *in vivo* distribution of bone marrow-derived mesenchymal stem cells after infusion. *Cells Tissues Organs,* 169, 12–20.

Garot, J., Unterseeh, T., Teiger, E. et al. 2003. Magnetic resonance imaging of targeted catheter-based implantation of myogenic precursor cells into infarcted left ventricular myocardium. *J Am Coll Cardiol,* 41, 1841–6.

Gimi, B., Leoni, L., Oberholzer, J. et al. 2006. Functional MR microimaging of pancreatic beta-cell activation. *Cell Transplant,* 15, 195–203.

Guzman, R., Uchida, N., Bliss, T. M. et al. 2007. Long-term monitoring of transplanted human neural stem cells in developmental and pathological contexts with MRI. *Proc Natl Acad Sci USA,* 104, 10211–6.

Hamano, K., Nishida, M., Hirata, K. et al. 2001. Local implantation of autologous bone marrow cells for therapeutic angiogenesis in patients with ischemic heart disease: Clinical trial and preliminary results. *Jpn Circ J,* 65, 845–7.

Harisinghani, M. G., Barentsz, J., Hahn, P. F. et al. 2003. Noninvasive detection of clinically occult lymph-node metastases in prostate cancer. *New Engl J Med,* 348, 2491–9.

Herrera, D. G., Garcia-Verdugo, J. M., and Alvarez-Buylla, A. 1999. Adult-derived neural precursors transplanted into multiple regions in the adult brain. *Ann Neurol,* 46, 867–77.

Heyn, C., Bowen, C. V., Rutt, B. K., and Foster, P. J. 2005. Detection threshold of single Spio-labeled cells with FIESTA. *Magn Reson Med,* 53, 312–20.

Hill, J. M., Dick, A. J., Raman, V. K. et al. 2003. Serial cardiac magnetic resonance imaging of injected mesenchymal stem cells. *Circulation,* 108, 1009–14.

Hoehn, M., Kustermann, E., Blunk, J. et al. 2002. Monitoring of implanted stem cell migration in vivo: A highly resolved *in vivo* magnetic resonance imaging investigation of experimental stroke in rat. *Proc Natl Acad Sci USA,* 99, 16267–72.

Jackson, J., Chapon, C., Jones, W., Hirani, E., Qassim, A., and Bhakoo, K. 2009. *In vivo* multimodal imaging of stem cell transplantation in a rodent model of Parkinson's disease. *J Neurosci Methods,* 183, 141–8.

Jain, M., Dersimonian, H., Brenner, D. A. et al. 2001. Cell therapy attenuates deleterious ventricular remodeling and improves cardiac performance after myocardial infarction. *Circulation,* 103, 1920–7.

Jaquet, K., Krause, K. T., Denschel, J. et al. 2005. Reduction of myocardial scar size after implantation of mesenchymal stem cells in rats: What is the mechanism? *Stem Cells Dev,* 14, 299–309.

Jendelova, P., Herynek, V., Urdzikova, L. et al. 2004. Magnetic resonance tracking of transplanted bone marrow and embryonic stem cells labeled by iron oxide nanoparticles in rat brain and spinal cord. *J Neurosci Res,* 76, 232–43.

Jiang, Y., Jahagirdar, B. N., Reinhardt, R. L. et al. 2002. Pluripotency of mesenchymal stem cells derived from adult marrow. *Nature,* 418, 41–9.

Josephson, L., Tung, C. H., Moore, A., and Weissleder, R. 1999. High-efficiency intracellular magnetic labeling with novel superparamagnetic-Tat peptide conjugates. *Bioconjug Chem,* 10, 186–91.

Kawamoto, A., Gwon, H. C., Iwaguro, H. et al. 2001. Therapeutic potential of *ex vivo* expanded endothelial progenitor cells for myocardial ischemia. *Circulation,* 103, 634–7.

Kawasaki, H., Mizuseki, K., Nishikawa, S. et al. 2000. Induction of midbrain dopaminergic neurons from ES cells by stromal cell-derived inducing activity. *Neuron,* 28, 31–40.

Koblas, T., Girman, P., Berkova, Z. et al. 2005. Magnetic resonance imaging of intrahepatically transplanted islets using paramagnetic beads. *Transplant Proc,* 37, 3493–5.

Kraitchman, D. L., Heldman, A. W., Atalar, E. et al. 2003. *In vivo* magnetic resonance imaging of mesenchymal stem cells in myocardial infarction. *Circulation,* 107, 2290–3.

Lagasse, E., Connors, H., AL-Dhalimy, M. et al. 2000. Purified hematopoietic stem cells can differentiate into hepatocytes in vivo. *Nat Med,* 6, 1229–34.

Lee, S. H., Lumelsky, N., Studer, L., Auerbach, J. M., and Mckay, R. D. 2000. Efficient generation of midbrain and hindbrain neurons from mouse embryonic stem cells. *Nat Biotechnol,* 18, 675–9.

Lewin, M., Carlesso, N., Tung, C. H. et al. 2000. Tat peptide-derivatized magnetic nanoparticles allow *in vivo* tracking and recovery of progenitor cells. *Nat Biotechnol,* 18, 410–4.

Liu, S., Qu, Y., Stewart, T. J. et al. 2000. Embryonic stem cells differentiate into oligodendrocytes and myelinate in culture and after spinal cord transplantation. *Proc Natl Acad Sci USA,* 97, 6126–31.

Louie, A. Y., Huber, M. M., Ahrens, E. T. et al. 2000. *In vivo* visualization of gene expression using magnetic resonance imaging. *Nat Biotechnol,* 18, 321–5.

Lu, Y., Dang, H., Middleton, B. et al. 2004. Bioluminescent monitoring of islet graft survival after transplantation. *Mol Ther,* 9, 428–35.

Lu, Y., Dang, H., Middleton, B. et al. 2006. Noninvasive imaging of islet grafts using positron-emission tomography. *Proc Natl Acad Sci USA,* 103, 11294–9.

Makino, S., Fukuda, K., Miyoshi, S. et al. 1999. Cardiomyocytes can be generated from marrow stromal cells in vitro. *J Clin Invest,* 103, 697–705.

Massoud, T. F. and Gambhir, S. S. 2003. Molecular imaging in living subjects: Seeing fundamental biological processes in a new light. *Genes Dev,* 17, 545–80.

Merbach, A. E. and Toth, E. 2001. *The Chemistry of Contrast Agents in Medical Magnetic Resonance Imaging.* New York: John Wiley and Sons.

Min, J. Y., Yang, Y., Converso, K. L. et al. 2002. Transplantation of embryonic stem cells improves cardiac function in postinfarcted rats. *J Appl Physiol,* 92, 288–96.

Modo, M., Beech, J. S., Meade, T. J., Williams, S. C., and Price, J. 2009. A chronic 1 year assessment of MRI contrast agent-labelled neural stem cell transplants in stroke. *Neuroimage,* 47(Suppl 2), T133–42.

Modo, M., Mellodew, K., Cash, D. et al. 2004. Mapping transplanted stem cell migration after a stroke: A serial, *in vivo* magnetic resonance imaging study. *Neuroimage,* 21, 311–7.

Noble, M. and Mayer-Proschel, M. 2001. Glial-restricted progenitors. *In:* Rao, M. S. (ed.) *Stem Cells and CNS Development.* Totowa, NJ: Humana Press.

Ourednik, J., Ourednik, V., Lynch, W. P., Schachner, M., and Snyder, E. Y. 2002. Neural stem cells display an inherent mechanism for rescuing dysfunctional neurons. *Nat Biotechnol,* 20, 1103–10.

Pfeffer, M. A. 1995. Left ventricular remodeling after acute myocardial infarction. *Annu Rev Med,* 46, 455–66.

Piccini, P., Brooks, D. J., Bjorklund, A. et al. 1999. Dopamine release from nigral transplants visualized *in vivo* in a Parkinson's patient. *Nat Neurosci,* 2, 1137–40.

Reimer, K. A., Lowe, J. E., Rasmussen, M. M., and Jennings, R. B. 1977. The wavefront phenomenon of ischemic cell death. 1. Myocardial infarct size vs. duration of coronary occlusion in dogs. *Circulation,* 56, 786–94.

Reubinoff, B. E., Itsykson, P., Turetsky, T. et al. 2001. Neural progenitors from human embryonic stem cells. *Nat Biotechnol,* 19, 1134–40.

Ryan, T. J., Antman, E. M., Brooks, N. H. et al. 1999. 1999 update: ACC/AHA guidelines for the management of patients with acute myocardial infarction. A report of the American College of Cardiology/American Heart Association Task Force on Practice Guidelines (Committee on Management of Acute Myocardial Infarction). *J Am Coll Cardiol,* 34, 890–911.

Sato, A., Klaunberg, B., and Tolwani, R. 2004. *In vivo* bioluminescence imaging. *Comp Med,* 54, 631–4.

Schoepf, U., Marecos, E. M., Melder, R. J., Jain, R. K., and Weissleder, R. 1998. Intracellular magnetic labeling of lymphocytes for *in vivo* trafficking studies. *Biotechniques,* 24, 642–6, 648–51.

Shake, J. G., Gruber, P. J., Baumgartner, W. A. et al. 2002. Mesenchymal stem cell implantation in a swine myo-cardial infarct model: Engraftment and functional effects. *Ann Thorac Surg,* 73, 1919–25; discussion 1926.

Shapiro, M. G., Atanasijevic, T., Faas, H., Westmeyer, G. G., and Jasanoff, A. 2006. Dynamic imaging with MRI contrast agents: Quantitative considerations. *Magn Reson Imaging,* 24, 449–62.

Shapiro, A. M., Lakey, J. R., Ryan, E. A. et al. 2000. Islet transplantation in seven patients with type 1 diabetes mellitus using a glucocorticoid-free immunosuppressive regimen. *N Engl J Med,* 343, 230–8.

Shapiro, A. M., Ricordi, C., and Hering, B. 2003. Edmonton's islet success has indeed been replicated elsewhere. *Lancet,* 362, 1242.

Shibuya, S., Miyamoto, O., Auer, R. N., Itano, T., Mori, S., and Norimatsu, H. 2002. Embryonic intermediate filament, nestin, expression following traumatic spinal cord injury in adult rats. *Neuroscience,* 114, 905–16.

Silva, A. C., Lee, J. H., Aoki, I., and Koretsky, A. P. 2004. Manganese-enhanced magnetic resonance imaging (MeMRI): Methodological and practical considerations. *NMR Biomed,* 17, 532–43.

Simonova, M., Shtanko, O., Sergeyev, N., Weissleder, R., and Bogdanov, A. 2003. Engineering of technetium-99m-binding artificial receptors for imaging gene expression. *J Gene Med,* 5, 1056–66.

Simpson, N. R., Souza, F., Witkowski, P. et al. 2006. Visualizing pancreatic beta-cell mass with [11C]Dtbz. *Nucl Med Biol,* 33, 855–64.

Sipe, J. C., Filippi, M., Martino, G. et al. 1999. Method for intracellular magnetic labeling of human mono-nuclear cells using approved iron contrast agents. *Magn Reson Imaging,* 17, 1521–3.

Souza, F., Simpson, N., Raffo, A. et al. 2006. Longitudinal noninvasive PET-based beta cell mass estimates in a spontaneous diabetes rat model. *J Clin Invest,* 116, 1506–13.

Srinivas, M., Aarntzen, E. H., Bulte, J. W. et al. 2010. Imaging of cellular therapies. *Adv Drug Deliv Rev.,* 62, 1080–93.

Stavridis, M. P. and Smith, A. G. 2003. Neural differentiation of mouse embryonic stem cells. *Biochem Soc Trans,* 31, 45–9.

Stuckey, D. J., Carr, C. A., Martin-Rendon, E. et al. 2006. Iron particles for noninvasive monitoring of bone marrow stromal cell engraftment into, and isolation of viable engrafted donor cells from, the heart. *Stem Cells,* 24, 1968–75.

Tai, J. H., Foster, P., Rosales, A. et al. 2006. Imaging islets labeled with magnetic nanoparticles at 1.5 Tesla. *Diabetes,* 55, 2931–8.

Tang, Y., Shah, K., Messerli, S. M., Snyder, E., Breakefield, X., and Weissleder, R. 2003. *In vivo* tracking of neural progenitor cell migration to glioblastomas. *Hum Gene Ther,* 14, 1247–54.

Taupitz, M., Schmitz, S., and Hamm, B. 2003. [Superparamagnetic iron oxide particles: Current state and future development]. *Rofo Fortschr Geb Rontgenstr Neuen Bildgeb Verfahr,* 175, 752–65.

Tomita, S., Li, R. K., Weisel, R. D. et al. 1999. Autologous transplantation of bone marrow cells improves damaged heart function. *Circulation,* 100, II247–56.

Wang, X., Rosol, M., GE, S. et al. 2003. Dynamic tracking of human hematopoietic stem cell engraftment using *in vivo* bioluminescence imaging. *Blood,* 102, 3478–82.

Weber, D. A. and Ivanovic, M. 1999. Ultra-high-resolution imaging of small animals: Implications for preclinical and research studies. *J Nucl Cardiol,* 6, 332–44.

Weber, A., Pedrosa, I., Kawamoto, A. et al. 2004. Magnetic resonance mapping of transplanted endothelial progenitor cells for therapeutic neovascularization in ischemic heart disease. *Eur J Cardiothorac Surg,* 26, 137–43.

Weimann, J. M., Charlton, C. A., Brazelton, T. R., Hackman, R. C., and Blau, H. M. 2003. Contribution of transplanted bone marrow cells to Purkinje neurons in human adult brains. *Proc Natl Acad Sci USA,* 100, 2088–93.

Weissleder, R., Cheng, H. C., Bogdanova, A., and Bogdanov, A., JR. 1997. Magnetically labeled cells can be detected by MR imaging. *J Magn Reson Imaging,* 7, 258–63.

Wunderbaldinger, P., Josephson, L., and Weissleder, R. 2002. Crosslinked iron oxides (CLIO): A new platform for the development of targeted MR contrast agents. *Acad Radiol,* 9(Suppl 2), S304–6.

Weissleder, R. and Ntziachristos, V. 2003. Shedding light onto live molecular targets. *Nat Med,* 9, 123–8.

Yamamoto, S., Yamamoto, N., Kitamura, T., Nakamura, K., and Nakafuku, M. 2001. Proliferation of parenchymal neural progenitors in response to injury in the adult rat spinal cord. *Exp Neurol,* 172, 115–27.

Yeh, T. C., Zhang, W., Ildstad, S. T., and Ho, C. 1993. Intracellular labeling of T-cells with superparamagnetic contrast agents. *Magn Reson Med,* 30, 617–25.

Yeh, T. C., Zhang, W., Ildstad, S. T., and Ho, C. 1995. *In vivo* dynamic MRI tracking of rat T-cells labeled with superparamagnetic iron-oxide particles. *Magn Reson Med,* 33, 200–8.

Zacharakis, G., Kambara, H., Shih, H. et al. 2005. Volumetric tomography of fluorescent proteins through small animals in vivo. *Proc Natl Acad Sci USA,* 102, 18252–7.

Zhang, R. L., Zhang, L., Zhang, Z. G. et al. 2003. Migration and differentiation of adult rat subventricular zone progenitor cells transplanted into the adult rat striatum. *Neuroscience,* 116, 373–82.

8

Computer Tomography and Micro-CT for Tissue Engineering Applications

Zhang Zhiyong
Shanghai Jiao Tong University
National Tissue Engineering Center of China

Jerry K. Y. Chan
KK Women's and Children's Hospital
National University of Singapore
Duke-NUS Graduate Medical School

Teoh Swee Hin
Nanyang Technological University

8.1 Introduction to CT/Micro-CT

Tissue engineering (TE) and regenerative medicine has been generally regarded as one of the most promising approaches to eventually address the ever-pressing clinical problem of organ and tissue shortage with the global aging in this century. The multi-disciplinary research efforts in this field mainly concentrate on generating functionally effective tissue-engineered (TE) constructs, which can be implanted, restore, and repair the damaged organ or tissue. The combinational use of three-dimensional (3D) scaffolds, stem cells, growth factors and culture conditioning to fabricate TE constructs has constituted the basic tenets in the field of TE and regenerative medicine. To facilitate the translational research of TE and regenerative medicine, it is essentially important and desirable to develop examination techniques, which can quantitatively and qualitatively analyze the spatial development of cell performance and tissue regeneration in the TE constructs both *in vitro* before implantation and *in vivo* after implantation.

Microscopy-based techniques are the most commonly used tool to examine the cellular and tissue profile of TE constructs (Liu et al., 2007, Dadsetan et al., 2008). Histological section of TE constructs is usually required prior to the microscopy examination and histological analysis using microscopy remains as the gold standard method for analyzing the performance of TE construct *in vitro* and *in vivo*. It provides a powerful tool to distinguish different tissue formation and study the fine structure of cell and ECM in a very high resolution. However, the information acquired is usually limited to two-dimensional, and in order to obtain 3D information for the spatial development of tissue regeneration, multiple histological sections are required, which is highly tedious experimental process and cannot fully cover all the information throughout the whole TE constructs. Furthermore, although histomorphometric analysis can be used, it only provides semi-quantitative information. This technique is further limited by its destructive and tedious sample processing, which does not allow the repeated utilization of samples and *in vivo* live application. An improved microscopy technique involves the use of fluorescence probes, which may simplify the sample preparation without histological sectioning and can be more quantitative when high-throughput approaches are applied (Simon et al., 2005, Dragunow, 2008). Furthermore, confocal laser fluorescence microscopy can yield 3D images for spatial information through Z-stack scanning (Liu et al., 2007, Zhang et al., 2009a,b). However, the information acquired in this approach, is only limited to a few hundred microns on the surface of samples, thus cannot image the interior structure of TE constructs (Dorsey et al., 2009).

Another common technique to analyze TE constructs involves the indirect colorimetric and fluoro-metric soluble assays of a variety of components in the cells or ECM, including the enzymes (dehydrogenase) (Mosmann, 1983, Simon et al., 2002), protein (BCA) (Smith et al., 1985, Popat et al., 2005) or DNA (Picogreen) (Zhang et al., 2009a,b, 2010b). These soluble assays are highly quantitative, accurate and sensitive, for example, the Picogreen dsDNA assay can detect the DNA concentrations as low as 10 ng/mL DNA, thus providing a powerful tool to investigate the cell proliferation and function (Dorsey et al., 2009). However, like histology, it requires destructive sample preparation procedure to release the cellular components and furthermore it does not yield information on the spatial distribution of cell and tissue within the TE constructs.

On the contrary, computed tomography (CT) imaging technique, especially the micro-CT with much higher resolution, provides a powerful imaging approach to investigate the *in vitro* and *in vivo* performance of TE constructs with both qualitative and spatial information. In addition, it is a rapid and nondestructive imaging technique, which is easy to perform and usually does not require any sample preparation. Over the past decades, it has become more important and favorable measuring tool in the field of TE and regenerative medicine, as indicated by the rising annual number of publications employing CT/micro-CT (Figure 8.1). The increased use of CT, especially micro-CT modalities in TE and regenerative medicine can be explained by the following advantages compared with other measurement techniques: (1) CT/micro-CT is noninvasive and offers ease of processing, thus eliminating the need for complicated, destructive, and potentially harmful preparation steps; (2) CT especially micro-CT can provide images with very high resolution with submicron scale, which can depict the fine cellular structure as conventional microscopy; (3) CT/micro-CT can provide both quantitative and qualitative information about the samples with high accuracy; (4) 3D imaging can be reconstructed from CT/micro-CT dataset with spatial information of the tissue and cellular distribution, allowing viewing in

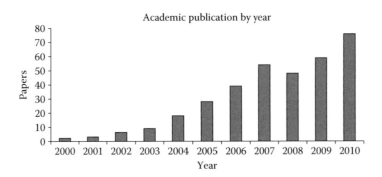

FIGURE 8.1 The rising number of annual publications regarding CT/micro-CT in the research field of TE and regenerative medicine indicates the importance and popularity of this technique. (This graph is based on a simple query of the public database PubMed by using the following mesh terms: "Computed tomography" or "CT" or "micro-CT" or "micro CT" and "Tissue Engineering" or "Regenerative Medicine.")

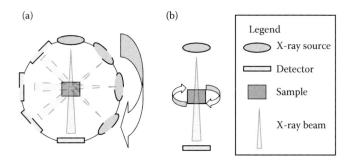

FIGURE 8.2 CT/micro-CT construction principles. (a) The samples to be imaged remain still while the x-ray source and detector rotate; (b) x-ray source and detector remain still while the samples rotate.

any cross section of samples, including the interior morphology; and (5) CT/micro-CT can be used for *in vivo* scanning without sacrificing the animals, making it possible for repetitive follow-up study using the same animals, which is highly favorable in animal experimentation (Ho and Hutmacher, 2006).

The invention of CT has been considered to be the greatest innovation in the field of radiology since the discovery of x-rays. G.N. Hounsfield conceived the idea about the design of CT in the mid-1960s, followed by the development of the first prototype at the EMI Central Research Laboratories, and the first clinical patient was scanned in 1971 (Beckmann, 2006). Owing to their contribution in this technological breakthrough, G.N. Hounsfield and A.M. Cormack were awarded the Nobel Prize in medicine in 1979 (Beckmann, 2006). The first description of micro-CT use in preclinical research emerged in the late 1970s and the early 1980s and ever since then micro-CT has been increasingly used in academic research especially in animal experimentations, as evidenced by the rapid increase in publications over the past three decades (Kujoory et al., 1980, Seguin et al., 1985, Schambach et al., 2010). Although providing much higher resolution, micro-CT is identical to the medical CT in its basic principles. Briefly, an x-ray beam projects through the sample to be imaged on a detector array, the relative rotation movement is applied between the sample and x-ray source coupled with detector array so that projection images of sample in different rotation angle can be acquired (Figure 8.2). According to the detector measurements, x-ray paths are calculated and the attenuation coefficients are derived. A 2D pixel map is created from these computations and each pixel is denoted by a threshold value which corresponds to the attenuation coefficient measured at a similar location within the sample. The process is repeated, generating a series of two-dimensional images through the length of the specimen. 3D modeling software can use computer algorithms to reconstruct the internal and external structure of the sample from 2D image (Hedberg et al., 2005, Ho and Hutmacher, 2006).

8.1.1 CT/Micro-CT Construction Principles

Generally speaking, the CT/micro-CT scanners can be divided into two categories according to their construction principles, especially the movement styles between samples and x-ray source detector. The first type of CT/micro-CT scanners had the samples that remained still during scanning, while the x-ray source-detector gantry rotates around them (Figure 8.2a). These scanners usually possess a fixed x-ray source–detector distance (SDD) as well as a determined imaging magnification level, which are conditioned during the manufacture of machines. This construction principle is usually found in clinical CT and some high-resolution *in vivo* micro-CT scanner, for example, Skyscan 1076 *in vivo* micro-CT, where the examined samples are usually too big and heavy to rotate without movement blurring during the scanning. Due to the fixed SDD and thus the magnification level, the resolution of this type of CT/micro-CT scanners depends mainly on the pixel pitch and matrix size of the detector used (Schambach et al., 2010).

On the contrary, in the second type of CT/micro-CT scanners, x-ray source and detector are fixed in a position, while the examined sample is rotated within the pathway of the x-ray beam (Figure 8.2b). This construction principle allows free adjustment of source–object distance (SOD) and object–detector distance (ODD), thereby allowing SDD adjustment. Free adjustment of SOD and ODD facilitates optimization of the geometric magnification level, depending on the signal-to-noise ratio (SNR) and the penumbra blurring (Badea et al., 2008, Schambach et al., 2010). Compared with the first type of scanners, these scanners can allow more flexibility during scanning and usually achieve higher resolution when short SOD is used to scan the small samples. Moreover, different rotation modes of samples have been used in these scanners, including the horizontal rotation, vertical rotation, or orthogonal rotation to the path of x-ray beam (Badea et al., 2004, Schambach et al., 2009, 2010).

8.1.2 X-Ray Sources

X-ray sources are used to generate x-ray beams to project through the samples. Generally, two different x-ray sources can be used for the CT/micro-CT scanners. X-ray tubes are most widely used in the self-contained CT/micro-CT scanner, with the resolution up to 1 µm. Alternatively, x-ray beam derived from synchrotron facilities can be used as a high-quality x-ray source for the CT/micro-CT scanners as well, providing much higher resolution and better phase contrast image (Betz et al., 2007, Westneat et al., 2008, Metscher, 2009a). High-brilliance x-ray beams can be generated from synchrotron's electron stream with narrow bandwidths at chosen energies, which can be manipulated using diffractive zone plates, just like focusing a light beam with refractive lenses in an optical microscope. Compared to the broadband x-ray beam generated from the conventional x-ray tubes, narrowband x-ray beam from synchrotron facilities can not only allow a finer imaging resolution, up to 60 nm for imaging subcellular structure (Le Gros et al., 2005, Parkinson et al., 2008, McDermott et al., 2009), but also provide better phase contrast to distinguish the subtle difference in soft-tissue imaging (Betz et al., 2007, Westneat et al., 2008, Metscher, 2009a).

8.1.3 X-Ray Detectors

X-ray detector is used to record x-ray intensity after the x-ray beam projects through the samples. Charge-coupled device (CCD) photo detectors, which are coupled to a scintillator by tapered glass fibers, are most widely used x-ray detector in CT/micro-CT scanners. These detectors are highly photosensitive with high signal yield, due to the lack of readout electronics covering the photosensitive layer, however, readout is facilitated by shift register readout with one signal output over the entire detector, leading to longer readout times (Schambach et al., 2010).

The second type of x-ray detector can be the active matrix flat panel imagers (AMFPI) that consist of an array of photodiodes connected by a matrix of thin film transistors (TFTs) such as TFT liquid crystal image displays. Compared to CCD detector, AMFPI detector possesses a feature of faster readout times, because the information can be read out row by row, however, the readout electronics (TFTs) can cover up to 50% of the photosensitive layers, thus reduce photosensitivity of the detectors (Schambach et al., 2010).

CT/micro-CT has become an increasingly popular imaging technique for biomedical research, especially in TE and regenerative medicine. They have been utilized to evaluate and improve the design of 3D scaffolds for TE applications. They are proven to be one of the most powerful imaging techniques to assess the mineralized hard tissue, especially bone tissue; therefore, they have been widely used to measure the new bone formation in the bone TE field. Last but not least, with the use of proper x-ray contrast agents, CT/micro-CT have been demonstrated their great potential to qualitatively and quantitatively evaluate a variety of nonmineralized tissue, including the vascular network, cells cultured on the 3D scaffolds, brain, kidney, and other animal soft tissues.

8.2 Applications of CT/Micro-CT to Image Tissue Engineering Scaffolds

3D scaffold is an essential component for TE. The 3D scaffold provides the necessary support for cells to attach, grow, and differentiate and defines the overall shape of the TE transplant (Langer and Vacanti, 1993). The structural and architectural features of scaffolds, including porosity, pore size, surface area-to-volume ratio, interconnectivity, anisotrophy and strut thickness, and so on are the crucial factors, which have the profound influence on the mechanical and biological functionality of scaffolds and the success of clinical therapeutic applications (Lin et al., 2003). Thus, effective examination techniques to assess those critical properties of scaffolds are highly essential and beneficial for the design and development of scaffolds with favorable characteristics for TE applications.

Several scaffold assessment techniques have been developed, including theoretical calculation, scanning electron microscopy (SEM), and mercury intrusion porosimetry. Theoretical calculation is quite frequently used for the estimation of scaffold porosity. The porosity of scaffolds can be estimated easily using the following equation (Woodfield et al., 2004, Ho and Hutmacher, 2006):

$$\text{Porosity} = (1 - \text{Vg} \times \text{Va}) \times 100\%$$

where Vg is the volume of scaffold material (= mass of scaffold/density of scaffold material), Va the apparent scaffold cube volume (which can be calculated by the direct measurement of the overall dimensions of the scaffold).

Other theoretical calculation methods include the Archimedes method (Ma and Zhang, 2001, Chu et al., 2002) and the liquid displacement technique (Zhang and Ma, 1999, Ramay and Zhang, 2003), where the porous scaffold is submerged under the liquid with the measurement of the changes of weight or volume. Theoretical calculation is easy and useful technique to assess the porosity of scaffolds; however, they are incapable to examine many other parameters of scaffolds such as pore size, surface area-to-volume ratio, interconnectivity, anisotropy, and strut/wall thickness.

SEM imaging is another popular scaffold assessment method, which can provide high magnification of microscopic image, depicting the surface morphology of scaffolds. The pore size and strut/wall thickness can be accurately measured and interconnectivity, cross-section area, and anisotropy can also be visually estimated (Ho and Hutmacher, 2006). However, SEM imaging can only provide two-dimensional qualitative information by imaging and analyzing the superficial layer of the scaffolds.

Mercury intrusion porosimetry is another well-established measurement tool to analyze scaffolds. Scaffolds are submerged in the mercury liquid, which usually does not intrude into the pore and cavity space of scaffold until sufficient pressure is applied, because of its nonwetting physical property. The changes of mercury volume and pressure are recorded, from which a number of parameters can be determined, including apparent scaffold volume, total pore volume, porosity, and pore tortuosity. However, this method failed to assess the information of interconnectivity, strut/wall thickness, and anisotropy. Some other drawbacks of this technique include that it does not account for the closed pore and the pores with the diameter less than 0.0018 μm as well as the high pressure applied to intrude the mercury volume into pore may compress and affect the whole scaffold architecture. Furthermore, mercury porosimetry measurement is a destructive process due to the use of toxic mercury liquid (Ho and Hutmacher, 2006).

More recently, micro-CT has been explored in the assessment of scaffold with a number of advantages over the conventional techniques. It provides 3D quantitative information about the external and internal structure of scaffolds. Comprehensive information about the scaffolds can be assessed, including the porosity, pore size, surface area-to-volume ratio, anisotropy, and strut/wall thickness (Table 8.1). Furthermore, micro-CT can be used to quantitatively analyze pore interconnectivity within the scaffolds, which is an important parameter influencing cell migration and tissue ingrowth but generally neglected by the previous assessment methods. Several approaches with different

TABLE 8.1 Comparison of Different Techniques for Scaffold Assessment

	Porosity	Pore Size	Pore Interconnectivity	Surface Area/ Volume	Strut/Wall Thickness	Anisotropy
Theoretical methods	+	−	−	−	−	−
SEM	Qualitative	+	−	−	+	Qualitative
Mercury intrusion porosimetry	+	+	−	−	−	−
Micro-CT	+	+	+	+	+	+

analyzing algorithms have been developed to use micro-CT for the quantitative evaluation of the pore interconnectivity (Moore et al., 2004, Otsuki et al., 2006, Jones et al., 2007). Other advantages to use micro-CT would include that scaffolds with intricate interior structures and closed pore can be analyzed. Additionally, micro-CT is a non-destructive method, where the scaffold samples can be re-used for other experiment. Last but not least, the data obtained from micro-CT on precise 3D structural and architectural information of scaffolds can subsequently be used in finite element modeling (FEM). These simulations can then be used to investigate the mechanical properties and fluid-diffusion patterns in scaffold via computational modeling methods (Teo et al., 2006, Singh et al., 2007, Hutmacher and Singh, 2008).

We have developed a 3D polycaprolactone (PCL) scaffold using fused deposition modeling rapid prototyping technique (Hutmacher et al., 2001, Zein et al., 2002). Micro-CT was used to analyze the architecture of this 3D PCL scaffold (Figure 8.3) nondestructively, which can be imaged from different view angles (Figure 8.3a and b). Radom digital section can be done at any planar to exhibit different geometric layouts at different spatial location (Figure 8.3c). Moreover, the 3D digital model of scaffold allows the zoom-in view of any specific location within scaffolds. For example, the cubic region of scaffold in Figure 8.3c (indicated by dotted line) can be zoomed into Figure 8.3d and the pore interconnectivity of the scaffold can be visualized and quantitatively analyzed by inverting the threshold (Figure 8.3e). Figure 8.3f showed the merged image of Figure 8.3d and e. The 3D model of the scaffold created by micro-CT can be further imported to other application software for other analysis, for instance, the permeability and fluid dynamic analysis of this 3D PCL scaffolds cultured in a biaxial rotating bioreactor using computational fluid dynamics (CFD) software (Singh et al., 2007, Hutmacher and Singh, 2008).

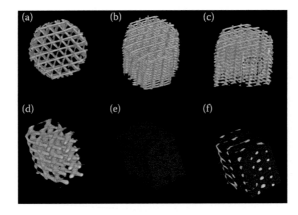

FIGURE 8.3 Micro-CT image of 3D PCL scaffold. (a, b) Different views of 3D PCL scaffold; (c) the digital section view of scaffold at certain planar; (d) the close-up view of scaffolds, the cubic region as indicated by dotted line in (c); (e) the inverted image showing the interconnected pore; and (f) the merged image of (d) and (e).

8.3 Applications of CT/Micro-CT to Image Hard Mineralized Tissue

Hard mineralized tissue such as bone tissue attenuates x-ray beams more significantly than the surrounding soft tissue, so it can easily be distinguished from the bone marrow and muscle tissue. Therefore, CT have been firstly and most widely used for the evaluation of bone morphology and microstructure, and in 1980s, Feldkamp et al. pioneered the use of micro-CT to examine the bone tissue in a much higher resolution, with a landmark study detailing the noninvasive analysis of trabecular bone (Feldkamp et al., 1989, Thomsen et al., 2005). CT/micro-CT have been explored and reported to investigate the quality of bone tissue in numerous applications, including assessing the bone density, osteogenesis, ovariectomy and osteoporosis, bone resorption, bone remodeling, bone neoplasm, and other topics (Schambach et al., 2010). Micro-CT also demonstrated as a powerful nondestructive imaging technique to monitor the dynamic change of bone morphology and microstructure under the influence of diseases or therapeutic treatments (Jiang et al., 2000, Patel et al., 2003). In addition, the 3D data, obtained by micro-CT has been verified against conventional 2D histological sections, considered to be the "gold standard" for studying bone tissue. The studies revealed high correlations between measures of bone structure obtained from conventional 2D sections and 3D micro-CT data, validating the application of micro-CT in bone tissue assessment (Müller et al., 1996, Thomsen et al., 2005).

Because of their advantages in analyzing the bone tissue, CT/micro-CT have become an increasingly popular imaging tool in the research field of bone TE. Together with conventional radiographic imaging, CT/micro-CT can be used to visualize the 3D morphology and structure of new bone formation and bony defect regeneration. For example, we managed to use micro-CT scanner to monitor the orthotopic and ectopic bone formation of the human fetal mesenchymal stem cells (hfMSC) mediated 3D scaffold constructs after *in vivo* implantation. Figure 8.4a showed the ectopic bone formation (white color) on the PCL scaffolds (dark color) after the implantation of hfMSC. The ectopic bone volume can then be quantified and statistically analyzed (Zhang et al., 2009b). Moreover, micro-CT can be used to monitor the new bone regeneration on the femoral defects in rats treated with the implantation of hfMSC cellular scaffold (Figure 8.4b).

Another favorable application of CT/micro-CT in the TE research field is their capability to conduct the live scanning of the animal without sacrificing the animals, allowing follow-up studies in the animal experiments, because the data-acquisition process of micro-CT is noninvasive. This *in vivo* scanning application will reduce the number of animals used in the experiment and minimize the experiment variation by repetitive follow-up scanning on the same animals. Figure 8.5 demonstrated the technical feasibility of repetitive live scanning of the femoral area in rats, where bone tissue can be clearly visualized and distinguished from surrounding tissues using a threshold of 100. *In vivo* micro-CT scanner (1076, SkyScan, Belgium) can be used for the live scanning, rats were anesthetized first, then placed in

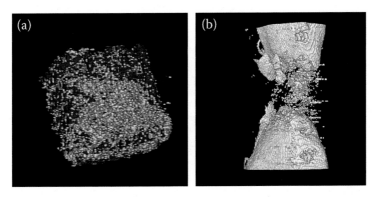

FIGURE 8.4   Micro-CT analysis of ectopic bone formation (a) and the orthotopic bone formation (b).

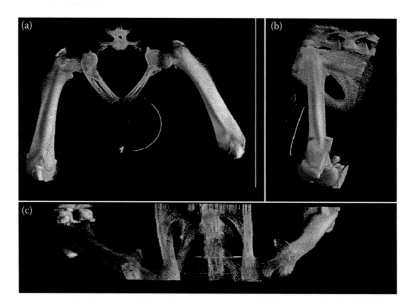

FIGURE 8.5 *In vivo* micro-CT scanning of rat's femora: (a) frontal plane view; (b) sagittal plane view; (c) axial plane view. Threshold = 100.

the animal holder with the hindlimbs stretched as straight as possible. A spatial resolution of 35 μm, an averaging of 4 and 1 mm aluminum filter were applied during the scanning. The scanning frequency of once per month, we found that rats survive healthily without any difference from other unscanned rats. Furthermore, because the femoral area is far away from the chest, the influence of rhythmical chest movement on the micro-CT scanning can be ignored. Thus, the *in vivo* live micro-CT scanning on the rat's femoral area can be carried out under normal anesthetization, in the absence of synchronizing scanning with the breath of rats.

Besides their capability to image bone tissue, CT/micro-CT can be used to assess mineralization deposition within a variety of porous biomaterial scaffolds either by *in vitro* cell culture or by the chemical processing (Cartmell et al., 2004, Jones et al., 2004, Oliveira et al., 2007, Porter et al., 2007, Guldberg et al., 2008). Porter et al. reported the use of micro-CT imaging-based technique to continuously monitor 3D mineralization over time in a perfusion bioreactor noninvasively and specifically assess mechanisms of construct mineralization by quantifying a number of parameters such as number, size, and distribution of mineralized particle formation within constructs varying in thickness (Porter et al., 2007). Cartmell et al. utilized the micro-CT to nondestructively and quantitatively monitor the pattern and amount of mineral deposition within 3D polymeric scaffolds and demineralized trabecular bone matrix (DTBM) by rat stromal cells under osteogenic inductive condition (Cartmell et al., 2004). Moreover, in the same study, Cartmell et al. found that the micro-CT was able to detect the partial remineralization of the DTBM when exposed to high phosphate osteogenic media (Cartmell et al., 2004). This application is line with the report of Oliveira et al., who managed to use micro-CT to assess the apatite mineralization layer onto 3D polymeric scaffolds by dynamic biomimetic coating (Oliveira et al., 2007). In one of our recent studies, micro-CT was used to qualitatively and quantitatively analyze the *in vitro* bio-mineralization of the large 3D cellular scaffolds under different culture conditions, and the result showed that a significantly higher amount of *in vitro* mineralization was deposited on the cellular scaffolds cultured in a unique biaxial rotating bioreactor compared to the ones under traditional static culture condition (Figure 8.6). This finding is consistent with our other quantitative assays, including calcium deposition assays, validating their application to assess *in vitro* mineralization of cellular scaffold constructs (Zhang et al., 2009b).

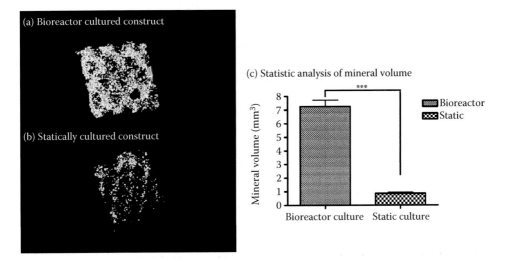

FIGURE 8.6 Micro-CT imaging of the bio-mineralization of cultured cellular scaffold constructs. (a) The bio-mineralization by the bioreactor cultured cellular construct; (b) the bio-mineralization by the statically cultured cellular construct; (c) the statistic analysis of mineral volume.

8.4 Applications of CT/Micro-CT to Image Nonmineralized Tissue

Unlike mineralized hard tissues, nonmineralized tissues possess much lower x-ray absorption capability with low inherent x-ray contrast, resulting in the difficulty for CT/micro-CT imaging. However, with the development of x-ray contrast agents and their staining techniques for contrast enhancement, CT/micro-CT have been increasingly applied for the assessment of nonmineralized tissues, including vascular network, animal soft tissue and cells cultured in 3D scaffolds.

8.4.1 Imaging Vascular Network

Vasculature plays a pivotal role in the functionality of tissue and organs, and has been under extensive investigation for decades. Various assessment techniques to study the morphology and structure of the vascular network have been developed to facilitate the relevant research. As early as 1842, vascular filling techniques have been utilized to study the vascular architecture by Bowman, who injected wax into the vasculature and corroded the tissue away to expose the resultant cast or injected various precipitates into the vasculature to study the finer vessels of the kidney (Bentley et al., 2002). Since then, vascular filing techniques have became a popular and primary approach to study the vascular network, which is usually filled with opaque substances such as India ink, radio-opaque contrast material, or various polymers. Subsequently the filled vascular tissue is either cleared, x-rayed, or corroded so that the filled vasculature may be studied by light microscopy and SEM with proper histological section or microangiography if radio-opaque contrast materials was used (Bentley et al., 2002). Besides vascular filling techniques, immunohistochemistry is another helpful method to study the vascular network, where histological section slices can be immunostained with blood vessel-specific markers (Lu et al., 2006). However, it should be noted that both of these two conventional assessment techniques involve the tedious procedure of sample preparation and histological section; furthermore, they are highly dependent on 2D assessment of micro-vascular density within representative histological sections without elaboration of the true 3D vascular tree (Young et al., 2008).

More recently, CT/micro-CT-based assessment technique with the aid of perfusable x-ray contrast agents such as barium sulfate or lead chromate has been explored to qualitatively and quantitatively evaluate the vasculature (Bentley et al., 2002, Guldberg et al., 2003). Using this technique, a resolution of 15–30 μm can be reached for the 3D images of vasculature. The advantages of this approach over 2D conventional methods include that the 3D connectivity of the vascular network can be reconstructed and quantitatively studied and the assessment is nondestructive with less complicated sample preparation (Guldberg et al., 2003). So far, CT/micro-CT have been reported to be capable to evaluate the vasculature of different organ systems, both in healthy and in diseased conditions, including renal vasculature, hepatic vasculature and portal hypertension, cerebral vasculature, coronary arteries, and ocular vasculature (Bentley et al., 2002, Schambach et al., 2010).

Recently, this contrast agent-enhanced micro-CT technique has been introduced to analyze the angiogenesis and neo-vascularization in TE research, especially in bone TE. Micro-CT becomes especially helpful to study the bone fracture healing, by not only measuring the new bone formation, but also quantifying the degree of neo-vascularization. It is well known that vascularization is one of the essential elements for bone regeneration. The rapid establishment of an effective vascular network is critical for maintenance and restoration of the bone tissue, by transporting elements such as growth factors and cells. However, it is the least studied in the context of bone TE perhaps due in part to the difficulty in quantifying vessel formation within 3D constructs. By quantifying both vascularization and bone formation, the degree of neo-vascularization can be correlated to the new bone formation with the potential to unfold the mechanism between neo-vascularization and the bone regeneration. Rai et al. have applied micro-CT to quantitatively study the vascularization and new bone formation in the rat femoral segmental defects, which were implanted with platelet-rich plasma (PRP) loaded 3D scaffolds, after the perfusion of Microfil via abdominal aorta (Rai et al., 2007). The authors found the PRP loaded scaffolds can increase the vascularization degree significantly with higher amount of new bone regeneration at week 3. This result is in line with our recent experimental finding, where significantly higher amount of vascular infiltration into the rat femoral defect region as illustrated by micro-CT analysis (Figure 8.7) have led to the successful healing of the critical sized defects (Zhang et al., 2010a). In our study, Microfil was perfused through the vascular network of rat via the left ventricle, and then the entire legs were analyzed by micro-CT, which can visualize both bone tissue and the vascular in-growth in the defect region (Figure 8.7a), correlating well with the macroscopic observation (Figure 8.7b). Furthermore, rat femurs were decalcified in acetic acid for 1 month, leading to a much clearer view of vasculature (Figure 8.7c), and the accurate quantification of vascular infiltration into the scaffolds (as indicated by a circle in cross-section view, Figure 8.7d). Together with the new bone formation quantification using micro-CT at late time point, the early rapid vascularization at initial stage has been shown to be crucial for the new bone formation in the later stage of fracture healing (Zhang et al., 2010a).

8.4.2 Imaging Other Nonmineralized Soft Tissue and Cells Cultured in 3D Scaffolds

Besides vascular tissue, micro-CT is attracting increasing research interest for their use to analyze the anatomic structure of other nonmineralized soft tissue, because of their capability to nondestructively generate accurate and quantitative 3D images with rapid and simple sample preparation (Metscher, 2009a,b). Compared to magnetic resonance microscopy (micro-MRI), another popular imaging tool for nondestructive analysis of soft tissue, micro-CT provided much better images with both resolution and the delineation of targeted tissue margin closely matching to the traditional histological analysis (deCrespigny et al., 2008, Faraj et al., 2009). Nonetheless, like vascular tissue, in order to achieve satisfactory image quality, proper contrast enhancement of soft tissue is required. To date, several contrast enhancement techniques have been developed to improve the image quality of micro-CT, by involving the use of various x-ray contrast agents.

The most successful contrast agent used for micro-CT imaging of soft tissues is osmium tetroxide (Johnson et al., 2006, Metscher, 2009b). Osmium is a heavy metal with K-shell energy of 73.9 keV

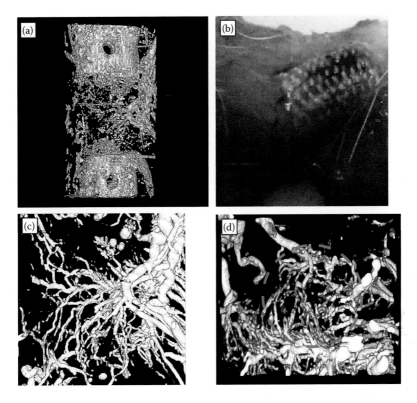

FIGURE 8.7 Micro-CT imaging of vasculature and bone tissue. (a) the micro-CT 3D image of both vasculature and bone tissue; (b) macroscopic image of sample at defect region; (c) micro-CT 3D image of the vasculature without bone tissue; and (d) the infiltration of blood vessel into the scaffold region as indicated by the circle at the cross-section view of micro-CT 3D image. (Reprinted from *Biomaterials*, 31, Zhang, Z.-Y. et al. Neo-vascularization and bone formation mediby fetal stem cell tissue-engineered bone grafts in critical-size femoral defects. 608–620. Copyright 2010a, with permission from Elsevier.)

and it can bind to the cell membrane, which can scatter x-ray and make the cell and tissue visible by micro-CT (Dorsey et al., 2009, Metscher, 2009b). Johnson and his colleagues reported the use of osmium tetroxide staining enhanced micro-CT imaging technique for the phenotypic assessment of mouse embryos and they managed to use this technique to conduct the 3D analysis of developmental patterning defects attributed to genetically engineered mutations and chemically induced embryotoxicity with a high resolution, up to 8 μm (Johnson et al., 2006). Besides imaging the soft tissue of vertebrates, small organs or tissues of invertebrates can be analyzed as well. For example, in a recent report, Ribi et al. managed to use the micro-CT with aid of osmium staining for 3D visualization of different compartments of the honey bee brain (Ribi et al., 2008). Alternative contrast agents have been explored as well. De Crespigny et al. used a hydrophilic iodinated contrast agent to stain the brains of rabbit and mice before the micro-CT scanning, which can identify a wide range of anatomical structure; furthermore, in glioma bearing mouse brains, qualitative and quantitative analysis of tumor contour and tumor volume by micro-CT is closely matched to the results by histological analysis (deCrespigny et al., 2008). Mizutani et al. described the use of a reduced-silver nerve staining method to visualize the neural network embedded in the 3D structure of the nerve tissue in the Drosophila under micro-CT (Mizutani et al., 2007). Metscher et al. developed three different simple staining methods using inorganic iodine and phosphotungstic acid, which allow high-contrast imaging of nonmineralized soft tissues at histological resolutions using a commercial micro-CT system (Metscher, 2009a,b).

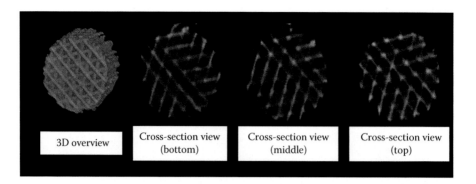

FIGURE 8.8 Micro-CT imaging of large cellular scaffolds without the use of contrast agent. More cell and ECM were found in the bottom region compared to middle and top regions. (Reprinted from *Biomaterials,* 31, Zhang, Z.-Y. et al. A comparison of bioreactors for culture of fetal mesenchycells for bone tissue engineering. 8684–8695. Copyright 2010b, with permission from Elsevier.)

Similarly, in the field of TE and regenerative medicine, cellular scaffold constructs, which are usually generated by the combinational use of scaffolds and cells *in vitro*, can be treated as a special type of non-mineralized soft tissue. Despite of their low intrinsic x-ray contrast, cellular scaffolds can be enhanced for their x-ray contrast with proper contrast agents staining technique and analyzed using micro-CT. This micro-CT-based imaging technique possesses several advantages over conventional methods to assess the cell adhesion and proliferation in the 3D scaffolds. On the one hand, compared to the traditional microscopy with histological section and fluorescence microscopy, micro-CT can see through the thick opaque scaffold constructs to generate quantitative information without the tedious sample preparation. On the other hand, it can provide the quantitative 3D information regarding the spatial distribution of the cells throughout the entire constructs, which is usually absent in common colorimetric and fluorometric soluble assays for enzymes, protein or DNA. For instance, Dorsey et al. employed the osmium tetroxide to stain the cellular scaffold constructs, which facilitated the visualization and quantitative and spatial analysis of cells cultured on the polymeric scaffold using micro-CT (Dorsey et al., 2009). However, it should be noted that the x-ray enhancement staining process involved the use of toxic chemicals such as osmium tetroxide, making the stained samples not applicable for other assays. As a result, we aim to develop a staining-free micro-CT imaging approach, which led to our successful visualization and quantitative analysis of cells and ECM cultured in the polymeric scaffolds without the use of toxic x-ray contrast agents (Figure 8.8) (Zhang et al., 2010b). In the different digital cross-section, we observed more cell and ECM proliferation at the bottom region compared to the middle and top regions (Figure 8.8). We believe that the visualization of the cell and ECM micro-CT under a relatively low thresholding may be likely to attribute to their increase of x-ray attenuation caused by the compact cellular structure and mineralized ECM deposited within the scaffolds. Without the use of toxic staining agent and any prestaining processing, this technique may provide a promising approach for the timely assessment of TE bone constructs prior to their transplantation into animal models or for clinical applications.

8.5 Conclusion

Thanks to their potentials to conduct nondestructive, quantitative, and 3D analysis with little requirement of sample preparation, CT/micro-CT especially micro-CT is becoming a more and more important imaging tool in the biomedical research, including the TE and regenerative medicine. It should be admitted that like other techniques, CT/micro-CT themselves are not perfect and have certain drawbacks, such as arbitrary thresholding, beam hardening as well as metal artifacts when scanning metal sample. However, we can foresee CT/micro-CT is becoming a more and more indispensable research technique in laboratories to complement or eventually replace the conventional assessment techniques to revolutionize and expedite the way of our research.

References

Badea, C. T., Drangova, M., Holdsworth, D. W., and Johnson, G. A. 2008. *In vivo* small-animal imaging using micro-CT and digital subtraction angiography. *Physics in Medicine and Biology,* 53, R319–R350.

Badea, C., Hedlund, L. W., and Johnson, G. A. 2004. Micro-CT with respiratory and cardiac gating. *Medical Physics,* 31, 3324–3329.

Beckmann, E. C. 2006. CT scanning the early days. *The British Journal of Radiology,* 79, 5–8.

Bentley, M. D., Ortiz, M. C., Ritman, E. L., and Romero, J. C. 2002. The use of microcomputed tomography to study microvasculature in small rodents. *American Journal of Physiology. Regulatory, Integrative and Comparative Physiology,* 282, R1267–R1279.

Betz, O., Wegst, U., Weide, D. et al. 2007. Imaging applications of synchrotron X-ray phase-contrast microtomography in biological morphology and biomaterials science. I. General aspects of the technique and its advantages in the analysis of millimetre-sized arthropod structure. *Journal of Microscopy,* 227, 51–71.

Cartmell, S., Huynh, K., Lin, A., Nagaraja, S., and Guldberg, R. 2004. Quantitative microcomputed tomography analysis of mineralization within three-dimensional scaffolds in vitro. *Journal of Biomedical Materials Research. Part A,* 69, 97–104.

Chu, T. M. G., Orton, D. G., Hollister, S. J., Feinberg, S. E., and Halloran, J. W. 2002. Mechanical and *in vivo* performance of hydroxyapatite implants with controlled architectures. *Biomaterials,* 23, 1283–1293.

Dadsetan, M., Hefferan, T. E., Szatkowski, J. P. et al. 2008. Effect of hydrogel porosity on marrow stromal cell phenotypic expression. *Biomaterials,* 29, 2193–2202.

De Crespigny, A., Bou-Reslan, H., Nishimura, M. C., Phillips, H., Carano, R. A. D., and Arceuil, H. E. 2008. 3D micro-CT imaging of the postmortem brain. *Journal of Neuroscience Methods,* 171, 207–213.

Dorsey, S. M., Lin-Gibson, S., and Simon, C. G. 2009. X-ray microcomputed tomography for the measurement of cell adhesion and proliferation in polymer scaffolds. *Biomaterials,* 30, 2967–2974.

Dragunow, M. 2008. High-content analysis in neuroscience. *Nature Reviews Neuroscience,* 9, 779–788.

Faraj, K. A., Cuijpers, V. M. J. I., Wismans, R. G. et al. 2009. Micro-computed tomographical imaging of soft biological materials using contrast techniques. *Tissue Engineering. Part C, Methods,* 15, 493–499.

Feldkamp, L. A., Goldstein, S. A., Parfitt, A. M., Jesion, G., and Kleerekoper, M. 1989. The direct examination of three-dimensional bone architecture *in vitro* by computed tomography. *Journal of Bone and Mineral Research: The Official Journal of the American Society for Bone and Mineral Research,* 4, 3–11.

Guldberg, R. E., Ballock, R. T., Boyan, B. D. et al. 2003. Analyzing bone, blood vessels, and biomaterials with microcomputed tomography. *IEEE Engineering in Medicine and Biology Magazine: The Quarterly Magazine of the Engineering in Medicine; Biology Society,* 22, 77–83.

Guldberg, R. E., Duvall, C. L., Peister, A. et al. 2008. 3D imaging of tissue integration with porous biomaterials. *Biomaterials,* 29, 3757–3761.

Hedberg, E. L., Kroese-Deutman, H. C., Shih, C. K. et al. 2005. Methods: A comparative analysis of radiography, microcomputed tomography, and histology for bone tissue engineering. *Tissue Engineering,* 11, 1356–1367.

Ho, S. T. and Hutmacher, D. W. 2006. A comparison of micro CT with other techniques used in the characterization of scaffolds. *Biomaterials,* 27, 1362–1376.

Hutmacher, D. W., Schantz, T., Zein, I., NG, K. W., Teoh, S. H., and Tan, K. C. 2001. Mechanical properties and cell cultural response of polycaprolactone scaffolds designed and fabricated via fused deposition modeling. *Journal of Biomedical Materials Research,* 55, 203–216.

Hutmacher, D. W. and Singh, H. 2008. Computational fluid dynamics for improved bioreactor design and 3D culture. *Trends in Biotechnology,* 26, 166–172.

Jiang, Y., Zhao, J., White, D. L., and Genant, H. K. 2000. Micro CT and Micro MR imaging of 3D architecture of animal skeleton. *Journal of Musculoskeletal Neuronal Interactions,* 1, 45–51.

Johnson, J. T., Hansen, M. S., WU, I. et al. 2006. Virtual histology of transgenic mouse embryos for high-throughput phenotyping. *PLoS Genetics,* 2, e61.

Jones, A. C., Milthorpe, B., Averdunk, H. et al. 2004. Analysis of 3D bone ingrowth into polymer scaffolds via micro-computed tomography imaging. *Biomaterials,* 25, 4947–4954.

Jones, J. R., Poologasundarampillai, G., Atwood, R. C., Bernard, D., and Lee, P. D. 2007. Non-destructive quantitative 3D analysis for the optimisation of tissue scaffolds. *Biomaterials,* 28, 1404–1413.

Kujoory, M. A., Hillman, B. J., and Barrett, H. H. 1980. High-resolution computed tomography of the normal rat nephrogram. *Investigative Radiology,* 15, 148–154.

Langer, R. and Vacanti, J. P. 1993. Tissue engineering. *Science,* 260, 920–926.

LE Gros, M. A., Mcdermott, G., and Larabell, C. A. 2005. X-ray tomography of whole cells. *Current Opinion in Structural Biology,* 15, 593–600.

Lin, A. S. P., Barrows, T. H., Cartmell, S. H., and Guldberg, R. E. 2003. Microarchitectural and mechanical characterization of oriented porous polymer scaffolds. *Biomaterials,* 24, 481–489.

Liu, E., Treiser, M. D., Johnson, P. A. et al. 2007. Quantitative biorelevant profiling of material microstructure within 3D porous scaffolds via multiphoton fluorescence microscopy. *Journal of Biomedical Materials Research. Part B, Applied Biomaterials,* 82, 284–297.

Lu, C., Marcucio, R., and Miclau, T. 2006. Assessing angiogenesis during fracture healing. *The Iowa Orthopaedic Journal,* 26, 17–26.

Ma, P. X. and Zhang, R. 2001. Microtubular architecture of biodegradable polymer scaffolds. *Journal of Biomedical Materials Research,* 56, 469–477.

Mcdermott, G., LE Gros, M. A., Knoechel, C. G., Uchida, M., and Larabell, C. A. 2009. Soft X-ray tomography and cryogenic light microscopy: The cool combination in cellular imaging. *Trends in Cellular Biology,* 19, 587–595.

Metscher, B. D. 2009a. MicroCT for comparative morphology: Simple staining methods allow high-contrast 3D imaging of diverse non-mineralized animal tissues. *BMC Physiology,* 9, 11.

Metscher, B. D. 2009b. MicroCT for developmental biology: A versatile tool for high-contrast 3D imaging at histological resolutions. *Developmental Dynamics: An Official Publication of the American Association of Anatomists,* 238, 632–640.

Mizutani, R., Takeuchi, A., Hara, T., Uesugi, K., and Suzuki, Y. 2007. Computed tomography imaging of the neuronal structure of Drosophila brain. *Journal of Synchrotron Radiation,* 14, 282–287.

Moore, M. J., Jabbari, E., Ritman, E. L. et al. 2004. Quantitative analysis of interconnectivity of porous biodegradable scaffolds with micro-computed tomography. *Journal of Biomedical Materials Research. Part A,* 71, 258–267.

Mosmann, T. 1983. Rapid colorimetric assay for cellular growth and survival: Application to proliferation and cytotoxicity assays. *Journal of Immunological Methods,* 65, 55–63.

Müller, R., Hahn, M., Vogel, M., Delling, G., and Rüegsegger, P. 1996. Morphometric analysis of noninvasively assessed bone biopsies: Comparison of high-resolution computed tomography and histologic sections. *Bone,* 18, 215–220.

Oliveira, A. L., Malafaya, P. B., Costa, S. A., Sousa, R. A., and Reis, R. L. 2007. Micro-computed tomography (micro-CT) as a potential tool to assess the effect of dynamic coating routes on the formation of biomimetic apatite layers on 3D-plotted biodegradable polymeric scaffolds. *Journal of Materials Science. Materials in Medicine,* 18, 211–223.

Otsuki, B., Takemoto, M., Fujibayashi, S., Neo, M., Kokubo, T., and Nakamura, T. 2006. Pore throat size and connectivity determine bone and tissue ingrowth into porous implants: Three-dimensional micro-CT based structural analyses of porous bioactive titanium implants. *Biomaterials,* 27, 5892–5900.

Parkinson, D. Y., Mcdermott, G., Etkin, L. D., Le Gros, M. A., and Larabell, C. A. 2008. Quantitative 3-D imaging of eukaryotic cells using soft X-ray tomography. *Journal of Structural Biology,* 162, 380–386.

Patel, V., Issever, A. S., Burghardt, A., Laib, A., Ries, M., and Majumdar, S. 2003. MicroCT evaluation of normal and osteoarthritic bone structure in human knee specimens. *Journal of Orthopaedic Research: Official Publication of the Orthopaedic Research Society,* 21, 6–13.

Popat, K. C., Leary Swan, E. E., Mukhatyar, V. et al. 2005. Influence of nanoporous alumina membranes on long-term osteoblast response. *Biomaterials, 26,* 4516–4522.

Porter, B. D., Lin, A. S. P., Peister, A., Hutmacher, D., and Guldberg, R. E. 2007. Noninvasive image analysis of 3D construct mineralization in a perfusion bioreactor. *Biomaterials, 28,* 2525–2533.

Rai, B., Oest, M. E., Dupont, K. M., Ho, K. H., Teoh, S. H., and Guldberg, R. E. 2007. Combination of platelet-rich plasma with polycaprolactone-tricalcium phosphate scaffolds for segmental bone defect repair. *Journal of Biomedical Materials Research Part A, 81,* 888–899.

Ramay, H. R. and Zhang, M. 2003. Preparation of porous hydroxyapatite scaffolds by combination of the gel-casting and polymer sponge methods. *Biomaterials, 24,* 3293–3302.

Ribi, W., Senden, T. J., Sakellariou, A., Limaye, A., and Zhang, S. 2008. Imaging honey bee brain anatomy with micro-X-ray-computed tomography. *Journal of Neuroscience Methods, 171,* 93–97.

Schambach, S. J., Bag, S., Steil, V. et al. 2009. Ultrafast high-resolution *in vivo* volume-CTA of mice cerebral vessels. *Stroke; a Journal of Cerebral Circulation, 40,* 1444–1450.

Schambach, S. J., Bag, S., Schilling, L., Groden, C., and Brockmann, M. A. 2010. Application of micro-CT in small animal imaging. *Methods (San Diego, Calif.), 50,* 2–13.

Seguin, F. H., Burstein, P., Bjorkholm, P. J., Homburger, F., and Adams, R. A. 1985. X-ray computed tomography with 50-Mum resolution. *Applied Optics, 24,* 4117.

Simon, C. G., Eidelman, N., Kennedy, S. B., Sehgal, A., Khatri, C. A., and Washburn, N. R. 2005. Combinatorial screening of cell proliferation on poly(L-lactic acid)/poly(D,L-lactic acid) blends. *Biomaterials, 26,* 6906–6915.

Simon, C. G., Khatri, C. A., Wight, S. A., and Wang, F. W. 2002. Preliminary report on the biocompatibility of a moldable, resorbable, composite bone graft consisting of calcium phosphate cement and poly(lactide-*co*-glycolide) microspheres. *Journal of Orthopaedic Research: Official Publication of the Orthopaedic Research Society, 20,* 473–482.

Singh, H., Ang, E. S., Lim, T. T., and Hutmacher, D. W. 2007. Flow modeling in a novel non-perfusion conical bioreactor. *Biotechnology and Bioengineering, 97,* 1291–1299.

Smith, P. K., Krohn, R. I., Hermanson, G. T. et al. 1985. Measurement of protein using bicinchoninic acid. *Analytical Biochemistry, 150,* 76–85.

Teo, J. C. M., SI-Hoe, K. M., Keh, J. E. L., and Teoh, S. H. 2006. Relationship between CT intensity, microarchitecture and mechanical properties of porcine vertebral cancellous bone. *Clinical Biomechanics (Bristol, Avon), 21,* 235–244.

Thomsen, J. S., Laib, A., Koller, B., Prohaska, S., Mosekilde, L., and Gowin, W. 2005. Stereological measures of trabecular bone structure: Comparison of 3D micro computed tomography with 2D histological sections in human proximal tibial bone biopsies. *Journal of Microscopy, 218,* 171–179.

Westneat, M. W., Socha, J. J., and Lee, W.-K. 2008. Advances in biological structure, function, and physiology using synchrotron X-ray imaging*. *Annual Review of Physiology, 70,* 119–142.

Woodfield, T. B. F., Malda, J., De Wijn, J., Péters, F., Riesle, J., and Van Blitterswijk, C. A. 2004. Design of porous scaffolds for cartilage tissue engineering using a three-dimensional fiber-deposition technique. *Biomaterials, 25,* 4149–4161.

Young, S., Kretlow, J. D., Nguyen, C. et al. 2008. Microcomputed tomography characterization of neovascularization in bone tissue engineering applications. *Tissue Engineering. Part B, Reviews, 14,* 295–306.

Zein, I., Hutmacher, D. W., Tan, K. C., and Teoh, S. H. 2002. Fused deposition modeling of novel scaffold architectures for tissue engineering applications. *Biomaterials, 23,* 1169–1185.

Zhang, R. and Ma, P. X. 1999. Poly(alpha-hydroxyl acids)/hydroxyapatite porous composites for bone-tissue engineering. I. Preparation and morphology. *Journal of Biomedical Materials Research, 44,* 446–455.

Zhang, Z.-Y., Teoh, S. H., Chong, M. S. K. et al. 2009a. Superior osteogenic capacity for bone tissue engineering of fetal compared with perinatal and adult mesenchymal stem cells. *Stem Cells, 27,* 126–137.

Zhang, Z.-Y., Teoh, S. H., Chong, W.-S. et al. 2009b. A biaxial rotating bioreactor for the culture of fetal mesenchymal stem cells for bone tissue engineering. *Biomaterials,* 30, 2694–2704.

Zhang, Z.-Y., Teoh, S. H., Chong, M. S. K. et al. 2010a. Neo-vascularization and bone formation mediated by fetal mesenchymal stem cell tissue-engineered bone grafts in critical-size femoral defects. *Biomaterials,* 31, 608–620.

Zhang, Z.-Y., Teoh, S. H., Teo, E. Y. et al. 2010b. A comparison of bioreactors for culture of fetal mesenchymal stem cells for bone tissue engineering. *Biomaterials,* 31, 8684–8695.

9

Intravascular Optical Coherence Tomography

Liu Linbo
*Nanyang Technological
University*

Atsushi Tanaka
*Wakayama Medical
University*

9.1 Introduction

Coronary artery disease (CAD) is the leading cause of death in the western world, resulting in close to a million deaths a year in the United States alone. CAD is a condition in which atherosclerotic plaque builds up inside the coronary arteries and narrows the arteries and reduces blood flow to downstream myocardium. A vulnerable atherosclerotic plaque that contains a necrotic lipid core and a thin fibrous cap can rupture, resulting in acute myocardial infarction (AMI). The rupture of a vulnerable atherosclerotic plaque causes the bloodstream to be exposed to procoagulant factors, forming thrombus, which can impede blood flow to downstream myocardium, trigging an acute coronary event. (Moreno et al., 1996).

Patients presenting with stenotic coronary lesions may be treated with stent implantation during percutaneous coronary intervention (PCI). The purpose of stent implantation is to restore and maintain blood flow through the artery. Stent implantation in human coronary arteries was initiated in 1986 by Sigwart et al. (1987). Coronary stents have been accepted as primary catheter-based therapy, and interventional cardiology has been revolutionized since the number of lesions treated with stents exceeds 85% of all interventional procedures (American Heart Association, 2003).

Although stents have been demonstrated to reduce restenosis compared to balloon angioplasty, in-stent restenosis has developed into a significant clinical problem with bare metal stents (BMS), leading to the need for a repeat PCI in a substantial number of patients (Hoffmann and Mintz, 2000). Drug-eluting stents (DES), coated with an agent designed to attenuate neointimal growth, reduce the need for repeat revascularization compared with BMS, also cause the problems of delayed arterial healing and rare cases of late stent thrombosis (Togni et al., 2005, Joner et al., 2006, Salam et al., 2006, Nakazawa et al., 2008, Katoh et al., 2009, Kubo et al., 2009, Barlis et al., 2010, Hassan et al., 2010). The potential risk of stent thrombosis in patients with DES mandates long-term administration of anticlotting drugs, repeat procedures with the attendant costs and potential for morbidity (Salam et al., 2006). Given these difficulties encountered with coronary stenting, there is a need for a tool to evaluate the stent healing process, which may be used to tailor antiplatelet regimen durations on an individual patient basis (Suter et al., 2010).

Angioscopy and intravascular ultrasound (IVUS) have been used as mainstream imaging modality for posttreatment monitoring of patients who have undergone PCI and stent implantation (Salam et al., 2006, Honda, 2009). Intracoronary optical coherence tomography (OCT) is an emerging investigation tool for the assessment of CAD (Brezinski et al., 1996a,b, Tearney et al., 1996a,b,c, 1997a,b, 2003, 2008a, Brezinski et al., 1997, Fujimoto et al., 1999, Jang et al., 2001, Bouma et al., 2003, Shite et al., 2006, Yun et al., 2006, Tanaka et al., 2008, 2009, Kubo et al., 2009, Villard et al., 2009a,b, Waxman et al., 2010) and evaluating the effects of coronary stenting (Brezinski et al., 1996a,b, 1997, Fujimoto et al., 1999, Jang et al., 2001, Bouma et al., 2003, Kume et al., 2005b, Shite et al., 2006, Yun et al., 2006, Matsumoto et al., 2007, Takano et al., 2007, Tearney et al., 1996a,b,c, 1997a,b, 2003, 2008a, Tanaka et al., 2008, 2009, Villard et al., 2009a,b, Moore et al., 2009, Otake et al., 2009, Toutouzas et al., 2009, Kubo et al., 2009, Barlis et al., 2010, Ozaki et al., 2010, Waxman et al., 2010). OCT is the only imaging modality with sufficient resolution to visualize the majority of the pathological features currently associated with the vulnerable plaque (Villard et al., 2009a). It has been demonstrated in cell culture and the human retina *in vivo* that cellular resolution can be achieved using ultrahigh resolution OCT technique (Drexler et al., 1999, Torti et al., 2009, Povazay et al., 2002), therefore, OCT holds the promise to perform cellular imaging in humans.

There had been some difficulties in conducting imaging *in vivo* with the first-generation OCT technology, namely time-domain OCT (TD-OCT), such as the need to remove blood from the imaging field to clearly visualize the artery wall, and 3-D imaging of long coronary segments. With the advent of second-generation Fourier-domain OCT (FD-OCT), which enables high-quality imaging at speeds up to 100× that of TD-OCT, 3-D imaging of long coronary segments during a brief transparent media flush is now possible (Suter et al., 2010). For detailed information on the media flushing technique and imaging speed issue, the readers are referred to a review by Suter et al. (2010). This chapter addresses the imaging principles of OCT that make it an ideal tool for interrogating coronary microstructure and assessing stent pathology. Finally, predictions of future OCT technologies are made to provide insight into what interventional cardiologists might expect in the future.

9.2 OCT Principles

9.2.1 Time-Domain Optical Coherence Tomography

Intravascular OCT is a structural imaging modality that is similar in principle to IVUS. In the axial direction, TD-OCT performs depth ranging by measuring echo time delay of the light, rather than acoustic waves, using low coherence interferometry (Huang et al., 1991, Hee et al., 1995, Puliafito et al., 1995, Tearney et al., 1997b, Drexler et al., 1999). While in the transverse direction, the optical beam is focused by the objective lens into a spot over a depth range defined by the confocal parameter of the focusing optics and the fiber pinhole. Therefore, OCT is essentially the combination of low coherence interferometry and confocal microscopy. A system configuration of a typical fiber-based TD-OCT is shown in Figure 9.1. The system is composed of a Michelson interferometer illuminated by a broadband light source, a moving reference mirror, sample arm optics with scanning mirror, and photo-detector with signal conditioning and acquisition electronics.

9.2.1.1 Low Coherence Interferometry

A beam splitter splits a broadband source field into a reference field E_r and sample field E_s. The sample field is focused into a small volume in the tissue. After scattering back from the tissue, the sample field modulated with the sample function E_s' mixes with E_r on the surface of the photodetector. The intensity that impinges on the photodetector is

$$I_d = \langle |E_d|^2 \rangle = 0.5(I_r + I_s') + \mathrm{Re}\left\{ \langle E_r^*(t+\tau)E_s'(t) \rangle \right\}, \tag{9.1}$$

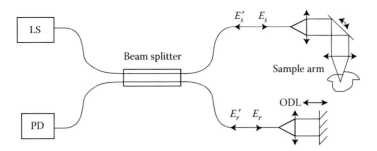

FIGURE 9.1 Schematics of a TD-OCT system. LS, light source; PD, photo detector; ODL, optical delay line.

where I_r and I_s are the DC intensities returning from the reference and sample arms of the interferometer. The second term in this equation, which depends on the optical time delay set by the position of the reference mirror, represents the amplitude of the interference fringes that carry information about the tissue structure. Time-domain A-scans was facilitated by moving the mirror in the reference path of the interferometer, so that the optical path length of the reflected light in the reference arm is scanned. The interference only occurs when the optical path length of the sample arm is matched to that of the reference arm (so that light of different wavelength is in phase), thus the interferometer functions as a cross-correlator and the amplitude of the interference signal generated after integration on the surface of the detector provides a measure of the cross-correlation amplitude (Schmitt, 1999). For detailed analytical expression of OCT signal, readers are referred to Schmitt (1999).

The axial resolution of low coherence interferometry is basically defined by the coherence length of the light signal detected by the sensor. Under Gaussian approximation of spectrum distribution, the free-space coherence length of the source, is given by Schmitt (1999).

$$l_c = \frac{2c\ln(2)}{\pi} \cdot \frac{1}{\Delta \upsilon} \approx 0.44 \frac{\lambda_0^2}{\Delta \lambda}, \tag{9.2}$$

where l_c is the full width of the coherence function at half-maximum measured in wavelength units, c is the speed of light, λ_0 is the center wavelength of the detected light. This equation provides important knowledge that sources with broad spectra are desirable because they produce interference patterns of short temporal (and spatial) extent.

9.2.1.2 Confocal Optics in the Sample Arm

OCT systems assume the confocal schematic in the sample arm optics. For example in Figure 9.1, the fiber tip in the sample arm serves physically as a confocal pinhole which actually creates a confocal volume in the sample space and rejects back-reflected or back-scattered photons from out-of-focus region. This confocal volume basically defined the three-dimensional range of an axial line (A-line). If the objective entrance pupil is larger than the Gaussian beam waist diameter d at the objective lens pupil plane, one can use the $1/e^2$-intensity beam waist radius in the focal plane to define the spot size or lateral resolution as

$$\Delta x = \lambda_0/(\pi\, NA) \tag{9.3}$$

NA is the numerical aperture of the OCT objective lens, which is inversely proportional to the lateral resolution. However, the depth of field of such an objective lens is given by the distance between the two points where the beam width expands to $\sqrt{2}$ times of the beam waist

$$\Delta z = 8\lambda_0\, (f/d)^2/\pi \approx 2\lambda_0/(\pi\, NA^2) \tag{9.4}$$

where f is the focal length of the objective lens.

In the low *NA* case, the confocal parameter is much larger than the coherence length of the light source, so that one of the apparent advantages of OCT is that the lateral resolution is completely decoupled from the axial resolution. Therefore, the optical design of the system can be optimized for lateral scanning, with no effect on the axial resolution. To provide a depth of focus more than 3 mm, the spot size of intra-coronary OCT system is usually 30–40 μm, a resolution 3–10 times poorer than the axial resolution. This compromise between lateral resolution and depth of focus is one of the major problem preventing intra-vascular OCT from imaging at cellular level resolution *in vivo*.

Shown in Figure 9.2, intracoronary OCT, which has a typical axial resolution of 10 μm and lateral reso-lution of 30–40 μm, is capable of characterizing the architectural morphology of plaque at resolution that is 10 times better than IVUS, the preceding technology for high-resolution imaging of the coronary wall. It had been demonstrated that intravascular OCT images provided superior resolution when compared to IVUS images obtained from the same locations in measuring cap thickness (Brezinski et al., 1997), intima-media thickness (Kume et al., 2005a), and enabled the visualization of features, such as the intima, includ-ing intimal flaps and defects, disruptions in the media, stent strut apposition that could not be identified by IVUS (Tearney et al., 2000), and tissue prolapse (Figure 9.3) (Jang et al., 2001). Afforded by the bandwidth of the light source and focusing optics, the resolution limits of OCT imaging systems have been pushed by researcher toward an axial resolution of 0.75–2 μm in tissue and a lateral resolution of 2–3 μm in tissue (Povazay et al., 2002, Fujimoto, 2003, Leitgeb et al., 2004, 2006, Fernandez et al., 2008, Torti et al., 2009, Villiger et al., 2009, Aguirre et al., 2010). An imaging tool that is capable of visualize cellular or even subcel-lular microstructure of the blood vessel wall will open up new opportunities for investigating CAD.

9.2.2 Fourier-Domain Optical Coherence Tomography

9.2.2.1 Spectral Interferometry

In TD-OCT, interference contrast is detected only if the object path length equals the reference path length. Therefore, the reference path has to be scanned through the depth range. Fourier-domain prin-ciples derive from the concept of "coherence radar" or "spectral radar" (Hausler and Lindner, 1998), which avoid scanning the reference through the depth range. These OCT obtain depth information by evaluating the spectrum of the interferogram. The Fourier transformation of the spectrum delivers the depth information. For this type of OCT, there are two approaches. For the basic implementation shown in Figure 9.4, the interferometer output is spectrally dispersed and the whole spectrum is detected by

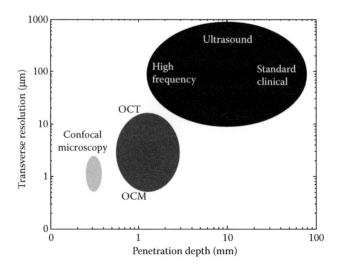

FIGURE 9.2   Transverse resolution and image penetration in OCT.

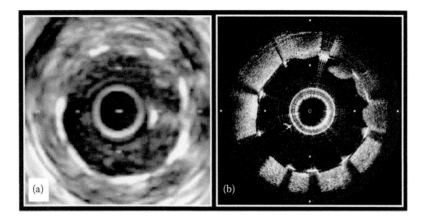

FIGURE 9.3 (a) IVUS and (b) OCT images of the stented right coronary artery are shown. Although IVUS showed a well-deployed stent, the detailed structure around the stent struts is not well visualized. In addition, OCT clearly visualized tissue prolapse between the stent struts (12 to 3 o'clock). The tissue prolapse occurred mainly in an area with lower OCT signal intensity, which is suggestive of a plaque with a large lipid content. (Reproduced with permission from Jang, I. K., Tearney, G., and Bouma, B. 2001. Visualization of tissue prolapse between coronary stent struts by optical coherence tomography—Comparison with intravascular ultrasound. *Circulation*, 104, 2754–2754.)

an array of photodiodes. Whereas those assume FD-OCT configuration require linear CCD or COMS as photodetectors with a typical rate of no more than ~100 kHz. In a different form of FD-OCT, named optical frequency domain imaging (OFDI) (Yun et al., 2006, Tearney et al., 2008b), the spectrum can be produced by a tunable laser or swept source and then be detected by a single photodiode, which allows A-line rates of more than 100 kHz through FDML technique (Huber et al., 2006a,b, Oh et al., 2010).

The measuring principle of FD-OCT is based on spectral interferometry. The signal from the object consists of many elementary waves emanating from different depths z. If the dispersion in the object is neglected, the scattering amplitude of the elementary waves versus depth is $a(z)$. The object signal is superimposed on the plane reference wave a_R. Detected by a spectrometer, the combined field is separated by wave numbers k, so that the interference signal $I(k)$ is (Hausler et al., 1998)

$$I(k) = S(k)\left| a_R \exp(ikl_r) + \int_0^\infty a(z) \times \exp\left\{ i2k\left[0.5l_r + n(z) \cdot z \right] \right\} dz \right|^2, \tag{9.5}$$

where the path length in the sample $l_s = l_r + 2n(z) \cdot z$; z_0 is the offset distance between the reference plane and object surface; n is the refractive index; a_R is the amplitude of the reference (for further investigations

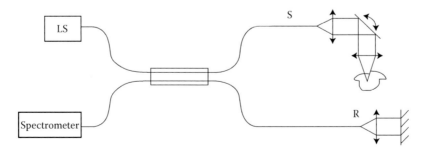

FIGURE 9.4 FDOCT setup with reference arm R, light source LS, sample S, and spectrometer.

set $a_R = 1$); $a(z)$ is the backscattering amplitude of the object signal; with regard to the offset z_0, and $a(z)$ is zero for $z < z_0$; and $S(k)$ is spectral intensity distribution of the light source. With these assumptions, the interference signal $I(k)$ can be written as (Hausler et al., 1998)

$$I(k) = S(k)\left|1 + \int_0^\infty a(z)\exp\{i2knz\}dz\right|^2$$

$$= S(k)\left[1 + 2\int_0^\infty a(z)\cos(2knz)dz + \int_0^\infty \int_0^\infty a(z)a(z')\exp[-i2kn \times (z - z')]dzdz'\right]. \quad (9.6)$$

It can be seen that $I(k)$ is the sum of the three terms. Besides a constant offset, the second term encodes the depth information of the object. It is a sum of cosine functions, where the amplitude of each cosine is proportional to the scattering amplitude $a(z)$. The depth z of the scattering event is encoded in the frequency $2nz$ of the cosine function. $a(z)$ can be acquired by a Fourier transformation of the interferogram. The third autocorrelation term describes the mutual interference of all elementary waves (Hausler et al., 1998). The measuring range ΔZ of the FD-OCT is limited by the resolution of the spectrometer (Hausler et al., 1998).

$$\Delta Z = \frac{1}{4n}\frac{\lambda^2}{\delta\lambda}, \quad (9.7)$$

9.2.2.2 Advantages of FD Method Over TD Method

FD-OCT has the following advantages over TD-OCT:

1. Much higher sensitivity (>30 dB): The increased sensitivity in FD-OCT compared with that of TD-OCT is based on the significant reduction of shot noise obtained by replacement of the single-element detector with a multi-element array detector. In a TD-OCT system, each wavelength is uniquely encoded as a frequency, and shot noise has a white-noise characteristic. In a single-detector TD-OCT system the shot noise generated by the power density at one specific wavelength is present at all frequencies and therefore adversely affects the sensitivity at all other wavelengths (Takada, 1998, Rollins and Izatt, 1999, de Boer et al., 2003, Choma et al., 2003, Leitgeb et al., 2003). For a detailed analysis of sensitivity of TD-OCT and FD-OCT, the reader is referred to references Choma et al. (2003), De Boer et al. (2003), and Leitgeb et al. (2003).
2. No moving parts are required to obtain axial scans.
3. Because of (1) and (2), up to ~100× increase in imaging speed compared with TD-OCT.

However, there are some problems with spectrometer-based FD-OCT and OFDI. The high-speed spectrometer required by FD-OCT is practically limited by the available spectral resolution, which ultimately limits the ranging depth and/or axial resolution. The spectral resolution of a high-speed, high-efficiency spectrometer is practically degraded by the aberration of the camera lens. To achieve theoretical resolution, the camera lens should be corrected for aberration, especially the field of curvature, chromatic aberration and astigmation within the field of view. Commercial camera lens are mostly designed for visible light and corrected for a much smaller field of view than high-speed line scan CCD used for FD-OCT. Normally, specially designed multi-element air-spaced camera lenses are used, but the performances are not consistent with the theoretical predictions, resulting in reduced bandwidth and spectral resolution, hence, reduced axial resolution and ranging depth (Cense et al., 2004, Wojtkowski et al., 2004). The degraded spectral resolution and effective bandwidth will also result in degraded sensitivity in spectroscopic FD-OCT which normally employs a broadband source. In contrast, TD-OCT and OFDI is free from these problems since a point detector, instead of a spectrometer, is used. This is probably one of reason why the highest axial resolution ever achieved is produced using TD systems.

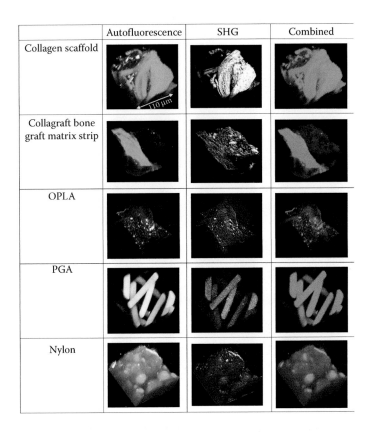

	Autofluorescence	SHG	Combined
Collagen scaffold			
Collagraft bone graft matrix strip			
OPLA			
PGA			
Nylon			

FIGURE 3.4

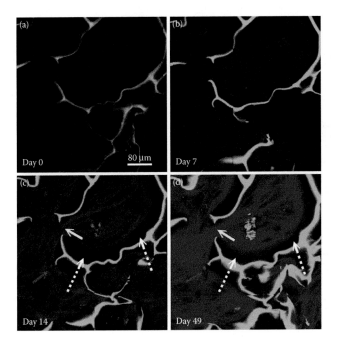

FIGURE 3.5

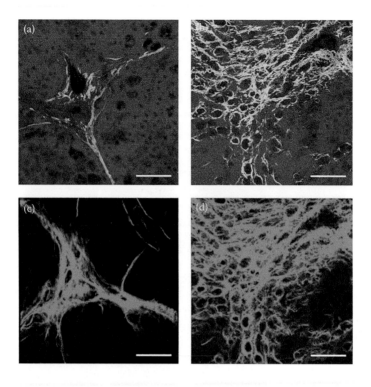

FIGURE 4.7

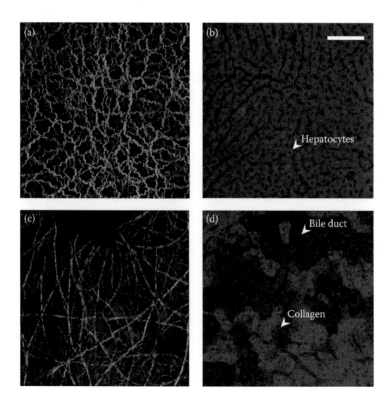

FIGURE 4.13

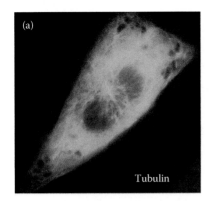

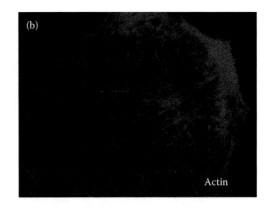

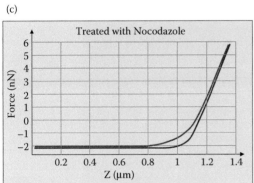

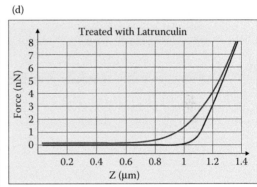

FIGURE 5.19

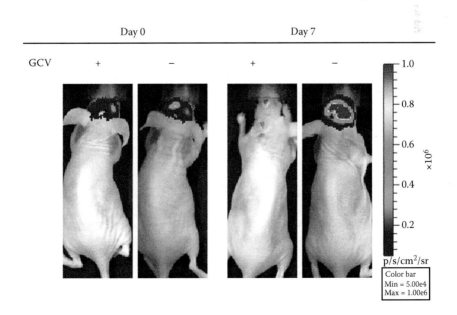

FIGURE 10.1

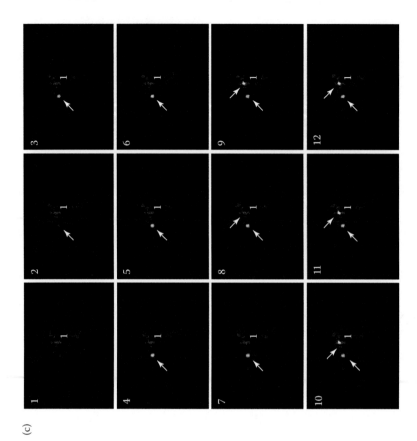

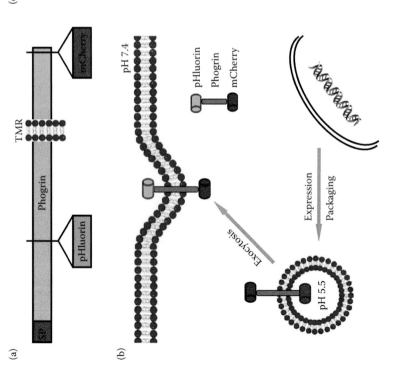

FIGURE 11.3

9.2.3 Intravascular OCT Systems

The intravascular OCT systems was composed of the imaging console, catheter, and console–catheter interface (Figure 9.5). OFDI-based intravascular OCT imaging catheters (0.8 mm diameter) is based on a design similar to that of commercial intracoronary ultrasound imaging catheters (Yun et al., 2006).

9.3 Evaluation Parameters

9.3.1 Normal Coronary Arterial Wall in OCT

The normal coronary artery wall usually shows a layered structure in OCT (Figure 9.6), including an intima as thin high signal layer, a media as low signal layer, and an adventitia for heterogeneous and usually high signal layer. Occasionally, the internal elastic lamina (IEL) and external elastic lamina (EEL), that are at the border of the intima-media or the media-adventia, respectively, can be visualized as highly backscattering thin structures.

9.3.2 Plaque Characterization by OCT

Although coronary angiography has been used as a gold standard tool for the evaluation of CAD, it can present only luminal stenosis. In other words, the era of angiography had directed toward stenosis-conscious cardiology. Over the last decade, IVUS has introduced area-conscious cardiology into CAD. For example, IVUS confirmed the concept of vascular remodeling and established its clinical significance. IVUS, however, cannot assess plaque characterization. Therefore, plaque characterization has

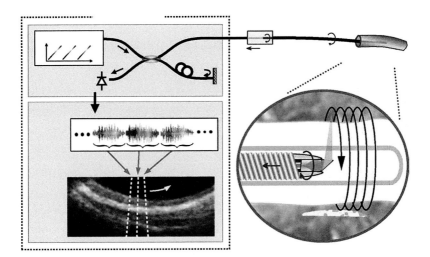

FIGURE 9.5 Principles of comprehensive optical frequency-domain imaging (OFDI). Minimally invasive catheters or endoscopes provide for access of the optical fiber to the organ or system of interest. An optical beam is focused into the tissue, and the echo-time delay and amplitude of the light reflected from the tissue microstructure at different depths are determined by detecting spectrally resolved interference between the tissue sample and a reference, as the source laser wavelength is rapidly varied from 1264 to 1376 nm. A Fourier transform of this signal forms image data along the axial line (A-line), which is determined by the optical beam emitted from the probe. A-lines are continuously acquired as the probe is actuated to provide spatial scanning of the beam in two directions that are orthogonal to the axial line (rotational and pull-back motion by a rotary junction). The resulting three-dimensional data sets can be rendered and viewed in arbitrary orientations for gross screening, and individual high-resolution cross-sections can be displayed at specific locations of interest. (Reproduced from Yun, S. H. et al. 2006. *Nature Medicine*, 12, 1429–1433. With permission.)

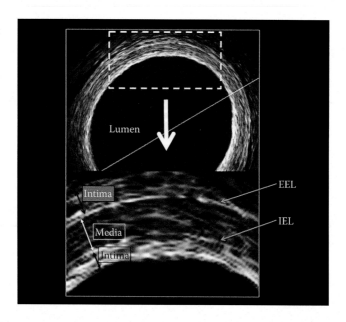

FIGURE 9.6 Human coronary artery. The normal coronary artery wall shows three-layered structure, high signal-low signal-high signal bands in the cross-sectional OCT image. These layers correspond to intimal, media, and adventia. In this case, the internal elastic lamina (IEL) and external elastic lamina (EEL), that are at the border of the intima-media or the media-adventia, respectively, can be visualized as highly backscattering thin structures.

been one of the dreams of cardiologists because it should contribute to improve clinical outcomes. In contrast to IVUS or angiography, OCT can evaluate the plaque characteristics like pathology with high-resolution cross-sectional image (Figure 9.7). *In vivo* plaque characterization by OCT may open the door to pathology-conscious cardiology.

9.3.2.1 Fibrous Plaque

A fibrous plaque by OCT is synonymous with a collagen-rich pathological intimal thickening. The fibrous plaque shows high backscattering and a homogeneous OCT signal (Yabushita et al., 2002). Usually, fibrous plaque is considered as a stable plaque for future cardiac events.

9.3.2.2 Fibrocalcific Plaque

In the OCT images, calcifications appear as a signal-poor and/or heterogeneous region with a sharply delineated border (Yabushita et al., 2002). A fibrocalcific plaque by OCT contains fibrous tissue and calcifications. The clinical significance of calcification is still controversial. An IVUS study reported that spotty calcification might be a sign of plaque instability. However, recent MDCT study cannot find the relationship between the future cardiac events and spotty calcification. For PCI, plaque containing a large amount of calcifications is likely to resist being expanded by balloon or stent implantation.

9.3.2.3 Lipid-Rich Plaque

A lipid-rich plaque by OCT is defined as a plaque containing a signal-poor region with poorly delineated border (Yabushita et al., 2002). This is equivalent to a term of lipid pool or necrotic core used in pathology.

9.3.2.4 Thin Cap Fibroatheroma

A fibrous cap is a tissue layer overlying the lipid components. Pathologically, fibrous cap thickness is considered as one of the morphological determinants for plaque vulnerability. Thin cap fibroatheroma

(TCFA) is defined as the necrotic core covered with less than 65 μm thickness fibrous cap in pathology. Since pathological studies reported that 95% of the culprit plaque ruptures showed less than 65 μm thickness fibrous cap, TCFA is recognized as a vulnerable plaque (Virmani et al., 2000).

9.3.2.5 Plaque Rupture

Plaque rupture is characterized as disruption of intima with creating new cavities in the OCT (Tanaka et al., 2008). Pathological studies have reported that the intimal disruption frequently occurs at the plaque shoulder (Virmani et al., 2000).

9.3.2.6 Thrombus

Thrombus appears as a mass attached to the luminal surface or floating within the lumen. OCT can identify the moving thrombus in some cases. A study reported that OCT could discriminate between the red and white thrombus using attenuation pattern (Kume et al., 2006a).

9.3.2.7 Macrophage Accumulations

Several studies reported that macrophages could be seen by FD-OCT as signal-rich, distinct, or confluent punctate dots (Tearney et al., 2003).

9.3.3 Stent

Metal stent. Since stent struts reflect the excited light, OCT can image only the leading edge of stent struts and makes a surface reflection figure. Shadowing without struts figure or struts without shadowing can be seen in some cases.

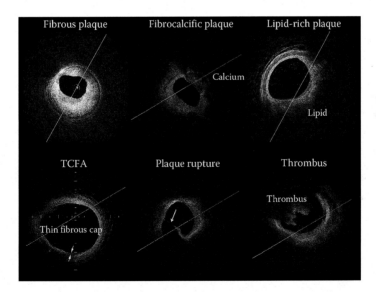

FIGURE 9.7 Plaque characterization by *in vivo* OCT. This figure shows various types of plaques that can be seen in human atherosclerotic plaque. This fibrous plaque shows eccentric accumulation of fibrous components and the media can be seen in circumference. This fibrocalcific plaque contains a few calcium depositions. Calcium depositions show a clear border. This large eccentric plaque contains a large amount of lipid component. The OCT signal rapidly decreases through the lipid components and the outer boundary of artery wall cannot be observed. Plaque disrupted at the plaque shoulder (white arrow). The broken-fibrous cap is very thin (<40 μm). Intracoronary thrombus can be observed and absorption of the light by thrombus made hard to observe the architectures behind the thrombus.

Absorbable stent. Recently, various types of bioabsorbable stent have been developed. Bioabsorbable stents show various stent reflection and shadowing figures depending on the material of struts. Recent study reported that some absorbable stent struts show a box-like appearance (Serruys et al., 2009).

Stent strut apposition. OCT can assess the stent struts apposition *in vivo* (Figure 9.8). Malapposition is diagnosed when the distance from the surface of stent struts to the luminal surface is greater than the strut thickness and polymer if present.

Tissue prolapse. Tissue prolapse is defined as the protrusion of tissue toward the lumen between stent struts after stent implantation (Figure 9.8). Tissue prolapse can be more frequently observed when the stent is placed over the lipid-rich plaque.

Neointimal coverage. Since neointimal coverage on stent strut is one of the powerful predictors for late stent thrombosis that may be rare but a serious complication of DES (Finn et al., 2007), assessment of reendothelization on stent struts is crucially important after DES implantation. Because OCT can provide a high-resolution coronary imaging, the assessment of stent coverage is expected as one of clinical applications of OCT. However, the current OCT cannot visualize the endothelium itself. A question has still remained whether neointimal coverage assessed by OCT has clinical significance.

9.3.4 Quantitative Assessment

OCT can also provide quantitative measurements for plaque and stent assessments. Some OCT measurements have been validated to morphometric analysis by histology.

Minimum and maximum lumen diameter: The shortest and longest diameter of the lumen.

Lumen area: The lumen area can be measured by tracing the leading edge of the intima.

Fibrous cap thickness: While an *ex vivo* study reported that OCT could directly measure the fibrous cap thickness (Kume et al., 2006b), it is usually difficult to determine the boundary of the cap and lipid contents *in vivo* OCT images.

Neointimal thickness: The neointimal thickness is measured as the distance from the leading edge of intima to the axial center of the stent strut surface reflection.

Stent profile: The stent profile (area) can be measured by tracing the axial center of the stent strut surface reflection.

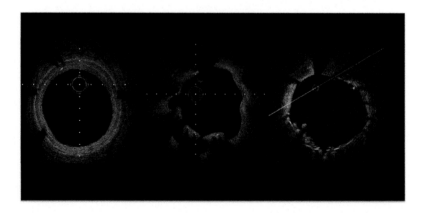

FIGURE 9.8 Stent imaging by OCT. The left panel shows that stent struts were completely covered with neo-intima. Usual metal stent struts show high surface reflection figure as high OCT signal with shadowing (11 or 3 o'clock). However, some struts does not show surface reflection figure (8 or 10 o'clock). The mid-panel shows stent malapposition (from 5 to 9 o'clock). This stent struts did not attach to the luminal surface but was covered with tissues. The right panel shows that tissue prolapsed between the stent struts. In this case, stent was deployed on the lipid-rich plaque.

Plaque area: The plaque area can be calculated by subtracting the lumen area from the area of the intima-media border. It is hard for OCT to assess this plaque area due to the shallow light penetration, especially in the lipid-rich plaque.

9.4 Implication for Stenting

9.4.1 Stent Imaging Technologies

Until recently, three imaging modalities, IVUS, angioscopy, and MDCT can be available for stent imaging in the clinical setting. It is widely believed that IVUS is an assist tool for PCI with stent because of an accumulation of clinical data and experience. However, many multicenter studies have failed to prove the fact that a routine IVUS usage could improve the outcome of PCI. Turning to the situation of angioscopy, the use of angioscopy is limited for the research purpose and no study proved any efficacies of it for PCI. Considering the noninvasive nature of MDCT, MDCT may be suitable for screening the lesions or making the strategy of PCI prior to the procedure. One of the limitations of MDCT is the fact that MDCT cannot use during the PCI procedure. While OCT is expected to participate with PCI in the DES era because of its high spatial resolution, it has still remained whether OCT could contribute to the patient care. Table 9.1 summarizes the features of each imaging modality. Each modality has both advantage and disadvantage for stent imaging. To take a look at this table, a combination of MDCT and OCT may be one of the promising solution for comprehensive assessment of stenting because they seem to be able to complement each other.

9.4.2 OCT for Stent Implantation: Procedure

Since OCT is an emerging imaging modality, there is not enough literature that proves the efficacy of OCT for stenting. However, the accumulation of clinical experience during PCI has suggested the possibility of OCT-guided PCI.

TABLE 9.1   Comparison between Imaging Modalities for Characterizing Atherosclerosis

	OCT	IVUS	Angioscopy	MDCT	Angiography
Invasive on non-invasive	Invasive	Invasive	Invasive	Non-invasive	Invasive
Spatial resolution (μm)	10	150–200	NA	500	500
Cross sectional image	○	○	×	○	×
3D image	○	△	×	@	×
Plaque characteristics	@	△	×	△	×
Thrombus	○	△	@	×	△
Positive remodeling	△	○	×	○	×
Luminal stenosis	○	○	×	○	○
Vessel dimension	△	○	×	○	×
Plaque area	△	○	×	○	×
Dissection	○	○	×	×	△
Malapposition	@	△	×	×	×
Neointimal coverage	@	△	○	×	×
Stent profile	@	○	×	△	△
Tissue prolapse	@	△	△	×	×
Stent fracture	NA	×	×	○	○

Note: NA = no data;   @ = best;   ○ = available;   △ = occasionally available;   × = not available.

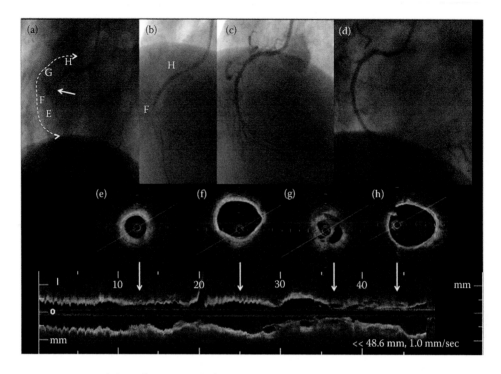

FIGURE 9.9 OCT-guided PCI for a case with chronic total occlusion. Base-line angiography (panel a) that shows chronic total occlusion at the proximal portion of the right coronary artery (white arrow). OCT could explore satisfactory length (55 mm) of the right coronary artery without any predilation by balloon (panels e–i, broken white line). OCT revealed the following: 1. The distal site had narrowed but not occluded lumen with fibrous plaque that might be stable (e). 2. The adjacent sites of the culprit lesion contained lipid-rich plaque that was likely to fall into restenosis or future cardiac events (f–h). 3. Culprit site shows honeycomb-like holes.Thus, the stent was implanted without any predilation to cover over the lipid-rich plaque (f–h) (panel b). While angiography just after stenting presented some narrowed sites in the distal portion of the right coronary artery (panel c), PCI had finished because OCT already revealed these sites had stable plaque. Follow-up angiography showed no restenosis in the stented site or no progression.

For example, smaller profile of the current TD-OCT probe enables us to explore the lesions that have never been acknowledged by other imaging modalities. Coronary angiography showed chronic total occlusion of the distal portion of the right coronary artery in the case of Figure 9.9. While IVUS could not pass the culprit lesion even after guide-wire crossing over the lesion, OCT imaging probe easily crossed over the lesion and could provide the detailed lesion morphology of chronic total occlusion.

9.4.3 No-Reflow Phenomenon

Recently published general consensus guidelines recommend an early invasive strategy with PCI for patients with acute coronary syndrome, particularly in high-risk patients (Braunwald et al., 2002). However, the patency of the culprit artery does not always guarantee salvage of the myocardium at risk of ischemia. The no-reflow phenomenon is defined as inadequate myocardial perfusion through a given segment of the coronary circulation without angiographic evidence of mechanical vessel obstruction (Kloner et al., 1974). It has also been reported that no-reflow is associated with poor functional and clinical patient outcomes when compared to patients with adequate reflow following reperfusion (Ito et al., 1992, 1996). Further, recent studies have revealed that distal embolization of thrombus and/or lipid plaque contents by PCI are one of the major causes of no-reflow (Tanaka et al., 2002, Sato et al., 2004). It may be critically important, therefore, to be able to predict which lesions are at high risk for myocardial no-reflow prior to beginning PCI. An OCT paper reported that the frequency of no-reflow increases according to the lipid arc at the culprit plaque (Tanaka et al., 2009). Considering the unique ability of OCT to detect lipid component and thrombus, the prediction for no-reflow is also one of the promising clinical applications for OCT.

9.4.4 Limitations and Future Development

9.4.4.1 Image Acquisition Methods

To acquire good-quality OCT images, it is necessary to displace red blood cells from viewing the area completely. Two image acquisition methods are postulated for OCT. One is a balloon-occlusion flushing method that is used for TD-OCT. Although the efficacy of this method has been reported, following critical limitations are noted: (1) a difficulty to observe the proximal lesion, (2) a provocation of myocardial ischemia, (3) a possibility to injure the coronary arterial wall, (4) an increase of the procedural complexity. To overcome these limitations, the other image acquisition method, namely a continuous-flushing method has been developed (Kataiwa et al., 2008). This method can be available for both TD-OCT and FD-OCT. Considering the switchover from TD-OCT to FD-OCT, this method would become the main image acquisition method for OCT.

9.4.4.2 Flush Medium

At current, viscous iso-osmolar x-ray contrast media or mixture of low molecular weight dextrose and lactated Ringer's solution has been successfully utilized for the continuous-flushing method. An intra-coronary OCT image session requires approximately 15–30 mL of flushing media. Therefore, repeated OCT imaging requires a certain amount of flushing media that might be harmful for patients with renal dysfunction or with heart failure. Alternative safe and efficient flushing media should be developed.

9.4.4.3 Functional Imaging

The current OCT provides the detail morphology of plaques that could contribute to our better understanding for pathobiology of atherosclerosis. In addition to this, the current intracoronary OCT engineering enables us to assess pathophysiology of coronary artery. As like a Doppler ultrasound technique, Doppler-OCT has a potential to assess blood flow.

9.4.4.4 PS-OCT

Several studies have demonstrated the possibility that polarization sensitive OCT (PS-OCT) may detect collagen contents in atherosclerotic plaques (Nadkarni et al., 2007). To measure birefringence may help us to identify the tissue components as like a special stain in histology.

References

American Heart Association. 2003. Heart and stroke statistical update.

Aguirre, A. D., Chen, Y., Bryan, B. et al. 2010. Cellular resolution ex vivo imaging of gastrointestinal tissues with optical coherence microscopy. *Journal of Biomedical Optics,* 15, 15(1), 016025.

Barlis, P., Regar, E., Serruys, P. W. et al. 2010. An optical coherence tomography study of a biodegradable vs. durable polymer-coated limus-eluting stent: A leaders trial sub-study. *European Heart Journal,* 31, 165–176.

Bouma, B. E., Tearney, G. J., Yabushita, H. et al. 2003. Evaluation of intracoronary stenting by intravascular optical coherence tomography. *Heart,* 89, 317–320.

Braunwald, E., Antman, E. M., Beasley, J. W. et al. 2002. ACC/AHA guideline update for the management of patients with unstable angina and non-ST-segment elevation myocardial infarction—2002: Summary article: A report of the American College of Cardiology/American Heart Association Task

Force on Practice Guidelines (Committee on the Management of Patients with Unstable Angina). *Circulation,* 106, 1893–900.

Brezinski, M. E., Tearney, G. J., Bouma, B. E. et al. 1996a. Imaging of coronary artery microstructure (*in vitro*) with optical coherence tomography. *American Journal of Cardiology,* 77, 92–93.

Brezinski, M. E., Tearney, G. J., Bouma, B. E. et al. 1996b. Optical coherence tomography for optical biopsy— Properties and demonstration of vascular pathology. *Circulation,* 93, 1206–1213.

Brezinski, M. E., Tearney, G. J., Weissman, N. J. et al. 1997. Assessing atherosclerotic plaque morphology: Comparison of optical coherence tomography and high frequency intravascular ultrasound. *Heart,* 77, 397–403.

Cense, B., Nassif, N. A., Chen, T. et al. 2004. Ultrahigh-resolution high-speed retinal imaging using spectral-domain optical coherence tomography. *Optics Express,* 12, 2435–2447.

Choma, M. A., Sarunic, M. V., Yang, C. H., and Izatt, J. A. 2003. Sensitivity advantage of swept source and Fourier domain optical coherence tomography. *Optics Express,* 11, 2183–2189.

De Boer, J. F., Cense, B., Park, B. H., Pierce, M. C., Tearney, G. J., and Bouma, B. E. 2003. Improved signal-to-noise ratio in spectral-domain compared with time-domain optical coherence tomography. *Optics Letters,* 28, 2067–2069.

Drexler, W., Morgner, U., Kartner, F. X. et al. 1999. *In vivo* ultrahigh-resolution optical coherence tomography. *Optics Letters,* 24, 1221–1223.

Fernandez, E. J., Hermann, B., Povazay, B. et al. 2008. Ultrahigh resolution optical coherence tomography and pancorrection for cellular imaging of the living human retina. *Optics Express,* 16, 11083–11094.

Finn, A. V., Joner, M., Nakazawa, G. et al. 2007. Pathological correlates of late drug-eluting stent thrombosis: Strut coverage as a marker of endothelialization. *Circulation,* 115, 2435–2441.

Fujimoto, J. G. 2003. Optical coherence tomography for ultrahigh resolution *in vivo* imaging. *Nature Biotechnology,* 21, 1361–1367.

Fujimoto, J. G., Boppart, S. A., Tearney, G. J., Bouma, B. E., Pitris, C., and Brezinski, M. E. 1999. High resolution *in vivo* intra-arterial imaging with optical coherence tomography. *Heart,* 82, 128–133.

Hassan, A. K. M., Bergheanu, S. C., Stijnen, T. et al. 2010. Late stent malapposition risk is higher after drug-eluting stent compared with bare-metal stent implantation and associates with late stent thrombosis. *European Heart Journal,* 31, 1172–1180.

Hausler, G. and Lindner, M. W. 1998. "Coherence radar" and "spectral radar"—new tools for dermatological diagnosis. *Journal of Biomedical Optics,* 3, 21–31.

Hee, M. R., Izatt, J. A., Swanson, E. A. et al. 1995. Optical coherence tomography of the human retina. *Archives of Ophthalmology,* 113, 325–332.

Hoffmann, R. and Mintz, G. S. 2000. Coronary in-stent restenosis—Predictors, treatment and prevention. *European Heart Journal,* 21, 1739–1749.

Honda, Y. 2009. Drug-eluting stents—Insights from invasive imaging technologies. *Circulation Journal,* 73, 1371–1380.

Huang, D., Swanson, E. A., Lin, C. P. et al. 1991. Optical coherence tomography. *Science,* 254, 1178–1181.

Huber, R., Adler, D. C., and Fujimoto, J. G. 2006a. Buffered Fourier domain mode locking: Unidirectional swept laser sources for optical coherence tomography imaging at 370,000 lines/s. *Optics Letters,* 31, 2975–2977.

Huber, R., Wojtkowski, M., and Fujimoto, J. G. 2006b. Fourier domain mode locking (Fdml): A new laser operating regime and applications for optical coherence tomography. *Optics Express,* 14, 3225–3237.

Ito, H., Maruyama, A., Iwakura, K. et al. 1996. Clinical implications of the 'no reflow' phenomenon. A predictor of complications and left ventricular remodeling in reperfused anterior wall myocardial infarction. *Circulation,* 93, 223–228.

Ito, H., Tomooka, T., Sakai, N. et al. 1992. Lack of myocardial perfusion immediately after successful thrombolysis. A predictor of poor recovery of left ventricular function in anterior myocardial infarction. *Circulation,* 85, 1699–705.

Jang, I. K., Tearney, G., and Bouma, B. 2001. Visualization of tissue prolapse between coronary stent struts by optical coherence tomography—Comparison with intravascular ultrasound. *Circulation,* 104, 2754.

Joner, M., Finn, A. V., Farb, A. et al. 2006. Pathology of drug-eluting stents in humans—Delayed healing and late thrombotic risk. *Journal of the American College of Cardiology,* 48, 193–202.

Kataiwa, H., Tanaka, A., Kitabata, H., Imanishi, T., and Akasaka, T. 2008. Safety and usefulness of non-occlusion image acquisition technique for optical coherence tomography. *Circulation Journal,* 72, 1536–1537.

Katoh, H., Shite, J., Shinke, T. et al. 2009. Delayed neointimalization on Sirolimus-eluting stents-6-month and 12-month follow up by optical coherence tomography. *Circulation Journal,* 73, 1033–1037.

Kloner, R. A., Ganote, C. E., and Jennings, R. B. 1974. The "no-reflow" phenomenon after temporary coronary occlusion in the dog. *Journal of Clinical Investigation,* 54, 1496–508.

Kubo, T., Bezera, H. G., Guagliumi, G. et al. 2009. Unhealed plaque ruptures after stenting in acute myocardial infarction assessed by intracoronary optical coherence tomography: Incidence, predictors, and clinical implication. *Circulation,* 120, S921–S921.

Kume, T., Akasaka, T., Kawamoto, T. et al. 2005a. Assessment of coronary intima-media thickness by optical coherence tomography—Comparison with intravascular ultrasound. *Circulation Journal,* 69, 903–907.

Kume, T., Akasaka, T., Kawamoto, T. et al. 2005b. Visualization of neointima formation by optical coherence tomography. *International Heart Journal,* 46, 1133–1136.

Kume, T., Akasaka, T., Kawamoto, T. et al. 2006a. Assessment of coronary arterial thrombus by optical coherence tomography. *American Journal of Cardiology,* 97, 1713–1717.

Kume, T., Akasaka, T., Kawamoto, T. et al. 2006b. Measurement of the thickness of the fibrous cap by optical coherence tomography. *American Heart Journal,* 152, 755 e1–e4.

Leitgeb, R. A., Drexler, W., Unterhuber, A. et al. 2004. Ultrahigh resolution Fourier domain optical coherence tomography. *Optics Express,* 12, 2156–2165.

Leitgeb, R., Hitzenberger, C. K., and Fercher, A. F. 2003. Performance of fourier domain vs. time domain optical coherence tomography. *Optics Express,* 11, 889–894.

Leitgeb, R. A., Villiger, M., Bachmann, A. H., Steinmann, L., and Lasser, T. 2006. Extended focus depth for Fourier domain optical coherence microscopy. *Optics Letters,* 31, 2450–2452.

Matsumoto, D., Shite, J., Shinke, T. et al. 2007. Neointimal coverage of sirolimus-eluting stents at 6-month follow-up: Evaluated by optical coherence tomography. *European Heart Journal,* 28, 961–967.

Moore, P., Barlis, P., Spiro, J. et al. 2009. A randomized optical coherence tomography study of coronary stent strut coverage and luminal protrusion with rapamycin-eluting stents. *JACC-Cardiovascular Interventions,* 2, 437–444.

Moreno, P. R., Bernardi, V. H., Lopezcuellar, J. et al. 1996. Macrophages, smooth muscle cells, and tissue factor in unstable angina—Implications for cell-mediated thrombogenicity in acute coronary syndromes. *Circulation,* 94, 3090–3097.

Nadkarni, S. K., Pierce, M. C., Park, B. H. et al. 2007. Measurement of collagen and smooth muscle cell content in atherosclerotic plaques using polarization-sensitive optical coherence tomography. *Journal of the American College of Cardiology,* 49, 1474–81.

Nakazawa, G., Finn, A. V., Joner, M. et al. 2008. Delayed arterial healing and increased late stent thrombosis at culprit sites after drug-eluting stent placement for acute myocardial infarction patients: An autopsy study. *Circulation,* 118, 1138–1145.

Oh, W. Y., Vakoc, B. J., Shishkov, M., Tearney, G. J., and Bouma, B. E. 2010. >400 kHz repetition rate wavelength-swept laser and application to high-speed optical frequency domain imaging. *Optics Letters,* 35, 2919–2921.

Otake, H., Shite, J., Ako, J. et al. 2009. Local determinants of thrombus formation following sirolimus-eluting stent implantation assessed by optical coherence tomography. *JACC-Cardiovascular Interventions,* 2, 459–466.

Ozaki, Y., Okumura, M., Ismail, T. F. et al. 2010. The fate of incomplete stent apposition with drug-eluting stents: An optical coherence tomography-based natural history study. *European Heart Journal,* 31, 1470–1476.

Povazay, B., Bizheva, K., Unterhuber, A. et al. 2002. Submicrometer axial resolution optical coherence tomography. *Optics Letters,* 27, 1800–1802.

Puliafito, C. A., Hee, M. R., Lin, C. P. et al. 1995. Imaging of macular diseases with optical coherence tomography. *Ophthalmology,* 102, 217–229.

Rollins, A. M. and Izatt, J. A. 1999. Optimal interferometer designs for optical coherence tomography. *Optics Letters,* 24, 1484–1486.

Salam, A. M., Al Suwaidi, J., and Holmes, D. R. 2006. Drug-eluting coronary stents. *Current Problems in Cardiology,* 31, 8–119.

Sato, H., Iida, H., Tanaka, A. et al. 2004. The decrease of plaque volume during percutaneous coronary intervention has a negative impact on coronary flow in acute myocardial infarction: A major role of percutaneous coronary intervention-induced embolization. *Journal of the American College of Cardiology,* 44, 300–4.

Schmitt, J. M. 1999. Optical coherence tomography (OCT): A review. *Selected Topics in Quantum Electronics, IEEE Journal of,* 5, 1205–1215.

Serruys, P. W., Ormiston, J. A., Onuma, Y. et al. 2009. A bioabsorbable everolimus-eluting coronary stent system (Absorb): 2-year outcomes and results from multiple imaging methods. *Lancet,* 373, 897–910.

Shite, J., Matsumoto, D., and Yokoyama, M. 2006. Sirolimus-eluting stent fracture with thrombus, visualization by optical coherence tomography. *European Heart Journal,* 27, 1389.

Sigwart, U., Puel, J., Mirkovitch, V., Joffre, F., and Kappenberger, L. 1987. Intravascular stents to prevent occlusion and restenosis after trans-luminal angioplasty. *New England Journal of Medicine,* 316, 701–706.

Suter, M. J., Tearney, G. J., Wang-Yuhl, O., and Bouma, B. E. 2010. Progress in intracoronary optical coherence tomography. *Selected Topics in Quantum Electronics, IEEE Journal of,* 16, 706–714.

Takada, K. 1998. Noise in optical low-coherence reflectometry. *IEEE Journal of Quantum Electronics,* 34, 1098–1108.

Takano, M., Inami, S., Jang, I.-K. et al. 2007. Evaluation by optical coherence tomography of neointimal coverage of sirolimus-eluting stent three months after implantation. *The American Journal of Cardiology,* 99, 1033–1038.

Tanaka, A., Imanishi, T., Kitabata, H. et al. 2008. Morphology of exertion-triggered plaque rupture in patients with acute coronary syndrome an optical coherence tomography study. *Circulation,* 118, 2368–2373.

Tanaka, A., Imanishi, T., Kitabata, H. et al. 2009. Lipid-rich plaque and myocardial perfusion after successful stenting in patients with non-ST-segment elevation acute coronary syndrome: An optical coherence tomography study. *European Heart Journal,* 30, 1348–1355.

Tanaka, A., Kawarabayashi, T., Nishibori, Y. et al. 2002. No-reflow phenomenon and lesion morphology in patients with acute myocardial infarction. *Circulation,* 105, 2148–52.

Tearney, G. J., Boppart, S. A., Bouma, B. E. et al. 1996a. Scanning single-mode fiber optic catheter-endoscope for optical coherence tomography. *Optics Letters,* 21, 543–545.

Tearney, G. J., Bouma, B. E., Boppart, S. A., Golubovic, B., Swanson, E. A., and Fujimoto, J. G. 1996b. Rapid acquisition of *in vivo* biological images by use of optical coherence tomography. *Optics Letters,* 21, 1408–1410.

Tearney, G. J., Bouma, B. E., and Fujimoto, J. G. 1997a. High-speed phase- and group-delay scanning with a grating-based phase control delay line. *Optics Letters,* 22, 1811–1813.

Tearney, G. J., Brezinski, M. E., Boppart, S. A. et al. 1996c. Catheter-based optical imaging of a human coronary artery. *Circulation,* 94, 3013–3013.

Tearney, G. J., Brezinski, M. E., Bouma, B. E. et al. 1997b. *In vivo* endoscopic optical biopsy with optical coherence tomography. *Science,* 276, 2037–2039.

Tearney, G. J., Jang, I. K., Kang, D. H. et al. 2000. Porcine coronary imaging *in vivo* by optical coherence tomography. *Acta Cardiologica*, 55, 233–237.

Tearney, G. J., Waxman, S., Shishkov, M. et al. 2008a. Three-dimensional coronary artery microscopy by intracoronary optical frequency domain imaging. *Journal of the American College of Cardiology and Imaging*, 1, 752–761.

Tearney, G. J., Waxman, S., Shishkov, M. et al. 2008b. Three-dimensional coronary artery microscopy by intracoronary optical frequency domain imaging. *Journal of the American College of Cardiology: Cardiovascular Imaging*, 1, 752–761.

Tearney, G. J., Yabushita, H., Houser, S. L. et al. 2003. Quantification of macrophage content in atherosclerotic plaques by optical coherence tomography. *Circulation*, 107, 113–119.

Togni, M., Windecker, S., Cocchia, R. et al. 2005. Sirolimus-eluting stents associated with paradoxic coronary vasoconstriction. *Journal of the American College of Cardiology*, 46, 231–236.

Torti, C., Povazay, B., Hofer, B. et al. 2009. Adaptive optics optical coherence tomography at 120,000 depth scans/s for non-invasive cellular phenotyping of the living human retina. *Optics Express*, 17, 19382–19400.

Toutouzas, K., Vaina, S., Riga, M. I., and Stefanadis, C. 2009. Evaluation of dissection after coronary stent implantation by intravascular optical coherence tomography. *Clinical Cardiology*, 32, E47–E48.

Villard, J., Cheruku, K., and Feldman, M. 2009a. Applications of optical coherence tomography in cardiovascular medicine, part 1. *Journal of Nuclear Cardiology*, 16, 287–303.

Villard, J., Paranjape, A., Victor, D., and Feldman, M. 2009b. Applications of optical coherence tomography in cardiovascular medicine, Part 2. *Journal of Nuclear Cardiology*, 16, 620–639.

Villiger, M., Goulley, J., Friedrich, M. et al. 2009. *In vivo* imaging of murine endocrine islets of Langerhans with extended-focus optical coherence microscopy. *Diabetologia*, 52, 1599–1607.

Virmani, R., Kolodgie, F. D., Burke, A. P., Farb, A., and Schwartz, S. M. 2000. Lessons from sudden coronary death—A comprehensive morphological classification scheme for atherosclerotic lesions. *Arteriosclerosis Thrombosis and Vascular Biology*, 20, 1262–1275.

Waxman, S., Freilich, M. I., Suter, M. J. et al. 2010. A case of lipid core plaque progression and rupture at the edge of a coronary stent: Elucidating the mechanisms of drug-eluting stent failure. *Circulation–Cardiovascular Interventions*, 3, 193–196.

Wojtkowski, M., Srinivasan, V. J., Ko, T. H., Fujimoto, J. G., Kowalczyk, A., and Duker, J. S. 2004. Ultrahigh-resolution, high-speed, Fourier domain optical coherence tomography and methods for dispersion compensation. *Optics Express*, 12, 2404–2422.

Yabushita, H., Bouma, B. E., Houser, S. L. et al. 2002. Characterization of human atherosclerosis by optical coherence tomography. *Circulation*, 106, 1640–1645.

Yun, S. H., Tearney, G. J., Vakoc, B. J. et al. 2006. Comprehensive volumetric optical microscopy in vivo. *Nature Medicine*, 12, 1429–1433.

10

Imaging of Therapeutic Processes in Animals Using Optical Reporter Genes

Ying Zhao
Institute of Bioengineering and Nanotechnology

Jiakai Lin
Institute of Bioengineering and Nanotechnology

Shu Wang
National University of Singapore

Institute of Bioengineering and Nanotechnology

10.1 Introduction

To traverse the chasm from *in vitro* therapeutic studies to clinical applications in translational research, animal models that mimic various human diseases are employed to evaluate the efficacy of new therapeutics. Such evaluations in animal models require various imaging and histochemical methods. For conventional histochemical or immunohistochemical analysis, animals are sacrificed at multiple time points prior to performing tissue sectioning and staining to visualize the location of certain molecules or cells. To reduce experimental variability, number of animals required, and the duration and cost of animal studies, noninvasive imaging modalities have been developed that allow researchers to follow disease progression by monitoring events repeatedly in the same living animals.

Five main classes of noninvasive imaging techniques have emerged as robust whole-animal imaging modalities: radionuclide imaging (positron emission tomography [PET] and single photon emission computed tomography [SPECT]), optical imaging (fluorescence imaging, fluorescence-mediated tomography, and bioluminescence imaging), magnetic resonance imaging (MRI), computed tomography (CT), and ultrasound imaging. With the exception of optical imaging, most modalities are used routinely in diagnostic laboratories and hospitals to scan patients. Smaller instruments, such as microPET, have also been built to facilitate animal research. Noninvasive imaging requires specific probes, contrast agents, or reporter genes to provide quantitative information. In particular, reporter genes have been utilized in modalities, including MRI, PET, SPECT, fluorescence imaging, and bioluminescence imaging. For gene therapy imaging, the use of reporter genes is a requisite for monitoring transgene expression (Table 10.1).

With noninvasive whole-animal imaging techniques, data obtained from the same animal significantly increases the power of the statistical analysis. Expenditure on laboratory animals is brought down by imaging the same animal throughout an experiment instead of killing many animals for histological staining at different time points. To carry out histological studies, tissues must be removed from the body, fixed by chemical means, and observed under nonphysiological conditions. Furthermore, images from conventional microscopy methods are extremely difficult to quantify. Live imaging provides more meaningful quantitative data, in consideration of the intactness and the physiology of the experimental subject (Massoud and Gambhir, 2003). Unlike the two-dimensional information obtained from tissue sections, noninvasive imaging techniques are able to image reporter genes in living animals in two dimensions or three dimensions.

Compared to MRI, PET, and SPECT, optical imaging methods are simple, low cost, more convenient, easier to set up, and suitable for high-throughput screening. Additional advantages particularly for fluorescence imaging and fluorescence-mediated tomography (FMT) modalities include the fact that no ionizing radiation is required and fluorophores are relatively stable and do not decay as with isotopes. Hence, it is possible to study the same animal repetitively at different time points without any fear of overexposure to ionizing radiation. It is also plausible that multiwavelength imaging can be performed to obtain information from multiple targets. This may be important clinically as determining expression patterns of multiple molecules, rather than single molecules, may form the basis for future imaging techniques. Optical imaging is not without its limitations. Light scatter, light absorption by tissue, and autofluorescence are the pressing problems for optical imaging that leads to its limited use in small animals currently. As a result, signal intensities detected may not be reflective of true reporter gene expression in the inner organs (Kang and Chung, 2008).

This review focuses on the use of optical reporter genes suitable for optical imaging, such as fluorescence imaging and bioluminescence imaging, when applied to imaging therapeutic processes in animals. Such methods will allow (i) noninvasive monitoring of target tissues that may benefit from the tested therapeutics, (ii) guided delivery of therapeutics, (iii) assessment of location, magnitude, and duration of therapeutic actions, and (iv) monitoring of experimental therapeutic response during and after treatment in animal disease models (Jacobs et al., 2007, Miletic et al., 2007).

10.2 Optical Reporter Genes

For optical imaging in live animals, reporter genes encoding proteins that are luminescent or fluorescent are commonly used. Widely used fluorescent proteins include the green or red fluorescent proteins while luminescent proteins originally isolated from the firefly, click beetle, sea pansy, or copepod have also proven popular with researchers. Photons from these luminescent or fluorescent proteins can be detected by several *in vivo* imaging systems based on sensitive charge-coupled device (CCD) cameras that are now commercially available.

Green fluorescent protein (GFP) and its variants have been extensively used as reporter genes in molecular and cell biology. GFP is a 27 kDa protein isolated from jellyfish *Aequorea victoria*, emitting green fluorescence at 510 nm when excited with blue light at 488 nm. GFP can be easily expressed in cells using standard transfection or transduction methods, and can be imaged in real time from live or fixed cells. Variants of GFP with shifted emissions, such as blue fluorescent protein (BFP), cyan fluorescent protein (CFP), and yellow fluorescent protein (YFP), have been engineered using mutagenesis to alter sequences in the region of the chromophore. Variants of red fluorescent proteins have also been discovered, including *Discosoma* red fluorescent protein (DsRed) and *Heteractis crispa* red fluorescent protein (HcRed). DsRed emits fluorescence at 583 nm and HcRed at 618 nm. Owing to tissue absorption

TABLE 10.1 Overview of Imaging Technologies Available

Imaging Modality	Signal Measured	Spatial Resolution	Tissue Depth	Inherently Quantitative	Strengths	Weaknesses	Applicable to Human Imaging	Cost of System
Single photon emission computed tomography (SPECT)	Gamma rays	1–2 mm	No limit	Yes	Can distinguish between different radionuclides, image multiple probes simultaneously	Radiation exposure	Yes	High
Positron emission tomography (PET)	Positrons	1–2 mm	No limit	Yes	High sensitivity	PET cyclotron needed; may only use one radionuclide at a time; radiation exposure	Yes	Very high
Fluorescence imaging	Visible light	2–3 mm	<1 cm	No	High sensitivity, ease of use, high throughput	Poor depth penetration, low spatial resolution, strong autofluorescence	Yes, with limitations	Low
Fluorescence-mediated tomography (FMT)	Near-infrared light	1–3 mm	<10 cm	Yes	Stronger depth penetration than fluorescence imaging	Stronger absorption of fluorescence in kidney or liver complicates fluorescence propagation	In development	High
Bioluminescence imaging	Visible light	≈ Depth of signal (several mm)	<2 cm	No	High sensitivity, ease of use, high throughput, no autoluminescence	Poor depth penetration, low spatial resolution	Yes, with limitations	High
Magnetic resonance imaging (MRI)	Changes in magnetic field	10–100 μm	No limit	Yes	Perform morphological, functional, and anatomical imaging	Long scan time, low sensitivity, big amount of probes needed	Yes	Very high
Computed tomography (CT)	X-rays	50–200 μm	No limit	Yes	Cross-sectional imaging	Poor resolution of soft tissues	Yes	High

of visible light, red-shifted fluorescent proteins such as DsRed and HcRed fare better than BFP or GFP in the transmission of light through tissues.

Luciferases are the large family of enzymes that catalyze the conversion of their substrate luciferins into oxyluciferins and in the process, emit photons. While oxygen is a prerequisite for the activity of all luciferases discovered so far, different luciferases require different cofactors for catalysis. Adenosine triphosphate (ATP) is one such cofactor that is required by the firefly luciferase (Fluc) and click beetle luciferase (CBRed). Fluc comes from the North-American firefly (*Photinus pyralis*) while CBRed comes from the click beetle (*Pyrophorus plagiophtalamus*). Both Fluc and CBRed utilize D-luciferin as a substrate with ATP as a cofactor. Another group of luciferases does not utilize ATP as a cofactor and uses coelenterazine as the substrate. These are marine luciferases, including *Renilla* luciferase (Rluc) derived from the sea pansy (*Renilla reniformis*) and *Gaussia* luciferase (Gluc) derived from the copepod (*Gaussia princeps*). The bacterial Lux luciferase can be derived from the genera *Photobacterium* and *Vibrio*. The bacterial luciferase operon LuxCDABE comes from *Photorhabdus luminescens* and is capable of expressing two luciferase monomers and substrate-regenerating enzymes. Autoluminescence occurs upon the expression of the operon in bacteria as no exogenous substrate (decanal) is needed.

Among different types of luciferases, Fluc and Rluc are the most commonly used luciferases in biomedical research (Gross and Piwnica-Worms, 2005, Wilson and Hastings, 1998). Fluc is a monomeric 61 kDA protein that catalyzes D-luciferin to oxyluciferin in the presence of oxygen, cofactors, Mg^{2+}, and ATP to release visible light peaking at 562 nm (at subphysiological temperature of 22°C) or 612 nm (37°C) (Min and Gambhir, 2004, Zhao et al., 2005). Rluc is a monomeric 36 kDA protein that catalyzes the oxidation of coelenterazine in the presence of oxygen, to produce blue luminescence at 482 nm (Min and Gambhir, 2004). This blue-shifted light limits its use in *in vivo* imaging and a codon-optimized Rluc was made that gave higher sensitivity than the native Rluc.

In certain cases, a therapeutic gene may serve as a reporter gene (e.g., herpes simplex virus thymidine kinase [HSV-TK] in PET imaging). In most cases, therapeutic genes are not capable of being monitored *in vivo* and the coupling of a reporter gene to the therapeutic gene is required for assessing levels of therapeutic gene delivery. The coexpression of reporter and therapeutic genes is possible with the modular design of an expression cassette. An expression cassette comprises of a promoter to confer the desired level of expression, complementary DNA (cDNA) for gene expression, and enhancer elements that boost transcriptional activity. Promoters are short stretches of DNA that recruit RNA polymerase to initiate gene transcription, while enhancers are DNA-binding sites for proteins that boost gene transcription. Researchers have flexibility in designing their reporter constructs according to the needs of their studies. For example, high level of transgene expression can be achieved with constitutively active promoters. Promoters found in viruses such as cytomegalovirus (CMV) promoter and long terminal repeat (LTR) are good candidates. Other potential highly active promoters to consider are elongation factor 1α (EF1α) promoter, phosphoglycerate kinase (PGK) promoter, and ubiquitin C promoter. Tissue-specific promoters, on the other hand, potentially restrict gene expression to target tissues and avoid adverse effects due to expression in nontargeted tissues. For the purpose of imaging gene delivery to tissues in whole animals, a promoter that offers high expression of reporter gene in target tissues is desired for optimum sensitivity during imaging.

Coexpression of two genes can be achieved by gene fusion, two-vector administration, dual-promoter, bidirectional transcription, and bicistronic approach using internal ribosomal entry site (IRES). In the case of fusion genes, a hybrid gene is formed from the therapeutic gene and reporter gene to ensure a firm correlation between reporter gene imaging and therapeutic gene expression level. The association of therapeutic effect with reporter gene expression level has been demonstrated with the creation of fusion proteins, HSV-TK with GFP (Loimas et al., 1998, Steffens et al., 2000), and HSV-TK with Fluc and neomycin resistance gene (Strathdee et al., 2000). However, gene fusion may affect the functions of both therapeutic and reporter genes, and this option is not applicable to many other combinations of therapeutic genes and reporter genes. Two-vector administration refers to the concurrent systemic administration of two gene delivery vectors that differ in the coding region (one coding for therapeutic gene and the other for reporter gene) but are otherwise identical. A critical assumption here is that both vectors have equal probability of delivering

TABLE 10.2 Comparison of Currently Available Gene Delivery Vectors

Delivery Vector	Advantages	Disadvantages
Retrovirus	Stable reporter gene expression Low immunogenicity	Insertional mutagenesis Limited to dividing cells
Lentivirus	Stable reporter gene expression Board cellular transduction Low immunogenicity	Insertional mutagenesis
Adenovirus	Board cellular transduction High titers	Transient transgene expression Toxicity
HSV-1	Capacity: 30 kb Persistent reporter gene expression Easily express multiple transgenes Efficient transduction on neurons	Neurovirulence Recombination with wild type High immunogenicity
Adeno-associated virus (AAV)	Persistent reporter gene expression Board cellular transduction Site-specific integration Low immunogenicity	Low transgene capacity: 4.5 kb
Baculovirus	Large insert capacity: 38 kb No viral gene expression or replication in human cells No preexisting immune response Efficient in board cell types	Loss of activity in blood
Liposomes	Low immunogenicity	Inefficient delivery *in vivo*
Stem cells	Homing on tumor Repair damage Virus free *Ex vivo* transduction	Immunogenicity: suppression/activation

their respective genes to the cells; hence, approximately equal expression patterns of both genes will result. Therapeutic gene and reporter gene can also be delivered in a single vector, driving by multiple promoters toward the same direction or in opposite directions. Since therapeutic gene and reporter gene are translated from two transcripts, reporter gene expression may not correlate exactly with therapeutic gene expression in some cases. Bicistronic approach involves the use of IRES in between coding regions of therapeutic gene and reporter gene, allowing the expression of two proteins from a common transcript. Several studies have utilized bicistronic vectors in reporter gene imaging and have shown that bicistronic vectors can be used to monitor the correlated expression of a therapeutic gene and a reporter gene (Herschman, 2004). With the range of techniques available for coupling reporter gene and therapeutic gene expression, the magnitude, location, and duration of therapeutic gene expression can be determined with reporter gene imaging.

Various gene delivery vectors have been used to deliver transgenes *in vivo*, including viral vectors, chemical vectors, and cellular vectors. Viral vectors are the most effective *in vivo* gene delivery reagents and have been well studied in clinical trials. Adenovirus, retrovirus, herpes simplex virus type 1 (HSV-1), and adeno-associated virus are the most commonly used viral vectors in gene therapy. In clinical trials, chemical vectors, such as liposome, have shown much lower transfection efficiency *in vivo* although it brings along fewer safety concerns than viral vectors. Recently discovered cellular vectors such as stem cells have tumor "homing" property that enables delivery of therapeutic gene to remote tumor sites. The strengths and weaknesses of the currently available gene delivery vectors make them suitable for different applications (Table 10.2).

10.3 Fluorescence Imaging

In fluorescence imaging, an external source of light (visible light ranging from 395 to 600 nm) excites a fluorophore at a particular wavelength, and light energy is emitted at a longer wavelength with lower energy when the fluorophore returns to its ground state. This emitted light is detected by a CCD camera

(Weissleder et al., 1999) to obtain two-dimensional or planar images. Suitable optical filters corresponding to the excitation and emission wavelength of the fluorophore have to be used. The behavior of normal cells or cancer cells can be observed *in vivo* after labeling with a fluorescent dye, fluorescence-labeled antibodies, or inducible expression of fluorescent proteins (Weissleder, 2002). 1000 to 10,000 fluorescently labeled metastatic gastric tumor cells can be imaged peritoneally (Kaneko et al., 2001). It is also possible to simultaneously image multiple targets using different fluorescent proteins, each with a unique emission wavelength in small animals (Chudakov et al., 2005). Transmission of light through mammalian tissues is hindered by strong light absorption from tissues and strong autofluorescence, resulting in low signal-to-background ratios (Troy et al., 2004). As such, fluorescence imaging systems have limited resolution beyond depths of 5 mm and are not inherently quantitative. A skin flap is required to visualize the fluorescently labeled tissues or organs under investigation in many cases (Bouvet et al., 2002, Yang et al., 2002).

The major absorbers of visible light and infrared light are hemoglobin and water, respectively, and their absorption coefficients are lowest in the near-infrared (NIR) region (Massoud and Gambhir, 2003). Consequently, imaging in NIR spectrum (700–900 nm) permits maximum tissue penetration and minimum autofluorescence from unlabeled cells and tissues. Several NIR fluorochromes with improved brightness, solubility, and photostability are now commercially available (Hilderbrand and Weissleder, 2009). One technique to circumvent the problem in fluorescence imaging of deep-seated tissues in whole-animal imaging is with the use of fluorescence-mediated tomography (FMT) (Ntziachristos et al., 2002, Weissleder and Ntziachristos, 2003). FMT is a volumetric imaging technique that accounts for the diffusive propagation of photons through mammalian tissues. The subject being imaged is subjected to continuous wave or pulsed light exposure. Fluorescence is detected by detectors arranged in a fixed spatial manner to reconstitute a tomographic image (Weissleder, 2002). FMT is able to quantify and resolve pico-to-femto moles of NIR probes as well as achieving several centimeters of tissue penetration.

Fluorescence imaging has found its applications in monitoring engrafted tumor growth (Yang et al., 2000a, Hoffman, 2001), metastasis formation (Moore et al., 1998), and gene expression (Yang et al., 2000b). GFP offers a method to quantify transduction efficiency, gene expression levels, and vector distribution. *In vivo* imaging of GFP has been used to track gene expression and distribution of adenovirus in nude mice or normal mice at a variety of sites (Yang et al., 2000a). In cancer therapy, GFP could help to image xenograft tumors in living animals. By imaging tumor cells stably express GFP, tumor growth, metastasis, and response to therapies in nude mice have been monitored (Bouvet et al., 2002, Ito et al., 2001). Although many strategies are being developed to improve the detection and visualization of fluorescent proteins *in vivo*, restrictions in tissue penetration and autofluorescence from tissues are likely to limit their imaging utility only to small animals.

10.4 Bioluminescence Imaging

Unlike fluorescence imaging, external light source is not required for bioluminescence imaging. Cells expressing luciferase are not visible until the luciferin substrate is administered by intraperitonealor intravenous injection. D-Luciferin circulates throughout many body compartments, readily crosses the blood–brain barrier, and rapidly enters many cells and is hence very suitable for *in vivo* imaging purposes (Contag et al., 1997). However, due considerations have to be taken when using the Rluc system. Coelenterazine was discovered to be a substrate for multidrug resistance P-glycoprotein and can be pumped out of cells (Pichler et al., 2004). Differential expression of P-glycoprotein in cells over time in longitudinal studies can affect signal intensities leading to misinterpretation of Rluc expression levels. Coelenterazine instability in serum is also a concern as it autooxidizes in the absence of Rluc, causing significant background signals (Zhao et al., 2004).

In the presence of oxygen and luciferins, luciferase enzymes such as Fluc and Rluc can catalyze the production of photons in the visible range of the spectrum with a short half-life in cells (Raty et al., 2007b). Dual luciferase imaging is possible by sequentially and systemically administering D-luciferin and coelenterazine that are specific substrates for Fluc and Rluc, respectively. Owing to the substrate specificity of Fluc and Rluc, there is no cross-reaction between these luciferases and substrates. Therefore, dual bioluminescence imaging has been applied to monitor two separate biological processes in the same animal over time (Bhaumik and Gambhir, 2002, Shah et al., 2003, 2005). The method is highly sensitive and specific because of the absence of autoluminescence. It has been estimated that the sensitivity of luminescence imaging is in the 10^{-15}–10^{-17} mol/L range at limited depths of not more than 2 cm (Massoud et al., 2008). The sensitivity is evident from the clear detection of 1000 luciferase-positive tumor cells in the peritoneal cavity (Contag et al., 2000). Nonetheless, bioluminescence imaging in live animals is constrained by issues pertaining to tissue absorption and scatter of light. Just like fluorescence imaging, bioluminescence imaging is likely to be feasible only in small animals where the source of signal is unlikely to be more than 1–2 cm beneath the skin.

Unlike autofluorescence, autoluminescence is almost nonexistent in living animals because luciferase is not found in mammalian organisms. The absence of autoluminescence makes bioluminescence imaging a highly sensitive and specific technique. Luciferases have been used for tracking tumor cells, stem cells, immune cells, and bacteria as well as for imaging gene expression (Weissleder and Ntziachristos, 2003). Cells stably expressing luciferase can be easily generated through plasmid transfection or viral transduction and cell fate *in vivo* can be monitored by noninvasive luminescence imaging. The migration of Fluc-expressing neural stem cells has been observed to cross the cerebral hemispheres toward tumor site in a brain tumor mouse model (Muller et al., 2006). The growth kinetics of brain tumor cells stably expressing Fluc, U87-Fluc, in the mouse brain can be monitored over time (Zhao and Wang, 2010). Gene transfers mediated by viral vectors can also be investigated by imaging the expression of luciferase reporter gene in living animals (De et al., 2003, Honigman et al., 2001, Lipshutz et al., 2001, Wu et al., 2001).

To improve the sensitivity and utility of bioluminescence in animal imaging, far-red or near-infrared bioluminescence imaging has been developed by conjugating luciferase protein to fluorochromes that emit light at long wavelengths in the far-red or NIR. This technology is based on the principle of bioluminescence resonance energy transfer (BRET). BRET is a natural phenomenon in marine organisms whereby resonance energy is transferred from a light-emitting protein (such as luciferase protein) to a fluorescent protein (So et al., 2006). In far-red or near-infrared bioluminescence imaging, the luciferase enzyme acts as the energy donor upon metabolizing luciferin. Resonance energy is further transferred to far-red or near-infrared fluorochromes, acting as energy acceptors, resulting in emission of light with enhanced penetration properties. Self-illuminating quantum dot luciferase conjugates have shown to greatly enhance sensitivity in small animal imaging (So et al., 2006). Another far-red bioluminescent probe was developed by conjugating a far-red fluorescent indocyanine derivative to biotinylated *Cypridina* luciferase and the accumulation of this probe in tumor-bearing mice was monitored by optical imaging (Wu et al., 2009).

10.5 Equipments for Optical Imaging

In vivo imaging systems for optical imaging are equipped with a cooled CCD camera to acquire light emitted from the organism. Ultra-cooled CCD camera (as low as −120°C) is highly sensitive to light and can detect very weak light sources within the organism. The wavelengths captured the range from visible light to near infrared. CCD cameras are normally mounted in a light-tight specimen chamber that holds anesthetized animals. Data acquisition and analysis is done by a camera controller linked to a computer. Data acquisition is often a straightforward process that does not require particular expertise. After administration of the luciferase substrate (D-luciferin for Fluc or coelenterazine for Rluc), the animals are placed on a heated stage and are ready to be imaged. Under a flash of weak illumination, a gray-scale photograph is captured as a reference image for anatomical information. Subsequently, a bioluminescent or fluorescent image is captured possibly over a range of user-preset integration times. Total integration times depend on the imaging system (fluorescence, Fluc, or Rluc-based luminescence) used. Owing to the short half-life of coelenterazine in serum, integration time for imaging Rluc should be less than 2 min while that for Fluc can be as long as 5 min. The software then produces an overlay of the fluorescent or bioluminescent image arising from reporter gene activity and a gray-scale light photographic image of the animal. The location of the optical signals with regard to the animal's anatomical information can be inferred from the overlaid image (Massoud et al., 2008). For interpretation of signal intensities, a pseudocolor scale bar is used. Blue typically represents the lowest intensity while red represents the highest intensity. Quantification of various signal intensities is made by demarcating a region of interest over the area of interest. Reading can be recorded as counts, photons, or efficiency. As signal intensities in bioluminescence imaging are depth dependent, quantification of serial data from the same animal is possible only if the animal is imaged in the exact same position and image acquisition parameters are kept identical throughout the time course of the experiment. Measurements are highly reproducible if the exposure conditions, for example, including time, f/stop, height of the sample shelf, binding ratio, and time after injection of the optical substrate, are kept identical (Massoud and Gambhir, 2003).

The major absorbers of visible light and infrared light are hemoglobin and water, respectively and their absorption coefficients are lowest in the NIR region (Massoud and Gambhir, 2003). Consequently, imaging in NIR spectrum (700–900 nm) permits maximum tissue penetration and minimum autofluorescence from unlabeled cells and tissues. Several NIR fluorochromes with improved brightness, solubility, and photostability are now commercially available (Hilderbrand and Weissleder, 2009). One technique to circumvent the problem in fluorescence imaging of deep-seated tissues in whole-animal imaging is with the use of FMT (Ntziachristos et al., 2002, Weissleder and Ntziachristos, 2003). FMT is a volumetric imaging technique that accounts for the diffusive propagation of photons through mammalian tissues. The subject being imaged is subjected to continuous wave or pulsed light exposure. Fluorescence is detected by detectors arranged in a fixed spatial manner to reconstitute a tomographic image (Weissleder, 2002). FMT is able to quantify and resolve pico-to-femto moles of NIR probes as well as achieving several centimeters of tissue penetration.

10.6 Applications of Optical Imaging

10.6.1 Monitoring Biodistribution of Gene Delivery Vectors

With a better understanding on genetic contributions to disease development, gene therapy has emerged as a promising treatment strategy. Gene therapy targets the causes of diseases and corrects them at their most basic levels, offering an obvious advantage over conventional therapeutic approaches such as chemotherapy and radiotherapy (Penuelas et al., 2005). This therapeutic intervention utilizes gene transfer vectors to deliver functional genes into target cells to replace dysfunctional gene variants, introduce a therapeutic gene into diseased tissues, or to knock out a disease-causing gene variant, thereby improving patient outcomes.

Preclinical studies have demonstrated the profound effectiveness of gene therapy strategy in treating central nervous system degeneration, cancer, cardiovascular diseases, and immunodeficiency diseases. However, in clinical trials, gene therapy has met with limited success. While adverse reactions caused by viral or nonviral vectors have limited the pervasive application of gene therapy technique on patients, failures in gene therapy clinical studies may be caused by low levels of *in vivo* transgene expression in target cells, therapeutic gene expression in nontarget tissues, and inability to dynamically monitor the location, magnitude, and duration of gene expression in the body (Shah, 2005). Gene delivery strategies also require tight control of both spatial and temporal expression of the therapeutic transgene (Rome et al., 2007). As one of the major hurdles encountered in clinical gene therapy trials, the inability to monitor gene transfection or transduction and transgene expression in the target tissue has been noted (Raty et al., 2007a, Waerzeggers et al., 2009). Thus, biopsies or autopsy samples are currently used for laboratory analysis to understand the distribution of pharmacokinetics of gene therapy reagents, which are unable to provide adequate information on temporal and spatial distribution of the applied gene therapy vectors.

As the first demonstration for the feasibility of imaging the location, magnitude, and time course of cardiac reporter gene expression in living animals, Wu and coworkers injected, into the hearts of rats, adenoviral particles expressing Fluc under the control of a constitutively active CMV promoter (Wu et al., 2002). The authors consequently detected hepatic expression of luciferase under whole-animal bioluminescent imaging, confirming leakage of adenoviral vectors into the systemic circulation. While most work investigating biodistribution of viral gene transfer vectors concentrate on detecting cells producing luciferase subsequent to viral infection, a novel fusion of HSV type 1 thymidine kinase and Fluc to an adenoviral capsid protein, pIX, also enabled direct imaging of the adenoviral particles (Ad-pIX-TK-Luc) (Matthews et al., 2006). Matthews and coworkers demonstrated that the intratumoral intensities of bioluminescent signals from Ad-pIX-TK-Luc compared favorably with bioluminescent intensities arising from adenoviral transduction of tumors and expressing Fluc under a constitutively active CMV promoter. The presence of a thymidine kinase protein also permits the imaging of adenoviral particle biodistribution using PET-based imaging methods. An earlier piece of work by Meulenbroek and coworkers demonstrated the feasibility of appending GFP to the pIX capsid protein and the authors were able to visualize adenoviral distribution around muscle fibers after intramuscular administration in mice (Meulenbroek et al., 2004). Hence, modification of adenoviral capsid protein, pIX, with optical reporter genes appears to be a feasible way of monitoring adenovirus trafficking *in vivo*.

10.6.2 Monitoring Growth of Tumor and Response to Therapy

Noninvasive imaging of reporter gene expression in tumor cells offers a great platform to temporally monitor cancer progression, metastasis, and response to therapy in whole animals. Cancer cells stably expressing imaging reporter genes can be engrafted into various sites within an animal and each animal can be monitored for tumor burden at the primary inoculation sites. If those tumor cells are capable of metastasizing, the development of metastases can be examined as well. Consequently, individual response to therapy can be followed in a noninvasive fashion that permits the collection of multiple data points from one individual, eliminating unnecessary sacrifice of tumor-bearing animals for invasive mid-study biopsies. Optical imaging methods that detect fluorescent proteins, near-infrared fluorochromes, or luciferases have been extensively utilized in several cancer therapy and drug screening studies.

The use of fluorescent proteins such as GFP in whole-animal imaging is hampered by extensive tissue attenuation of both the exciting and emitted light. In some cases, imaging of tumor xenografts expressing GFP is only possible at subcutaneous positions in nude mice while imaging of deep-seated tumors or organs require surgical procedures such as creating a skin flap for fluorescence observation (Yang et al., 2002). In spite of those limitations, studies that utilized tumor xenografts stably expressing GFP have proven to be experimentally feasible and useful (Bouvet et al., 2002, Ito et al., 2001, Zhou et al., 2002). GFP-expressing tumors are especially valuable for intravital microscopy as they provide a clear distinction between the host and tumor cells (Bogdanov et al., 2002, Fukumura et al., 1998).

Near-infrared fluorescence (NIRF) imaging provides enhanced tissue penetration while minimizing tissue autofluorescence henceforth, partially addressing restrictions imposed by GFP imaging in tumors *in vivo*. NIRF probes conjugated to protease-sensitive quenching peptides are well suited for imaging tumors that have upregulated levels of proteases such as cathepsins (Weissleder et al., 1999). Cathepsin cleavage of the quenching peptides increases the fluorescence over several hundred folds (Tung et al., 1999) with no detectable toxicity (Weissleder and Mahmood, 2001). While this strategy was described in the context of cathepsin-expressing tumor cells, NIRF probes can also be used for *in vivo* monitoring of tumor cells transduced by viral vectors delivering protease genes (Shah et al., 2004). In that study, glioma cells transduced by a HSV amplicon vector expressing a human immunodeficiency virus type 1 (HIV-1) protease were more reliably detected than control vector-transduced tumors in the abdomen of nude mice. That study demonstrated that a viral protease expressed within tumor cells can be imaged in live animals, availing itself as a transgene marker in tumor therapy and for testing the efficacy of HIV-1

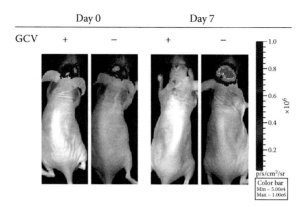

FIGURE 10.1   (See color insert.) Luminescence imaging. An example of monitoring tumor xenografts with biolu-minescence. Luc-labeled U87 glioablastoma cells were injected into the striatum to form brain tumor xenografts. A single injection of baculovirus-expressing HSVtk suicide gene was given to each of human U87 glioblastoma xeno-grafts in the mouse brain 7 days after tumor inoculation, followed by intraperitoneal injection of ganciclovir (GCV), or vehicle control (−GCV) for 7 days. Bioluminescence images are shown 0 and 7 days after GCV injection. (Unpublished data.)

protease inhibitors *in vivo*. In the same light, there is a need for direct and noninvasive assays to detect or image matrix metalloproteinase (MMP) activity *in vivo* for preclinical drug screening. Bremer and coworkers reported the synthesis of NIRF probes that can be used as activatable probes to sense MMP activity in intact tumors in nude mice (Bremer et al., 2001). Similarly, Lee and coworkers conjugated an MMP-sensitive gold nanoparticle that efficiently quenches NIRF in the absence of MMP activation (Lee et al., 2008). Cleavage of the MMP-sensitive quenching peptide by MMP-expressing tumor cells permits the detection of NIR dyes and using this system, nano-molar amounts of MMPs can be detected both *in vitro* and *in vivo*. The ability to detect MMP activity *in vivo* within hours after MMP inhibitor treat-ment makes it a useful drug-screening platform.

FMT obliterates the tissue penetration limitation associated with fluorescent proteins and lucif-erase reporter genes and is the preferred imaging method for deep-seated organs and target tissues. Animals are injected with NIRF probes and subjected to continuous wave or pulsed light exposure. Fluorescence is detected by detectors arranged in a fixed spatial manner to reconstitute a tomographic image (Weissleder, 2002). Cathepsin B, a protease implicated in tumor invasion (Demchik et al., 1999), from gliosarcomas transplanted into mice brains was successfully imaged and its activity was quanti-fied using FMT (Ntziachristos et al., 2002). FMT imaging is now used as an imaging modality in the research fields of oncology, inflammatory disease, skeletal disease, pulmonary disease, and cardiovas-cular disease.

The use of luciferase as an optical reporter gene in tumor xenografts is favored by researchers in recent years due to its ease of implementation and detection in small animals and lack of tissue autofluores-cence. Luminescence imaging can be used to monitor growth of a variety of tumors and their response to experimental therapies (Figure 10.1) (Brakenhielm et al., 2007, Contag et al., 2000, Honigman et al., 2001, Kanerva et al., 2003, Shah et al., 2003, Zhao and Wang, 2010). The ability to detect as little as 1000 luciferase-expressing cells intraperitoneally provides a great opportunity to follow tumor growth and metastatic progression (Contag et al., 2000).

10.6.3 Monitoring Systemic Distribution of Stem Cells and Differentiated Cells in Cell-Based Therapy

Stem cells are the cell source of all organs in our body, including brain, heart, liver, bones, and muscles. Under the situation that diseased or damaged tissue that would not ordinarily regenerate, transplants of stem cells can be used to trigger healing in the patient. To this date, stem cell therapies have been demonstrated, although in trials or in the laboratory, to heal broken bones, bad burns, blindness, deafness, and heart damage.

Cell motility and behavior *in vivo* have been investigated in many oncology and immunology studies to monitor the recruitment and time of arrival or departure of specific cells at a specific location within the body. Various adult stem cells or progenitor cells such as neural stem cells or mesenchymal stem cells have been reported to possess tumor-targeting properties (Nakamura et al., 2004, Shah et al., 2003). Verifying the engraftment, migration, and proliferation of progenitor cells after engraftment is paramount to understanding the efficacy and safety of this approach. Shah and coworkers engineered mouse neural progenitor cells (NPCs) to express Fluc concurrently with a therapeutic protein, a secreted form of tumor necrosis factor-related apoptosis-inducing ligand (sTRAIL) (Shah et al., 2003). They simultaneously engineered human glioma cells to express Rluc, hence permitting monitoring the activity and localities of two different luciferases intracranially. Creating a human glioma xenograft mouse model, Shah and coworkers demonstrated the migration of Rluc-expressing NPCs into Fluc-expressing glioma xenografts. The antitumor effect of sTRAIL was evident from the temporal reduction in Fluc intensities after migration of NPCs to glioma sites. Dual luciferase imaging in this work enabled the monitoring of both NPC migration and tumor burden over time. Bioluminescence imaging however does not provide inherent quantitative information as mentioned earlier. Consequently, researchers will not able to directly infer the number of cells giving rise to a particular luminescent signal intensity (Weissleder and Pittet, 2008) although relative quantification is possible. Immune cells such as T-cells have been explored for cell therapy purposes and can be engineered, similarly as adult stem cells or progenitor cells, to express luciferase (Costa et al., 2001). Costa and coworkers detected the migration of CD4[+] T-cells transduced with luciferase to systemic lymph nodes, lungs, spleen, and spinal cord in an experimental autoimmune encephalomyelitis mouse model. Histological analysis confirmed long-term therapeutic transgene expression and correlated with significant reduction in clinical disease, affirming the use of bioluminescent imaging of cellular vectors in cell-based therapies.

In addition to cells, transplantation of tissues stably expressing Fluc may also be examined using bioluminescence imaging. Islet transplantation has been demonstrated to successfully improve long-term blood glucose instability and severe hypoglycemic complications in a subgroup of patients with type 1 diabetes mellitus (Shapiro et al., 2006). Conventional preclinical protocols involved monitoring blood glucose levels as a measure of islet cells' viability and functionality *in vivo* after transplantation. By the time poor blood glucose homeostasis manifests itself, the viability and mass of transplanted islets would have already reduced drastically. Chen and coworkers studied if bioluminescent islet cells would provide a sensitive method for tracking the fate of transplanted islet cells even before dysfunction of blood glucose homeostasis is evident (Chen et al., 2006). Islets from a transgenic mouse strain that constitutively expressed Fluc were transplanted into syngeneic wild-type mice or allogeneic diabetic mice. Bioluminescence signals could be detected from as few as 10 islets transplanted at the renal subcapsular space, intrahepatic, intra-abdominal, and subcutaneous locations. Bioluminescence intensities remained consistent for up to 18 months posttransplantation in isografts, reflecting the presence of stable engraftments. In allografts, initial homeostasis of blood glucose and stabilization of bioluminescence were achieved immediately after transplantation but both features progressively decreased hyperglycemia that eventually recurred with islet failure. Consequently, Chen and coworkers were able to administer antilymphocyte serum (ALS) to mice prior to detectable dysregulated blood glucose homeostasis and achieve a 40% incidental graft loss and a mean time to graft loss of 53.5 days (Chen et al., 2008), whereas mice that received ALS treatment only upon detectable hyperglycemia had a 100% incidental graft loss and a mean time to graft loss of 22.1 days. Hence, bioluminescence imaging of allogenic islet transplants enables early monitoring of transplantation success and testing of therapeutic strategies to enhance transplant survival.

10.6.4 Monitoring Treatment of Pathogenic Infections within Host Animal

The versatility of bioluminescence imaging is not restricted to mammalian cells. The association of viral and bacterial pathogens with a range of biological systems may be investigated using bioluminescence (Prosser et al., 1996, Francis et al., 2000). Conventional studies of pathogen infection in small animal models rely on the observation of clinical symptoms and the harvesting of organs for pathological and molecular analyses. The use of genetically engineered luciferase-expressing bacterial cells simplifies pathogen studies with respect to monitoring bacterial burden and locations (Contag et al., 1995, Wiles et al., 2006), and bacteria viability in response to antibiotic treatments (Contag et al., 1995). Contag and coworkers transformed three strains of *Salmonella* with bacterial luciferase and were able to monitor the specific tissues colonized by *Salmonella* in a mouse model (Contag et al., 1995). Noninvasive live bioluminescent imaging over time allowed the determination of progressive, abortive, or persistent patterns of infection within different mice while identifying the cecum to play a role in *Salmonella* pathogenesis. Critically for drug development, the authors were able to demonstrate the *in vivo* efficacy of an antibiotic through temporal monitoring of bacterial burden in live mice, demonstrating the viability of using bioluminescent pathogens for *in vivo* drug screening.

In a similar attempt at determining bacteria colonization dynamics *in vivo*, Wiles and coworkers engineered a bioluminescent strain of *Citrobacter rodentium* to determine the dynamic infection pattern of *C. rodentium* (Wiles et al., 2006), which is known to colonize murine mucosa (Mundy et al., 2005). Conventional belief held that the dominance of strictly anaerobic bacteria in the gastrointestinal tract imply a lack of oxygen in the colonic environment that will likely not favor the use of luciferases that are also oxygenases. However, Wiles and coworkers showed that *C. rodentium* colonization of the gastrointestinal tract could be followed in living mice, proving that sufficient oxygen, probably in the nano-molar range, exists in the murine gastrointestinal tract for bioluminescence imaging. The spread of pathogens to unexpected anatomical sites in an organism can be objectively detected as shown for *C. rodentium,* for which the murine rectum is shown to be a primary site of infection, a site not reported previously.

Bioluminescent tracking of virus infection is also possible *in vivo*. HSV-1 expressing Fluc was administered into mice and bioluminescent signals were detected in mouse footpads, peritoneal cavity, brain, and eyes (Luker et al., 2002). Luker and coworkers found that the magnitude of bioluminescence intensities correlated with the input titers of virus used for infection. Importantly, treatment of infected mice with valacyclovir, an inhibitor of HSV-1 replication, produced dose-dependent decreases in luciferase activity—an indication of decreased HSV-1 burden in valacyclovir-treated mice. This important piece of work validates the use of bioluminescent imaging for antiviral drug development. This bioluminescent HSV-1 has since been employed in various studies (Burgos et al., 2006, Luker et al., 2003, Summers and Leib, 2002). Noteworthy was Luker and coworkers discovery that the absence of interferon receptors type I and II in knockout mice rendered those mice susceptible to HSV-1 infection, demonstrating another utility of whole-animal bioluminescent imaging for viral pathogenesis studies (Luker et al., 2003).

Bioluminescent imaging of the unicellular protozoan *Plasmodium berghei* parasite within mice has also been described (Franke-Fayard et al., 2006). The infected erythrocytes (schizonts) adhere to endothelial cells of certain tissues and organs such as lungs, spleen, and adipose tissues, resulting in detectable bioluminescent signals at the schizont stage. Eventually, luciferase-expressing ring form of *P. berghei* is released from infected erythrocytes into the peripheral blood circulation, resulting in an intense bioluminescent signal from the whole mouse body.

A minor limitation in bioluminescence imaging for monitoring infection dynamics is that the requirement for ATP by luciferase excludes the detection of extracellular pathogens for example, virus found in blood (Cook and Griffin, 2003). With that taken into account, bioluminescence imaging has proven to be a valid methodology for noninvasive, real-time monitoring of pathogen infection and systemic colonization.

10.7 Perspectives

Optical imaging of reporter genes in animals provides a convenient method to quantify transgene expression level, localize the distribution of gene delivery vectors, and monitor the effect of novel therapies in live small animals. In the future, it will be important to develop new imaging systems with high resolution and accurate signal localization, to image luminescence and fluorescence in deep organs, not only in science laboratories but also in clinical applications. Further advancements in genetic engineering may generate new molecular probes with further red shift, to minimize light absorption, scattering, and autofluorescence. Novel imaging probes that can be activated by a specific protein target could be developed by combining reporter genes to fusion proteins. Imaging at multiple wavelengths may provide researchers opportunities to monitor multiple targets at the same time. Besides the development of better reporter genes, instruments may be equipped with more sensitive and precise detectors that will aid in better image reconstruction. One exciting research direction for small animal imaging is in the combination of optical imaging with traditional anatomical imaging modalities. Different levels of information are provided by images obtained from different imaging modalities. Synergizing different information of the same subject requires, for example, coregistration of images acquired from different systems and this technique works well for immobile organs such as the brain. However, organs located in the abdominal region are relatively more mobile and deformable, rendering the coregistration technique challenging for such organs. To address this issue, researchers are looking into the development of multiple imaging modalities within one system for simultaneous imaging of the same subject with minimal movement (Townsend, 2008). The combination of CT and PET was commercialized in 2001 while that of CT and SPECT was commercialized in 2004. Combining multiple imaging modalities into one system is not without its challenges. The high magnetic field in MRI can interfere with the detection of radioactivity in PET and SPECT imaging and the radiofrequencies of MRI and PET systems interfere with each other. While x-ray and optical imaging do not require the subject to be anesthetized for more than a few minutes at most, an MRI scan can typically take a few hours, thereby drastically reducing throughput and posing health risks to the animals. Despite all these difficulties, high levels of clarity and information in images are still much sought after, for example, the ability to coregister optical signals with x-ray anatomical information that allows researchers to easily identify where the signals are coming from. Consequently, the development of probes, reporter gene constructs, and instruments for multimodality imaging is accelerating to meet the clinical demands. Dual-imaging probes such as MRI-optical, PET-optical, and SPECT-optical probes have been developed for *in vivo* imaging (Lee and Chen, 2009, Jennings and Long, 2009). In particular, there appears to be a strong interest in the development of MRI-optical probes such as those involving gadolinium chelates, functionalized quantum dots, and iron oxide nanoparticles (Frullano and Meade, 2007). Various triple-modality fusion reporter genes for fluorescent, bioluminescent, and radionuclide imaging have also been reported (Ponomarev et al., 2004, Ray et al., 2004). We envision that the development of multimodality reporter genes will allow researchers and clinicians to make use of the strengths of each imaging platform to gather high-quality experimental data in laboratories and clinics that will positively impact disease diagnosis and management.

Acknowledgments

This work was supported by the Institute of Bioengineering and Nanotechnology, Biomedical Research Council, Agency for Science, Technology and Research (A*STAR) in Singapore and a grant from National Medical Research Council in Singapore (NMRC/1203/2009).

References

Bhaumik, S. and Gambhir, S. S. 2002. Optical imaging of Renilla luciferase reporter gene expression in living mice. *Proc Natl Acad Sci USA*, 99, 377–82.

Bogdanov, A. A. Jr., Lin, C. P., Simonova, M., Matuszewski, L., and Weissleder, R. 2002. Cellular activation of the self-quenched fluorescent reporter probe in tumor microenvironment. *Neoplasia,* 4, 228–36.

Bouvet, M., Wang, J., Nardin, S. R. et al. 2002. Real-time optical imaging of primary tumor growth and multiple metastatic events in a pancreatic cancer orthotopic model. *Cancer Res,* 62, 1534–40.

Brakenhielm, E., Burton, J. B., Johnson, M. et al. 2007. Modulating metastasis by a lymphangiogenic switch in prostate cancer. *Int J Cancer,* 121, 2153–61.

Bremer, C., Tung, C. H., and Weissleder, R. 2001. *In vivo* molecular target assessment of matrix metalloproteinase inhibition. *Nat Med,* 7, 743–8.

Burgos, J. S., Guzman-Sanchez, F., Sastre, I., Fillat, C., and Valdivieso, F. 2006. Non-invasive bioluminescence imaging for monitoring herpes simplex virus type 1 hematogenous infection. *Microbes Infect,* 8, 1330–8.

Chen, X., Zhang, X., Larson, C. S., Baker, M. S., and Kaufman, D. B. 2006. *In vivo* bioluminescence imaging of transplanted islets and early detection of graft rejection. *Transplantation,* 81, 1421–7.

Chen, X., Zhang, X., Larson, C., Xia, G., and Kaufman, D. B. 2008. Prolonging islet allograft survival using *in vivo* bioluminescence imaging to guide timing of antilymphocyte serum treatment of rejection. *Transplantation,* 85, 1246–52.

Chudakov, D. M., Lukyanov, S., and Lukyanov, K. A. 2005. Fluorescent proteins as a toolkit for *in vivo* imaging. *Trends Biotechnol,* 23, 605–13.

Contag, C. H., Contag, P. R., Mullins, J. I., Spilman, S. D., Stevenson, D. K., and Benaron, D. A. 1995. Photonic detection of bacterial pathogens in living hosts. *Mol Microbiol,* 18, 593–603.

Contag, C. H., Jenkins, D., Contag, P. R., and Negrin, R. S. 2000. Use of reporter genes for optical measurements of neoplastic disease *in vivo*. *Neoplasia,* 2, 41–52.

Contag, C. H., Spilman, S. D., Contag, P. R. et al. 1997. Visualizing gene expression in living mammals using a bioluminescent reporter. *Photochem Photobiol,* 66, 523–31.

Cook, S. H., and Griffin, D. E. 2003. Luciferase imaging of a neurotropic viral infection in intact animals. *J Virol,* 77, 5333–8.

Costa, G. L., Sandora, M. R., Nakajima, A. et al. 2001. Adoptive immunotherapy of experimental autoimmune encephalomyelitis via T cell delivery of the IL-12 p40 subunit. *J Immunol,* 167, 2379–87.

De, A., Lewis, X. Z., and Gambhir, S. S. 2003. Noninvasive imaging of lentiviral-mediated reporter gene expression in living mice. *Mol Ther,* 7, 681–91.

Demchik, L. L., Sameni, M., Nelson, K., Mikkelsen, T., and Sloane, B. F. 1999. Cathepsin B and glioma invasion. *Int J Dev Neurosci,* 17, 483–94.

Francis, K. P., Joh, D., Bellinger-Kawahara, C., Hawkinson, M. J., Purchio, T. F., and Contag, P. R. 2000. Monitoring bioluminescent *Staphylococcus aureus* infections in living mice using a novel luxABCDE construct. *Infect Immun,* 68, 3594–600.

Franke-Fayard, B., Waters, A. P., and Janse, C. J. 2006. Real-time *in vivo* imaging of transgenic bioluminescent blood stages of rodent malaria parasites in mice. *Nat Protoc,* 1, 476–85.

Frullano, L. and Meade, T. J. 2007. Multimodal MRI contrast agents. *J Biol Inorg Chem,* 12, 939–49.

Fukumura, D., Xavier, R., Sugiura, T. et al. 1998. Tumor induction of VEGF promoter activity in stromal cells. *Cell,* 94, 715–25.

Gross, S. and Piwnica-Worms, D. 2005. Spying on cancer: Molecular imaging *in vivo* with genetically encoded reporters. *Cancer Cell,* 7, 5–15.

Herschman, H. R. 2004. Noninvasive imaging of reporter gene expression in living subjects. *Adv Cancer Res,* 92, 29–80.

Hilderbrand, S. A. and Weissleder, R. 2009. Near-infrared fluorescence: Application to *in vivo* molecular imaging. *Curr Opin Chem Biol,* 14, 71–9.

Hoffman, R. M. 2001. Visualization of GFP-expressing tumors and metastasis *in vivo*. *Biotechniques,* 30, 1016–22, 1024–6.

Honigman, A., Zeira, E., Ohana, P. et al. 2001. Imaging transgene expression in live animals. *Mol Ther,* 4, 239–49.

Ito, S., Nakanishi, H., Ikehara, Y. et al. 2001. Real-time observation of micrometastasis formation in the living mouse liver using a green fluorescent protein gene-tagged rat tongue carcinoma cell line. *Int J Cancer,* 93, 212–7.

Jacobs, A. H., Rueger, M. A., Winkeler, A. et al. 2007. Imaging-guided gene therapy of experimental gliomas. *Cancer Res,* 67, 1706–15.

Jennings, L. E. and Long, N. J. 2009. Two is better than one—Probes for dual-modality molecular imaging. *Chem Commun (Camb),* 24, 3511–24.

Kaneko, K., Yano, M., Yamano, T. et al. 2001. Detection of peritoneal micrometastases of gastric carcinoma with green fluorescent protein and carcinoembryonic antigen promoter. *Cancer Res,* 61, 5570–4.

Kanerva, A., Zinn, K. R., Chaudhuri, T. R. et al. 2003. Enhanced therapeutic efficacy for ovarian cancer with a serotype 3 receptor-targeted oncolytic adenovirus. *Mol Ther,* 8, 449–58.

Kang, J. H. and Chung, J. K. 2008. Molecular-genetic imaging based on reporter gene expression. *J Nucl Med,* 49 Suppl 2, 164S–79S.

Lee, S., Cha, E. J., Park, K. et al. 2008. A near-infrared-fluorescence-quenched gold-nanoparticle imaging probe for *in vivo* drug screening and protease activity determination. *Angew Chem Int Ed Engl,* 47, 2804–7.

Lee, S. and Chen, X. 2009. Dual-modality probes for *in vivo* molecular imaging. *Mol Imaging,* 8, 87–100.

Lipshutz, G. S., Gruber, C. A., Cao, Y., Hardy, J., Contag, C. H., and Gaensler, K. M. 2001. *In utero* delivery of adeno-associated viral vectors: Intraperitoneal gene transfer produces long-term expression. *Mol Ther,* 3, 284–92.

Loimas, S., Wahlfors, J., and Janne, J. 1998. Herpes simplex virus thymidine kinase-green fluorescent protein fusion gene: New tool for gene transfer studies and gene therapy. *Biotechniques,* 24, 614–8.

Luker, G. D., Bardill, J. P., Prior, J. L., Pica, C. M., Piwnica-Worms, D., and Leib, D. A. 2002. Noninvasive bioluminescence imaging of herpes simplex virus type 1 infection and therapy in living mice. *J Virol,* 76, 12149–61.

Luker, G. D., Prior, J. L., Song, J., Pica, C. M., and Leib, D. A. 2003. Bioluminescence imaging reveals systemic dissemination of herpes simplex virus type 1 in the absence of interferon receptors. *J Virol,* 77, 11082–93.

Massoud, T. F. and Gambhir, S. S. 2003. Molecular imaging in living subjects: Seeing fundamental biological processes in a new light. *Genes Dev,* 17, 545–80.

Massoud, T. F., Singh, A., and Gambhir, S. S. 2008. Noninvasive molecular neuroimaging using reporter genes: Part I, principles revisited. *AJNR Am J Neuroradiol,* 29, 229–34.

Matthews, Q. L., Sibley, D. A., Wu, H. et al. 2006. Genetic incorporation of a herpes simplex virus type 1 thymidine kinase and firefly luciferase fusion into the adenovirus protein IX for functional display on the virion. *Mol Imaging,* 5, 510–9.

Meulenbroek, R. A., Sargent, K. L., Lunde, J., Jasmin, B. J., and Parks, R. J. 2004. Use of adenovirus protein IX (pIX) to display large polypeptides on the virion—Generation of fluorescent virus through the incorporation of pIX-GFP. *Mol Ther,* 9, 617–24.

Miletic, H., Fischer, Y. H., Giroglou, T. et al. 2007. Normal brain cells contribute to the bystander effect in suicide gene therapy of malignant glioma. *Clin Cancer Res,* 13, 6761–8.

Min, J. J. and Gambhir, S. S. 2004. Gene therapy progress and prospects: Noninvasive imaging of gene therapy in living subjects. *Gene Ther,* 11, 115–25.

Moore, A., Sergeyev, N., Bredow, S., and Weissleder, R. 1998. A model system to quantitate tumor burden in locoregional lymph nodes during cancer spread. *Invasion Metastasis,* 18, 192–7.

Muller, F. J., Snyder, E. Y., and Loring, J. F. 2006. Gene therapy: Can neural stem cells deliver? *Nat Rev Neurosci,* 7, 75–84.

Mundy, R., Macdonald, T. T., Dougan, G., Frankel, G., and Wiles, S. 2005. *Citrobacter rodentium* of mice and man. *Cell Microbiol,* 7, 1697–706.

Nakamura, K., Ito, Y., Kawano, Y. et al. 2004. Antitumor effect of genetically engineered mesenchymal stem cells in a rat glioma model. *Gene Ther,* 11, 1155–64.

Ntziachristos, V., Tung, C. H., Bremer, C., and Weissleder, R. 2002. Fluorescence molecular tomography resolves protease activity *in vivo. Nat Med,* 8, 757–60.

Penuelas, I., Haberkorn, U., Yaghoubi, S., and Gambhir, S. S. 2005. Gene therapy imaging in patients for oncological applications. *Eur J Nucl Med Mol Imaging,* 32 Suppl 2, S384–403.

Pichler, A., Prior, J. L., and Piwnica-Worms, D. 2004. Imaging reversal of multidrug resistance in living mice with bioluminescence: MDR1 P-glycoprotein transports coelenterazine. *Proc Natl Acad Sci USA,* 101, 1702–7.

Ponomarev, V., Doubrovin, M., Serganova, I. et al. 2004. A novel triple-modality reporter gene for whole-body fluorescent, bioluminescent, and nuclear noninvasive imaging. *Eur J Nucl Med Mol Imaging,* 31, 740–51.

Prosser, J. I., Killham, K., Glover, L. A., and Rattray, E. A. 1996. Luminescence-based systems for detection of bacteria in the environment. *Crit Rev Biotechnol,* 16, 157–83.

Raty, J. K., Liimatainen, T., Huhtala, T. et al. 2007a. SPECT/CT imaging of baculovirus biodistribution in rat. *Gene Ther,* 14, 930–8.

Raty, J. K., Liimatainen, T., Unelma Kaikkonen, M., Grohn, O., Airenne, K. J., and Yla-Herttuala, S. 2007b. Non-invasive imaging in gene therapy. *Mol Ther,* 15, 1579–86.

Ray, P., De, A., Min, J. J., Tsien, R. Y., and Gambhir, S. S. 2004. Imaging tri-fusion multimodality reporter gene expression in living subjects. *Cancer Res,* 64, 1323–30.

Rome, C., Couillaud, F., and Moonen, C. T. 2007. Gene expression and gene therapy imaging. *Eur Radiol,* 17, 305–19.

Shah, K. 2005. Current advances in molecular imaging of gene and cell therapy for cancer. *Cancer Biol Ther,* 4, 518–23.

Shah, K., Bureau, E., Kim, D. E. et al. 2005. Glioma therapy and real-time imaging of neural precursor cell migration and tumor regression. *Ann Neurol,* 57, 34–41.

Shah, K., Tang, Y., Breakefield, X., and Weissleder, R. 2003. Real-time imaging of TRAIL-induced apoptosis of glioma tumors *in vivo. Oncogene,* 22, 6865–72.

Shah, K., Tung, C. H., Chang, C. H. et al. 2004. *In vivo* imaging of HIV protease activity in amplicon vector-transduced gliomas. *Cancer Res,* 64, 273–8.

Shapiro, A. M., Ricordi, C., Hering, B. J. et al. 2006. International trial of the Edmonton protocol for islet transplantation. *N Engl J Med,* 355, 1318–30.

So, M. K., Xu, C., Loening, A. M., Gambhir, S. S., and Rao, J. 2006. Self-illuminating quantum dot conjugates for *in vivo* imaging. *Nat Biotechnol,* 24, 339–43.

Steffens, S., Frank, S., Fischer, U. et al. 2000. Enhanced green fluorescent protein fusion proteins of herpes simplex virus type 1 thymidine kinase and cytochrome P450 4B1: Applications for prodrug-activating gene therapy. *Cancer Gene Ther,* 7, 806–12.

Strathdee, C. A., Mcleod, M. R., and Underhill, T. M. 2000. Dominant positive and negative selection using luciferase, green fluorescent protein and beta-galactosidase reporter gene fusions. *Biotechniques,* 28, 210–2, 214.

Summers, B. C. and Leib, D. A. 2002. Herpes simplex virus type 1 origins of DNA replication play no role in the regulation of flanking promoters. *J Virol,* 76, 7020–9.

Townsend, D. W. 2008. Dual-modality imaging: Combining anatomy and function. *J Nucl Med,* 49, 938–55.

Troy, T., Jekic-Mcmullen, D., Sambucetti, L., and Rice, B. 2004. Quantitative comparison of the sensitivity of detection of fluorescent and bioluminescent reporters in animal models. *Mol Imaging,* 3, 9–23.

Tung, C. H., Bredow, S., Mahmood, U., and Weissleder, R. 1999. Preparation of a cathepsin D sensitive near-infrared fluorescence probe for imaging. *Bioconjug Chem,* 10, 892–6.

Waerzeggers, Y., Monfared, P., Viel, T., Winkeler, A., Voges, J., and Jacobs, A. H. 2009. Methods to monitor gene therapy with molecular imaging. *Methods,* 48, 146–60.

Weissleder, R. 2002. Scaling down imaging: Molecular mapping of cancer in mice. *Nat Rev Cancer,* 2, 11–8.

Weissleder, R. and Mahmood, U. 2001. Molecular imaging. *Radiology,* 219, 316–33.

Weissleder, R. and Ntziachristos, V. 2003. Shedding light onto live molecular targets. *Nat Med,* 9, 123–8.

Weissleder, R. and Pittet, M. J. 2008. Imaging in the era of molecular oncology. *Nature*, 452, 580–9.

Weissleder, R., Tung, C. H., Mahmood, U., and Bogdanov, A. A. Jr. 1999. *In vivo* imaging of tumors with protease-activated near-infrared fluorescent probes. *Nat Biotechnol*, 17, 375–8.

Wiles, S., Pickard, K. M., Peng, K., Macdonald, T. T., and Frankel, G. 2006. *In vivo* bioluminescence imaging of the murine pathogen *Citrobacter rodentium*. *Infect Immun*, 74, 5391–6.

Wilson, T. and Hastings, J. W. 1998. Bioluminescence. *Annu Rev Cell Dev Biol*, 14, 197–230.

Wu, J. C., Inubushi, M., Sundaresan, G., Schelbert, H. R., and Gambhir, S. S. 2002. Optical imaging of cardiac reporter gene expression in living rats. *Circulation*, 105, 1631–4.

Wu, C., Mino, K., Akimoto, H. et al. 2009. *In vivo* far-red luminescence imaging of a biomarker based on BRET from *Cypridina* bioluminescence to an organic dye. *Proc Natl Acad Sci USA*, 106, 15599–603.

Wu, J. C., Sundaresan, G., Iyer, M., and Gambhir, S. S. 2001. Noninvasive optical imaging of firefly luciferase reporter gene expression in skeletal muscles of living mice. *Mol Ther*, 4, 297–306.

Yang, M., Baranov, E., Jiang, P. et al. 2000a. Whole-body optical imaging of green fluorescent protein-expressing tumors and metastases. *Proc Natl Acad Sci USA*, 97, 1206–11.

Yang, M., Baranov, E., Moossa, A. R., Penman, S., and Hoffman, R. M. 2000b. Visualizing gene expression by whole-body fluorescence imaging. *Proc Natl Acad Sci USA*, 97, 12278–82.

Yang, M., Baranov, E., Wang, J. W. et al. 2002. Direct external imaging of nascent cancer, tumor progression, angiogenesis, and metastasis on internal organs in the fluorescent orthotopic model. *Proc Natl Acad Sci USA*, 99, 3824–9.

Zhao, H., Doyle, T. C., Coquoz, O., Kalish, F., Rice, B. W., and Contag, C. H. 2005. Emission spectra of bioluminescent reporters and interaction with mammalian tissue determine the sensitivity of detection *in vivo*. *J Biomed Opt*, 10, 41210.

Zhao, H., Doyle, T. C., Wong, R. J. et al. 2004. Characterization of coelenterazine analogs for measurements of Renilla luciferase activity in live cells and living animals. *Mol Imaging*, 3, 43–54.

Zhao, Y. and Wang, S. 2010. Human NT2 neural precursor-derived tumor-infiltrating cells as delivery vehicles for treatment of glioblastoma. *Hum Gene Ther*, 21, 683–94.

Zhou, J. H., Rosser, C. J., Tanaka, M. et al. 2002. Visualizing superficial human bladder cancer cell growth *in vivo* by green fluorescent protein expression. *Cancer Gene Ther*, 9, 681–6.

11

Imaging in Metabolic Medicine

Jinling Lu
Singapore Bioimaging Consortium

Kishore Kumar Bhakoo
Singapore Bioimaging Consortium

Kai-Hsiang Chuang
Singapore Bioimaging Consortium

George K. Radda
Singapore Bioimaging Consortium

Philip W. Kuchel
Singapore Bioimaging Consortium

Weiping Han
Singapore Bioimaging Consortium

11.1 Introduction

Metabolic syndrome is a collection of risk factors that include obesity, hypertension, and dyslipidemia, which leads to increased incidents of coronary artery disease, stroke, and type 2 diabetes. It is a rapidly growing public health issue in almost all developed and many developing countries. The syndrome affects 20–30% of adults and is increasing in prevalence worldwide (Han et al., 2010, Meng et al., 2010).

Despite the intense efforts of biomedical scientists and clinical researchers, many fundamental questions regarding the etiology and development of metabolic syndrome remain unclear. This is partially due to the lack of suitable imaging technologies to assess β-cell mass and functions, and to detect and quantify lipid composition and distribution, *in vivo*. The development of these technologies requires close collaboration among life scientists, engineers, chemists, physicists, and clinicians. Such technologies not only impact on our understanding of the complexity of metabolic disorders such as obesity and diabetes but also aid in their diagnosis, drug development, and assessment of treatment efficacy. In this chapter, we discuss potential imaging strategies that may be applied in metabolic medicine, with a particular focus on metabolic diseases. We then present examples of visualization of insulin secretion, evaluation of β-cell function and pancreatic islet transplantation, and detection and quantification of ectopic fat deposition.

11.2 Imaging Modalities in Metabolic Medicine

A wide variety of imaging modalities have been proposed for medical applications in metabolic medicine, especially "molecular imaging." This covers a range of imaging technologies that aim to provide disease-specific molecular information along with traditional anatomical readouts (Cassidy and Radda, 2005, Jaffer and Weissleder, 2005). A range of molecular imaging techniques provide useful information for the diagnosis and therapy of metabolic diseases, including nuclear approaches (positron emission tomography [PET] and single photon emission computed tomography [SPECT]), magnetic resonance imaging (MRI), x-ray computed tomography (CT), optical imaging, and ultrasound (Table 11.1) (Lin et al., 2008, Moore, 2009, Delporte et al., 2009, Han et al., 2010).

In developing a molecular imaging strategy for delineating the metabolic syndrome, the following issues are need to be considered: (1) identification of specific molecular markers that are relevant in the development of metabolic diseases; (2) development of specific high-affinity ligands that bind to an identified molecular marker; (3) choice of an appropriate combination of molecular imaging modalities that provide a suitable spatial resolution, sensitivity, and penetration depth; (4) synthesis of molecular imaging agents with high specificity and affinity to the desired molecular target (Cassidy and Radda, 2005, Jaffer and Weissleder, 2005, Han et al., 2010).

The first, and probably the biggest, challenge in metabolic imaging is to identify the specific markers that are relevant to a specific metabolic process, and subsequently designing highly specific ligands that bind to these markers, which could be further combined with imaging reporter molecules. The specific markers can include transporter/receptor proteins that are expressed in a specific cell or tissue, or those that can be introduced by genetic engineering. For example, the vesicular monoamine transporter 2 (VMAT2), which is expressed in β-cells but is absent from the exocrine pancreas, was used as a specific marker for the *in vivo* estimation of β-cell mass by PET. VMAT2 was targeted by ^{18}F- or ^{11}C-labeled dihydrotetrabenazine (DTBZ), a compound that is already in use for PET imaging of patients with disorders of the central nervous system (CNS) (Souza et al., 2006, Moore, 2009). In general, the development of molecular markers is intimately dependent on the scientific understanding of the diseases. An increased understanding of the molecular mechanisms underlying different pathophysiological processes and stages of metabolic disease development will allow the development and testing of more specific markers.

In the isolation of specific ligands for identified molecular markers of a disease, the most straightforward method is to target the markers with specific antibodies. However, this approach is restricted to tissue or cell-surface antigens (not intracellular ones), and it is further complicated by the possibility of host-immune responses that are stimulated by repeated administration, which is required for longitudinal studies. Alternatively, high-affinity ligands can be identified by using a screening-based method that is cognizant of data derived from genomic, proteomic, and metabolomic analyses, and the application of these methods has the potential to greatly expand the number of candidate ligands that

TABLE 11.1 Imaging Modalities Used in Metabolic Medicine

Imaging Modality	Form of Energy Used	Depth of Penetration (mm)	Whether Used in Clinical Diagnosis	Application in Metabolic Medicine
CT	X-rays	>300	Yes	Body fat distribution β-Cell function Islets transplantation
MRI/MRS	Radio frequency waves	>300	Yes	Body fat distribution β-Cell function Islets transplantation
PET	Annihilation photons	>300	Yes	β-Cell function Islets transplantation
SPECT	γ-Photons	>300	Yes	β-Cell function Islets transplantation
Ultrasound	High-frequency sound waves	1–200	Yes	Body fat distribution
Optical imaging	Visible to infrared light	1–20	No	β-Cell function Insulin secretion

may ultimately serve as imaging agents (Moore, 2009). For example, the diversity-oriented fluorescence library approach (DOFLA) was recently introduced; it involves screening a diverse library of compounds in a high throughput manner, either with purified analytes as targets or against whole cells, tissues, or organisms (Lee et al., 2009b). The fluorescence molecules used in the screening can potentially be modified by attaching an MRI-, PET-, or SPECT-compatible functional group, or a single radioisotope, which then renders them suitable for deep tissue imaging in animals or for clinical studies (Figure 11.1). DOFLA has been successfully used in screening for α- and β-cell-specific probes (Lee et al., 2009a). These probes are presently being developed as contrast agents for *in vivo* imaging. Another example of such a screening-based approach was the isolation of two 20-mer peptides (RIP1 and RIP2) that bind to islets *ex vivo*, which were identified from a random phage-displayed 20-mer peptide library (Samli et al., 2005).

In addition to the screening approach to identify small molecules and peptides for the development of *in vivo* imaging probes, genetic introduction of cDNAs encoding proteins that bind to MRI-, PET-, or SPECT-compatible probes, or that can accumulate iron and form MR-detectable iron particles, such as ferritin (Cohen et al., 2005, 2007, Genove et al., 2005, Zurkiya et al., 2008, Goldhawk et al., 2009) has proven valuable. Genetic modification of specific protein targets with fluorescence proteins, such as green fluorescent protein (GFP), provides target-specific labeling platforms to study live systems. Although these methods are not applicable to a clinical setting, they can be used to generate useful preclinical animal models, or cell-based models for the evaluation of therapeutic compounds in their ability to promote β-cell growth and/or to preserve β-cell mass.

Another important issue is selecting an appropriate molecular imaging system that provides the required spatial resolution, sensitivity, and penetration depth to interrogate the disease of interest. Different modalities usually provide different but overlapping information. For example, MRI, ultrasound, and CT can provide high-quality anatomical images of the body, but they provide only limited functional information; however, rapid developments in MRI techniques are increasingly providing functional readouts (Koretsky and Silva, 2004, Silva et al., 2004, Massaad and Pautler, 2011). In comparison, SPECT, PET, and MRS can provide exquisite molecular and functional information, but with poor anatomical resolution. Optical imaging can provide real-time molecular information in live systems, but have very poor tissue penetration; thus limiting its use in clinical applications to the detection of surface targets, such as detection of tumors in the oral cavity (Han et al., 2010, Moore, 2009). More recently, the distinction between structural and functional imaging has become increasingly blurred with advanced CT and MRI technologies providing both functional and structural information. The development of contrast agents that specifically label defined cell populations on the basis of their specific protein expression contributed to the evolution of MRI into a modality that can provide not only anatomical and functional but also molecular information. The use of these new contrast agents has enabled researchers to monitor molecular processes and deliver image-guided therapeutic capabilities (Medarova and Moore, 2009).

Today's medical imaging technologies are expected to provide anatomical, physiological, molecular, and genomic information for better understanding of the biological basis of a disease, earlier and more accurate disease diagnosis, more accurate evaluation and prediction of treatment responses, and more rapid development of highly specific and sensitive drugs and imaging agents (Culver et al., 2008). However, no single imaging modality currently used in the clinic can provide a comprehensive overview of the disease. Multimodality imaging, combining two or more imaging modalities that provide complementary information, has become an attractive strategy for *in vivo* imaging studies owing to its ability to provide concurrent high-resolution anatomical as well as functional information. The combination of CT with PET was introduced commercially in 2001, followed by CT with SPECT in 2004, and PET with MRI in 2008. The potential applications of multimodality optical and SPECT/PET or MRI systems have also been proposed (Culver et al., 2008). One example in metabolic imaging is the use of MN-NIRF that consists of magnetic nanoparticles (MN) and a near-infrared fluorescent (NIRF) dye (Cy5.5), for the detection of labeled pancreatic islets by both MRI and near-infrared optical imaging

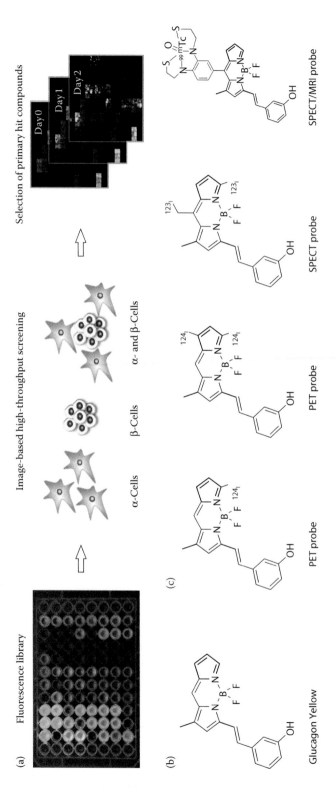

FIGURE 11.1 Schematic representation of DOFLA in α- and β-cell screening and its application in *in vivo* imaging-probe development. (a) Application of DOFLA in high-throughput screening of cell type-specific probes in α- and β-cells. (b) Structure of an α-cell selective probe Glucagon Yellow. (c) Potential modifications to convert Glucagon Yellow into PET or SPECT probes. The ⁹⁹ᵐTc ligand could be replaced by a Gd-chelate to form an MRI contrast agent. (Han, W. et al. Imaging metabolic syndrome. *EMBO Mol Med.* 2010. 2, 196–210. Copyright Wiley-VCH Verlag GmbH & Co. KGaA. Reproduced with permission.)

(Evgenov et al., 2006). Another example in metabolic imaging is a combined strategy of nuclear imaging and optical imaging in studying islet transplantation by introducing a herpes simplex virus type 1-thymidine kinase-green fluorescent protein (HSV1-thymidine kinase-GFP) fusion construct (tkgfp) into isolated islets (Tai et al., 2008).

11.3 Specific Considerations Related to Metabolic Diseases

11.3.1 Metabolic Syndrome and Diagnosis Criteria

Elevated blood glucose levels, high blood pressure, obesity, and dyslipidemia tend to occur together in some individuals. This cluster of clinical and biochemical indicators, previously called syndrome X, or insulin-resistance syndrome, is now commonly referred to as metabolic syndrome. More than a quarter of adults in industrialized countries are affected by this syndrome, making it a leading public health burden with significant impact on the productivity of the workforce and quality of life for the affected individuals and their families. Individuals with metabolic syndrome have increased risk of developing diseases, especially diabetes and cardiovascular disease.

There are two major diagnostic criteria defining the metabolic syndrome that are provided by the International Diabetes Federation (IDF) and the revised National Cholesterol Education Program (NCEP), respectively. It can be expected that many of the same individuals with metabolic syndrome can be classified under both criteria. The widely used clinical criteria from NCEP-Adult Treatment Panel III (NCEP-ATPIII) identify a person with metabolic syndrome when they present at least three out of five abnormal clinical observations in reference to increased waist circumference (WC), fasting plasma glucose, triglycerides, high blood pressure, and reduced HDL-cholesterol (Table 11.2).

Despite years of intense investigations by biomedical and clinical scientists, the exact causes of metabolic syndrome remain largely a mystery. Virtually all subjects with metabolic syndrome have insulin resistance, viz., a decreased ability of the body's tissues to respond to insulin. It is conceivable that insulin plays a central role in the development of metabolic disease as it does in diabetes; that is impairment in insulin secretion and/or function leads to diabetes. The NCEP-ATPIII criteria place its emphasis particularly on WC since there is evidence that ectopic adipose tissue increases the incidence of metabolic syndrome.

11.3.2 Insulin Secretion and β-Cell Mass in Metabolic Disease

In essence, diabetes is a disease of insulin deficiency due to either a lack of insulin production (type 1) because of autoimmunity-triggered β-cell death and diminishing β-cell mass, or deficient insulin secretion (type 2) that fails to overcome peripheral insensitivity to insulin (Porte, 1991). Insulin is released from pancreatic β-cells and is the major anabolic hormone involved in glucose homeostasis in the body.

TABLE 11.2 Clinical Identification of the Metabolic Syndrome: Any Three of the Following

Risk Factor	Defining Level
Abdominal obesity	Waist circumference
Men	>102 cm (>40 in)
Women	>88 cm (>35 in)
Triglycerides	≥150 mg dL^{-1}
HDL cholesterol	
Men	<40 mg dL^{-1}
Women	<50 mg dL^{-1}
Blood pressure	≥130/≥85 mmHg
Fasting glucose	≥110 mg dL^{-1}

Insulin secretion is a complex and highly regulated process, which is under the close control of the blood glucose concentration. Under physiological conditions, an elevation of blood glucose leads to its rapid uptake into the pancreatic β-cells, and induces exocytosis of insulin granules and the release of insulin into the bloodstream (Gustavsson et al., 2008). Insulin resistance refers to the impaired ability of tissues to respond to the action of insulin in promoting glucose transport from blood into the tissues, most notably skeletal muscle. Because of the central role that insulin secretion and insulin resistance play in the metabolic syndrome, many studies have focused on understanding the molecular mechanisms and regulation of insulin secretion, and cellular responses to insulin signaling. Understanding the mechanisms of insulin secretion and its regulation in the course of disease progression will be instrumental for identifying drug targets and for the evaluation of therapeutic strategies in the treatment of diabetes and related metabolic diseases.

Besides defective insulin secretion, reduced β-cell mass, with consequently decreased production and secretion of insulin, also contributes to the progression of diabetes. The amount of β-cells present in the pancreas is a major determinant of the quantity of insulin that can be secreted at a given time (Weir et al., 1990). Defects in glucose-triggered insulin secretion, possibly exacerbated by a decrease in β-cell mass, are ultimately responsible for the development of type 2 diabetes (Rutter, 2004). Pancreatic islet transplantation has been used in clinical trials on type 1 diabetic patients, and will almost certainly be used for late-stage type 2 diabetes patients in the near future. Considering the importance of β-cell mass in both types of diabetes, it is crucial to develop a means of noninvasive assessment of β-cell mass during diabetes development in a patient, and after islet transplantation.

11.3.3 Ectopic Fat and Metabolic Syndrome

In addition to serving as a repository for lipids, adipose tissue is also a remarkable endocrine organ releasing numerous hormones and cytokines, including leptin, adiponectin, and proinflammatory molecules, such as interleukin (IL)-6 and tumor necrosis factor-α (TNF-α) (Malaisse and Ladriere, 2001). Ectopic fat is defined as the deposition of triglycerides within cells of nonadipose tissue, which normally contains only small amount of fat, or the abnormal expansion of adipose tissue in other than subcutaneous sites. The most common ectopic fat deposition occurs as abdominal or visceral fat, which is central to the other components of the metabolic syndrome and has been found to be most predictive of cardiovascular risk (Delporte et al., 2009). Ectopic fat deposition also occurs in liver and skeletal muscle, two important metabolic organs. Fat accumulation in these organs appears to affect their ability to respond to insulin and to take up glucose. As such, one of the key questions in the field of metabolic disease research is how lipids are "directed" to accumulate ectopically, and which particular lipid species causes the most severe tissue damage leading to insulin and leptin resistance as diabetes and obesity develop. Imaging strategies that enable assessment of fat distribution and identification of lipid species within fat depots are needed to address these questions.

11.4 Imaging Insulin Secretion and Insulin Granule Exocytosis in Model Organisms

Understanding insulin secretion and its regulation mechanisms *in vivo*, and the ability to measure β-cell mass as the disease progresses is instrumental in identifying and validating drug targets, and for formulating therapeutic strategies. Since the discovery of insulin almost a century ago, most research efforts in the field have focused on understanding the mechanisms of insulin secretion and insulin signaling events, and more recently, attention has also been given to understanding the regulation of exocytosis of insulin granules (Gustavsson et al., 2008, Gustavsson and Han, 2009). These studies have established cellular mechanisms governing insulin secretion (Figure 11.2), but many of the details remain untested *in vivo* due to a lack of suitable tools to measure and quantify insulin secretion at high spatial and temporal resolution.

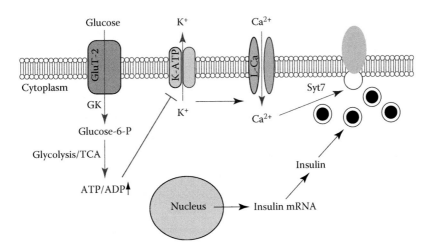

FIGURE 11.2 Cellular and molecular regulation of insulin secretion. The cellular events leading to insulin secretion start with a rise in blood glucose levels, which results in glucose uptake into pancreatic β-cells. Glucose in the cells then undergoes glycolysis and the constituent carbon atoms enter the tricarboxylic (TCA) cycle to produce ATP, resulting in an increased ATP/ADP ratio, and consequent closure of K_{ATP}-channels. Membrane depolarization due to K_{ATP}-channel closure opens L-type calcium channels, allowing calcium influx into the cells; this triggers exocytosis of insulin granules and the release of insulin into the bloodstream. GLUT2, glucose transporter 2; GK, glucokinase; TCA, tricarboxylic acid cycle; Syt7, synaptotagmin-7; K-ATP, ATP-sensitive potassium channel; L-Ca, L-type calcium channel.

Insulin secretion is typically measured by detecting the released hormone with enzyme-linked immunosorbent assay (ELISA) or radioimmunoassay (RIA). These traditional methods suffer from poor temporal resolution, a complete lack of cellular spatial resolution, and long processing time. In model organisms, optical imaging of genetically introduced exocytosis markers may provide an alternative approach, with high spatial and temporal resolution, along with instantaneous read-out. One such strategy, shown in Figure 11.3, allows visual monitoring of insulin granule movement and exocytosis, plus instantaneous quantification of the number of exocytotic events of insulin granules and hence implied insulin secretion. The genetically engineered optical probe for exocytosis can be introduced into mouse β-cells under the control of mouse insulin I promoter (MIP) (Lu et al., 2009) or other β-cell-specific promoters. Insulin granule exocytosis, in response to systemic glucose concentration changes or drugs, may be examined by intravital fluorescent microscopy of the transgenic mice after tail vein injections of reagents that affect insulin secretion.

An alternative strategy, which may be combined with the use of an optical probe, is to use surface-enhanced Raman spectroscopy (SERS) for the simultaneous detection of insulin and glucose (Figure 11.4). In this SERS-based approach, a SERS tag (e.g., an organic molecule immobilized on a gold nanoparticle) serves to label and track analytes such as insulin and glucose. Advantages such as high information content, multiplexing capability, lack of extensive sample preparation, high tissue penetration, and a highly sensitive detection limit down to the level of single molecules (Qian et al., 2008) have led to various applications of the SERS-based approach for biosensing in cell suspensions (Kim et al., 2006) and tissues (Zhang et al., 2008). A key requirement for the SERS-based method is that an analyte or its reporter must lie close to the nanoparticles or nano-roughened surface of noble metals (Table 11.1). The SERS technique has been used in mice for *in vivo* tumor detection (Keren et al., 2008, Qian et al., 2008, von Maltzahn et al., 2009, Xiao et al., 2009, Zavaleta et al., 2009). As for metabolic sensing, SERS may be used for the detection of glucose or various derivatives, and of insulin and other peptide hormones, using SERS-active nanoparticles attached to a fiber optic sensor (Zhang et al., 2007). Even though optical fiber-based SERS biosensing is still in its infancy (Shi et al., 2009), the technique will

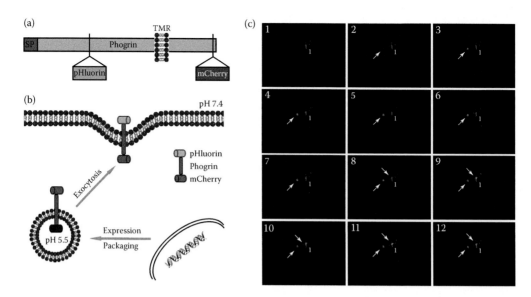

FIGURE 11.3   (See color insert.) Optical sensor for visualizing insulin granule exocytosis. (a) The optical sensor was based on a chimeric fusion protein consisting of a secretory granule resident protein, phogrin, and two fluorescent proteins: (1) highly pH-sensitive pHluorin localized inside the secretory granules and (2) a red mCherry protein localized in the cytoplasma. (b) Schematic representation of the design strategy for the optical sensor. pHluorin is inside the acidic lumen and remains nonfluorescent in the resting state. Upon exocytosis, pHluorin faces the extracellular fluid at neutral pH, and becomes highly fluorescent, while mCherry remains in the cytoplasm and serves as a label for granule tracking and a standard light-emitter for ratiometric quantification. (Adapted from Lu, J. et al. 2009. *J Innov Opt Health Sci*, 2, 397–405.) (c) A time-lapse movie showing exocytosis events in cultured insulin-secreting cells. Arrows indicate exocytosed insulin-granules. SP, signal peptide; TMR, transmembrane region. (Han, W. et al. Imaging metabolic syndrome. *EMBO Mol Med*. 2010. 2, 196–210. Copyright Wiley-VCH Verlag GmbH & Co. KGaA. Reproduced with permission.)

increasingly be used because the fibers can be readily configured for *in vivo* applications thus enabling SERS-based metabolic studies in living animals (Han et al., 2010).

11.5 *In Vivo* Imaging of β-Cell Mass and Islet Transplantation

As noted above, it is very important to measure, in a noninvasive manner, β-cell mass during diabetes development, and to track both endogenous and transplanted islet masses. It will also be important to develop imaging strategies to evaluate the function of residual and transplanted islets over time, and to characterize changes in islet vasculature (Leoni and Roman, 2010). The largest obstacle to imaging β-cell mass is their low abundance in the pancreas; the islets of Langerhans occupy only 1–2% of the total pancreatic tissue and are dispersed throughout the organ (Moore, 2009). Despite these challenges, progress has been made in imaging endogenous and transplanted islets, islet vasculature, islet apoptosis, and infiltration of immune cells during the progression of diabetes. Among the more "forthcoming" modalities are magnetic resonance, PET, SPECT, and fluorescence optical imaging. Current imaging strategies for determining β-cell mass and assessing islet vitality are summarized in Table 11.3.

11.5.1 MRI in Assessing β-Cell Mass and Islets Transplantation

In recent years, new methods have been developed to overcome the sensitivity limit of conventional MRI for the estimation of pancreatic β-cell mass. The most significant progress has been achieved in

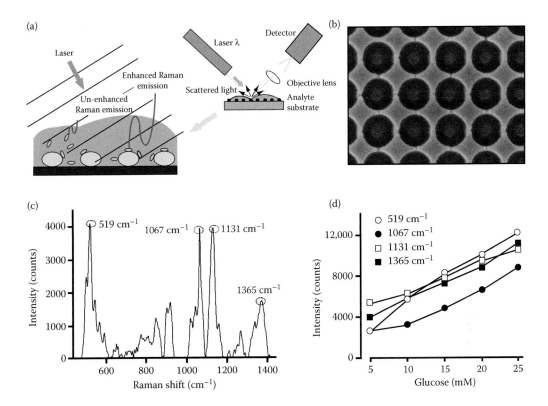

FIGURE 11.4 Basic concept and applications of surface-enhanced Raman spectroscopy (SERS). (a) Basic concept of SERS. Raman scattering occurs at very low intensities in the solution phase when molecules (represented by oval-shaped grains) are far from the metallic nanoparticles (gray spheres). When the molecule is close to the nanosurface, upon laser excitation, the intensity is enhanced by the interaction of the molecule with the surface electrons. (b) Scanning electron micrograph (EM) of silver and gold bimetallic SERS substrate by deep ultraviolet (UV) lithography on silicon wafers designed for glucose sensing. (c) Raman spectrum of glucose. Glucose sensing using a surface-functionalized bimetallic SERS substrate shown in (b) in the SERS mode. The peaks indicate different vibrational frequencies in the molecules. By virtue of the the narrowness of the peaks (high spectral resolution), structurally similar multiple analytes can be detected simultaneously. (d) Glucose quantification by SERS-based glucose sensing. Area-under-curves of the vibrational bands of a glucose Raman spectrum at wavelengths of 519, 1067, 1131, and 1365 cm^{-1} plotted against glucose concentration. (Han, W. et al. Imaging metabolic syndrome. *EMBO Mol Med.* 2010. 2, 196–210. Copyright Wiley-VCH Verlag GmbH & Co. KGaA. Reproduced with permission.)

detecting pancreatic islets that are prelabeled with either reporter gene or exogenous contrast agent, prior to transplantation. The success is largely ascribed to the simplicity of the prelabeling procedure (Moore, 2009).

Superparamagnetic iron oxide nanoparticles (SPIOs) are major exogenous contrast agents used in MRI. They have been used to tag islet transplants *in vivo* and enable longitudinal monitoring of the fate of the cells (Jirak et al., 2004, Tai et al., 2006, Kim et al., 2009, Marzola et al., 2009). For labeling with the contrast agents, isolated pancreatic islets are incubated in the presence of the contrast agent prior to transplantation. The labeled islets are then detected in an animal model by T_2-weighted MRI, thus following the distribution and fate of the transplanted tissue. Hence, MRI of SPIO-labeled islets can be used to detect noninvasively *in vivo* tolerance and rejection of islet transplants. In addition to iron oxide particles, the Gd-based T_1 contrast agent GdHP-DO3A, a neutral and very hydrophilic contrast agent for MRI that is currently used in clinical practice, has been used to record the fate of transplanted islets

TABLE 11.3   Image Strategies for Assessing β-Cell Mass and Islets Transplantation

Imaging Technology	Imaging Agent	Target	References
MRI	SPIO	Isolated pancreatic islets	Marzola et al. (2009), Jirak et al. (2004), Kim et al. (2009), Tai et al. (2006)
	GdHP-DO3A	Transplanted islets	Biancone et al. (2007)
	Manganese (Mn^{2+})	β-Cell via Ca^{2+} channels	Gimi et al. (2006), Antkowiak et al. (2009)
	Lanthanide complexes (EuDOTA-4AmBBA)	Transplanted islets	Woods et al. (2004)
	Transgene encoding MRI-compatible probes (e.g., ferritin and MagA)	β-Cell or isolated islets	To be demonstrated
Nuclear imaging (SPECT/PET)	^{111}In-DTPA-IC2 antibody	Unknown surface epitope of β-cell	Moore et al. (2001)
	^{125}I-R2D6	β-Cell surface gangliosides	Ladriere et al. (2001)
	^{125}I–DTPA-K14D10 and fragment	Surface epitope of β-cell	Hampe et al. (2005)
	^{18}F-FDG	Glucose handling machinery	Malaisse et al. (2000a,b)
	^{125}I-D-glucose	Glucose handling machinery	Malaisse et al. (2000e)
	[11]C/^{18}F/^{123}I-D-mannoheptulose	GLUT2	Malaisse (2005), Malaisse et al. (2009), Malaisse and Ladriere (2001), Sener et al. (2001)
	^{14}C-Alloxan	GLUT2	Malaisse et al. (2001b)
	^{18}F-Bromoglibenclamide	SUR1	Schneider et al. (2005)
	^{99m}Tc-Glibenclamide	SUR1	Schneider et al. (2007)
	^{18}F-Repaglinide	SUR1	Wangler et al. (2004b)
	^{11}C-Repaglinide	SUR1	Wangler et al. (2004a)
	^{18}F-FHBG	HSV1-TK	Lu et al. (2006b)
	^{131}I-FIAU	HSV1-TK	Tai et al. (2008)
	RIP1-phage clone	Unknown	Samli et al. (2005)
	^{111}In-DTPA-exendin-4	GLP-1R	Wicki et al. (2007), Wild et al. (2006)
	^{111}In-DOTA-exendin-4	GLP-1R	Christ et al. (2009)
	[11]C/^{18}F-DTBZ	VMAT2	Souza et al. (2006), Harris et al. (2008), Simpson et al. (2006), Kung et al. (2008a,b)
	^{18}F-FBT	Acetylcholine transporter	
	^{18}F-L-DOPA		de Lonlay et al. (2006), Otonkoski et al. (2006)
Optical imaging (BLI or FI)	MIP-luciferase	β-Cell	Park et al. (2005)
	MIP-GFP	β-Cell	Hara et al. (2003)
	RIP-GFP	β-Cell	Speier et al. (2008)
	Adenovirus- or lentivirus-mediated luciferase	Isolated islets	Fowler et al. (2005), Virostko et al. (2004), Lu et al. (2004)
	Cy5.5–streptozotocin	GLUT2	Ran et al. (2007)
	Cy5.5 with annexin V	Phosphatidylserine (PS)	Medarova et al. (2005)

Note: SPIO, superparamagnetic iron oxide nanoparticle; GLUT2, glucose transporter 2; SUR1, sulfonylurea receptor 1; BLI, bioluminence imaging; FI, fluorescence imaging; HSV1-TK, herpes simplex virus type 1 thymidine kinase; GLP-1R, glucagon-like peptide-1 receptor; VMAT2, vesicular monoamine transporter type 2.

(Biancone et al., 2007). In this study, the researchers detected the transplanted islets by MRI after *ex vivo* cell labeling with GdHP-DO3A, demonstrating sufficient levels of label, and retention within pancreatic islets, to permit effective monitoring of the islets after transplantation.

Although the two contrast agents can label isolated islets, they are not β-cell specific and cannot be used to assess β-cell function. Manganese (Mn^{2+}) is proving to be a useful MRI contrast agent that enters cells via Ca^{2+} channels; it has been applied in so-called manganese-enhanced MRI (MEMRI) for neuronal functional mapping (Kawai et al., 2010). Since β-cells are electrically excitable, and insulin secretion from them is Ca^{2+} dependent, MEMRI can be used to measure Ca^{2+} influx in β-cells after stimulation with glucose. Mn^{2+} accumulates in β-cells in proportion to the glucose concentration to which the cells are exposed, resulting in ~200% increase in MRI contrast enhancement when compared with nonactivated cells. MEMRI has been used to observe activity in isolated β-cells (Gimi et al., 2006, Leoni et al., 2010) and in mouse pancreas *in vivo* (Antkowiak et al., 2009). Although the toxicity of Mn^{2+} may limit its application in humans, this noninvasive technique complements optical fluorescent imaging approaches in assessing β-cell mass in animal models.

Glucose and glycogen in β-cells can be detected by MRI using the chemical exchange saturation transfer (CEST) technique and its derivative PARACEST, which is generated by specific paramagnetic lanthanide (e.g., Ln^{3+}) complexes (van Zijl et al., 2007). Glucose distribution in the liver has been mapped *ex vivo* in this manner (Ren et al., 2008). Encouraging results were obtained with the glucose-sensitive Europium-based PARACEST agent, EuDOTA-4AmBBA (Woods et al., 2004). Several recent studies have demonstrated the use of iron-binding proteins, including ferritin and the bacterial protein MagA with *in vitro* and *in vivo* imaging (Cohen et al., 2005, 2007, Genove et al., 2005, Zurkiya et al., 2008, Goldhawk et al., 2009). It is plausible that transgenic mouse lines with pancreatic β-cell-specific expression of one of these proteins could be generated and studied for the detection and quantification of β-cell mass in the animal.

Although MRI is sensitive to different physical and physiological parameters *in vivo,* it detects an average value of all the NMR parameters across a voxel, so it cannot differentiate between individual cellular components of a tissue. This limits the ability to image multiple cell or islet populations and/or biological processes in parallel with understanding their interactions and temporal changes. One potential strategy to overcome this limitation is to use frequency-shifting contrast agents that change the resonance frequency of the water instead of changing the T_1 and/or T_2 relaxation times (Zabow et al., 2008). This enables images to be generated for the different frequencies as can be achieved by using quantum dots in optical imaging (Wang et al., 2010). An alternative approach is to combine the use of conventional relaxation agents (e.g., iron oxide nanoparticles, or Mn^{2+}) and a CEST contrast agent (Gilad et al., 2009). This might be used to track transplanted β-cells with iron oxide (or a frequency-shifting agent) and to monitor their function using Mn^{2+}- or Zn^{2+}-sensitive agents, and/or the CEST effect.

11.5.2 Nuclear Imaging in Assessing β-Cell Mass and Islet Transplantation

In the past decade, significant progress has been made in nuclear imaging techniques for noninvasive visualization of pancreatic β-cells with the development of specific radioactive probes for β-cells (Figure 11.5). However, most of the probes have not been useful for *in vivo* determination of β-cell mass due to their poor uptake, and/or low specificity for pancreatic islets.

11.5.2.1 Isotope-Labeled Antibody Probes

Antibodies that are directed against surface antigens of β-cells are likely to provide highly specific targeting. In 2001, Ladrier et al. attempted to detect β-cells by PET, using a ^{125}I-labeled mouse monoclonal antibody (R2D6) that was directed against surface gangliosides (Ladriere et al., 2001). However, no significant differences were observed in the *in vitro* uptake of radioactivity by pancreatic slices or isolated islets from control and diabetic animals. Later, Hampe et al. used another existing β-cell-specific antibody (K14D10), and its Fab fragments, that were labeled with DTPA-^{125}I to assess β-cell mass *in vivo* (Hampe et al., 2005). The Fab was rapidly cleared from the blood, but its low binding specificity for β-cells did not

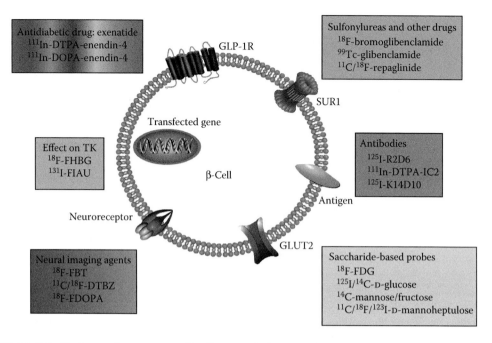

FIGURE 11.5 Nuclear probes in assessing β-cell mass and islets transplantation. Potential β-cell and islet imaging probes including radiolabeled antibodies, compounds targeting to glucagon-like peptide-1 receptor (GLP-1R) and SUR1, neural imaging agent, transfected gene, and the saccharide-based probes.

justify further *in vivo* evaluation. A more promising example is from the group of Moore (Moore et al., 2001); they estimated β-cell mass by nuclear imaging using a β-cell-specific monoclonal antibody IC2 that had been modified by linkage to a radioisotope chelator, viz., IC2-DTPA-^{111}In. IC2 is a rat–rat hybridoma autoantibody that is produced from the BB rat with an unknown antigen target (Aaen et al., 1990). Moore et al. showed that highly specific binding and accumulation in β-cells occurs after intravenous administration of the probe molecules, with virtually no extraneous binding to the exocrine pancreas or stromal tissues (Moore et al., 2001). However, the low extent of uptake of the antibody into the endocrine pancreas will probably hamper future applications *in vivo* for estimating β-cell mass.

11.5.2.2 Saccharide-Based Probes

β-Cells are sensitive to changes in blood glucose, and a glycogen pool exists in β-cells but not in acinar cells in situations of sustained hyperglycemia. On the basis of this principle, Malaisse's group tested many radiolabeled sugar analogs for imaging and quantification of the endocrine pancreas in animal models, including the most commonly used PET imaging probe, ^{18}F-FDG, ^{125}I and ^{14}C-labeled D-glucose, ^{14}C-labeled mannose and fructose, and radiolabeled D-mannoheptulose (Malaisse et al., 2000a,b,c,d,e,f, 2001a, Malaisse and Ladriere, 2001, Malaisse, 2005). However, most of these probes could not provide enough specific radioactive signals for imaging of β-cells. Only the D-mannoheptulose showed potential for *in vivo* imaging of the pancreatic β-cells. D-Mannoheptulose is transported into cells by the GLUT2 transporter, which is expressed mainly in hepatocytes and insulin-producing β-cells, but not in other cell types. The uptake of D-mannoheptulose coincides with the intracellular accumulation of acidic metabolites generated by phosphorylation of D-mannoheptulose (Malaisse, 2005, Malaisse et al., 2009). Hence, Malaisse and coworkers designed several nuclear imaging probes based on D-mannoheptulose, such as 1-^{11}C-labeled D-mannoheptulose, 3-deoxy-3-^{18}F-fluoro-D-mannoheptulose, and 7-deoxy-7-^{123}I-iodo-D-mannoheptulose (Malaisse and Ladriere, 2001, Sener et al., 2001, Malaisse, 2005, Malaisse et al., 2009). Mannoheptulose has been proposed as one of the more promising candidate tracers. However, because of the close proximity of liver and pancreas, it is not clear whether accurate quantitative information on β-cell mass can be obtained using D-mannoheptulose as the probe. On the basis of the specific expression of GLUT2 in pancreatic β-cells, Malaisse et al. also proposed the diabetogenic drug alloxan as a candidate ligand for β-cell nuclear imaging (Malaisse et al., 2001b). Since alloxan destroys insulin-producing β-cells, it is difficult to imagine alloxan as a probe to image β-cells in animals or humans.

11.5.2.3 Specific Receptors on Pancreatic Cells

Many compounds used in the treatment of type 2 diabetes are proven to be targeted to specific receptors on β-cell surface; therefore, it is reasonable to screen these compounds for β-cell targeting agents, following isotope incorporation, for nuclear imaging. The sulfonylurea receptor 1 (SUR1) is highly expressed at the internal face of the plasma membrane of the pancreatic islet cells, but not in the exocrine pancreas, and represents the target for the well-known antidiabetic agent sulfonylureas, such as glibenclamide and tolbutamide. Schneider's group evaluated 20 different fluorine-labeled glibenclamide derivatives and chose the 2-[^{18}F]-fluoroethoxy-5-bromoglibenclamide and ^{99m}Tc-glibenclamide for *in vivo* evaluation in humans (Schneider et al., 2005, 2007). The same group also examined a nonsulfonylurea antidiabetic drug repaglinide, which also targets SUR1. They synthesized and evaluated ^{18}F-labeled and ^{11}C-labeled repaglinide derivative in human PET study (Wangler et al., 2004a,b). However, these probes were taken up by various tissues in a nonspecific manner, and therefore would not be very useful in determining β-cell mass in clinical nuclear imaging.

Other antidiabetic drugs, such as exenatide, which targets glucagon-like peptide-1 receptor (GLP-1R), has been radiolabeled (^{111}In-DTPA-exendin-4 or ^{111}In-DOTA-exendin-4) and evaluated as an imaging candidate for GLP-1R targeting in Rip1-Tag2 mice (Wild et al., 2006, Wicki et al., 2007, Christ et al., 2009). In this model, multiple insulinomas were visible on SPECT, suggesting that it is a highly efficient radiotherapeutic for GLP-1R-targeted therapy for insulinoma. The same target is also under intense evaluation by Gotthardt and colleagues in determining β-cell mass by PET. Initial results are most encouraging (Gotthardt et al., 2006, Brom et al., 2010).

11.5.2.4 Neural Imaging Probes

The pancreas is an innervated peripheral organ with both parasympathetic and sympathetic neurons to tightly control its endocrine and exocrine functions (Lin et al., 2008). Many transporters that are expressed in neurons of the CNS are also expressed in pancreatic β-cells but not in the exocrine pancreas or other peripheral organs. Several PET ligands that are currently used in clinical brain imaging have been tested in assessing β-cell functions *in vivo*, including ^{11}C/^{18}F-DTBZ, ^{18}F-FBT, ^{18}F-L-DOPA (de Lonlay et al., 2006, Otonkoski et al., 2006). One of the most promising agents is the ^{11}C-dihydrotetrabenazine (^{11}C-DTBZ), which binds specifically to VMAT2. Harris and coworkers reported the use of ^{11}C-DTBZ in estimating β-cell mass in a rodent model of spontaneous type 1 diabetes (Souza et al., 2006, Harris et al., 2008, Simpson et al., 2006). Recently, Inabnet et al. used the ^{11}C-DTBZ PET method to assess β-cell performance after sleeve gastrectomy and duodenal–jejunal bypass in GotoKakizaki rats (Inabnet et al., 2010). To explore the potential clinical use of this probe over a longer period, Kung et al. synthesized the longer half-life ^{18}F-labeled analogs of DTBZ (Kung et al., 2008a,b) and assessed its β-cell accumulation. These studies suggested that PET-based quantitation of VMAT2 receptors provides a noninvasive measurement of β-cell mass in rat animal model. However, further whole-body biodistribution studies in baboons with human radiation dosimetry failed to show any significant accumulation of the tracer in the pancreas (Murthy et al., 2008).

11.5.2.5 Exogenous Gene Transduction

The method of reporter gene transfer has also been applied to nuclear imaging of β-cell mass and islet transplantations, for example, by expressing the herpes simplex virus type 1 thymidine kinase (*HSV1-TK*) gene and its mutant in pancreatic islets. Since cells that express the *HSV1-TK* gene could trap the nuclear imaging probes, ^{18}F-FHBG and ^{131}I-FIAU, by thymidine kinase-mediated monophosphory-lation, these compounds can be used for noninvasive nuclear imaging of the HSV1-TK infected cells or islets (Lu et al., 2006a, Tai et al., 2008).

11.5.3 Optical Imaging

There are two major approaches using optical imaging methods to study the biology and function of β-cells in the intact pancreas and in transplanted islets. The more frequently used method is genetic introduction of reporter genes that produce light-emitting proteins upon external light excitation (e.g., GFP and its mutants) or enzymes that act on substrates (e.g., luciferase and luciferin) to emit visible light in β-cells or islets. Recently, a number of groups have successfully applied these methods to image β-cell mass in animals.

In these methods, GFP or luciferase reporter genes were introduced into β-cells under the control of MIP (Hara et al., 2003, Park et al., 2005) or rat insulin I promoter (RIP) (Speier et al., 2008) to generate transgenic mice that express GFP or luciferase in β-cells. The luciferase-expressing β-cells can be read-ily visualized in living mice using whole-body bioluminescent imaging (BLI) and the GFP-expressing β-cells can be visualized by fluorescence optical imaging. These transgenic mice have normal glucose tolerance and pancreatic islet architecture and allow easier identification and separation of β-cells from other endocrine cell types than previous methods, and were suggested to be useful for monitoring changes in β-cell function or mass in living animals with normal or altered metabolic states (Park et al., 2005). Recently, Speier et al. reported the application of the RIP-GFP islets and laser-scanning micro-scopy for noninvasive and longitudinal studies of pancreatic islet cell biology *in vivo* (Speier et al., 2008). In this work, the RIP-GFP transgenic mouse islets were transplanted into the anterior chamber of the eye using injection through the cornea. Using the anterior chamber of the eye as a natural body window, the engrafted GFP-islets could be visualized under fluorescence imaging, allowing repetitive *in vivo* imaging of islet vascularization, β-cell function, and death at cellular resolution (Speier et al., 2008). Besides trans-genic mice, virus-mediated gene transfer and BLI have been utilized by various groups for the assess-ment of pancreatic β-cell mass after islet transplantation. Healthy murine or human islets are infected by adenovirus or lentivirus carrying the luciferase gene prior to transplantation and were imaged and quantified by *in vivo* BLI after transplantation (Fowler et al., 2005, Virostko et al., 2004, Lu et al., 2004).

The alternative method is to develop targeted molecular beacons that can emit suitable light for opti-cal imaging. Considering that the pancreas is deep within the abdomen, near-infrared radiation chro-mophores such as commercially available NIR dye, Cy5.5, are preferred for reasons of tissue penetration (Lin et al., 2008). By conjugating Cy5.5 with β-cell-specific ligand streptozotocin, which selectively tar-gets β-cells by specific interaction with the GLUT2 transporter, Ran and colleagues synthesized two near-infrared probes for imaging β-cells (Ran et al., 2007). The two probes labeled almost all insulin-producing INS-1 cells, but only 18–33% of control cells. By conjugating Cy5.5 with annexin V, Moore's group could demonstrate different levels of β-cell apoptosis between diabetic and control animals in the animal models of both type 1 and type 2 diabetes *in vitro* and *ex vivo* (Medarova et al., 2005).

11.6 Imaging Ectopic Fat Deposition

Ectopic fat deposition, that is, lipid accumulation in tissues/organs other than white adipose tissue (WAT), such as liver, heart, and muscle, is often associated with metabolic abnormalities, including insulin and leptin resistance (Muoio and Newgard, 2006). There is strong evidence that excessive vis-ceral fat plays an important role in the development of metabolic syndrome and may predispose even young children to adult diseases, such as heart disease and diabetes. Since ectopic fat accumulation in the metabolic organs may potentially be used as a marker for the diagnosis of metabolic syndrome, accurate estimates of the amount of fat and specific lipid species of the fat in these organs are especially critical (Han et al., 2010).

TABLE 11.4 Imaging Modalities in Body Fat Measurements

Method	Capability Measuring Total Body Fat	Capability Measuring Fat Distribution	Applicability in Large Population Studies
CT	Moderate	Very high	Low
MRI and MRS	High	Very high	Low
DXA	Very high	High	Moderate
Densitometry	Very high	Very low	Low
Dilution techniques	High	Very low	Moderate
Anthropometry (BMI, WC, HC, WHR)	Moderate	Very low	Very high

Note: CT, computed tomography; MRI, magnetic resonance imaging; MRS, magnetic resonance spectroscopy; DXA, dual-energy x-ray absorptiometry; BIA, bioelectrical impedance analysis; BMI, body mass index; WC, waist circumference; HC, hip circumference; WHR, waist-to-hip ratio.

Although there are a number of methods available to estimate the body fat, not all of them can measure the abnormal distribution of the body fat (Table 11.4) (Sener et al., 2001). Conventional anthropometry methods, such as body mass index (BMI) and WC, can be easily applied in epidemiological studies and routine clinical screening. However, these methods cannot distinguish between fat mass and lean (nonfat) mass, or report the distribution of fat over the body. Multicompartment models, such as underwater weighing, dilution techniques, and dual-energy x-ray absorptiometry (DEXA) are all reliable methods to allow accurate measures of total body fat, but cannot provide information on the distribution of body fat either. Sophisticated imaging techniques, such as MRI and CT, have proven to be the optimal techniques for the accurate assessment of ectopic fat deposition, making it possible to noninvasively quantify ectopic fat within skeletal muscle cells (intramyocellular lipids [IMCL]), liver (intrahepatocyte lipids [IHCL]), and other tissues (Han et al., 2010).

CT is an optimal technique for the accurate assessment of ectopic fat and has been used to measure abdominal fat distribution in humans for decades (Yoshizumi et al., 1999). Accurate imaging quantification of visceral fat can be important because clinical proxies for abdominal obesity such as WC or BMI cannot discriminate subcutaneous fat from visceral fat entities that exhibit vastly different metabolic profiles. Figure 11.6 shows the CT sections for the fat distribution of individuals with and without the metabolic syndrome (Meng et al., 2010). In patients with metabolic syndrome, the subcutaneous fat is often lacking or dysfunctional, which may be a cause of undesirable fat deposition ectopically in other organs. CT has also been widely used to measure the ectopic fat accumulation in liver (Ma et al., 2009) and myocardium (Tansey et al., 2005).

Although CT is highly accurate, the associated radiation dose makes it unsuitable for longitudinal studies, especially in younger subjects. MRI also allows ectopic fat assessment noninvasively, including fat distribution and total volume, but without delivering any radiation. Siegel et al. explored the use of T1-weighted MR imaging at a single-slice level as a fast, reproducible means of assessing fat distribution and volume in preadolescents and adolescents (Siegel et al., 2007). Cali and Caprio examined the impact of varying degrees of obesity on the prevalence of the metabolic syndrome and its relation to ectopic fat deposition in a large, multiethnic cohort of children and adolescents, and found that the prevalence of the metabolic syndrome is high in obese children and adolescents, and increases with worsening obesity. Obese adolescents with a high proportion of visceral fat and relatively low abdominal subcutaneous fat exhibit a phenotype reminiscent of partial lipodystrophy: hepatic steatosis, profound insulin resistance, and an increased risk of the metabolic syndrome (Cali and Caprio, 2009). MR spectroscopy, especially localized 2D correlation spectroscopy, has been successfully applied in the detection and quantification of fat in skeletal muscle. The method allows measurement of intra- and extra-myocellular fat and determination of the type and amount of different lipid species in each compartment (Han et al., 2010). This represents an exciting development in metabolic disease research, and its wide application in human imaging will offer insights in the development of metabolic syndrome in humans.

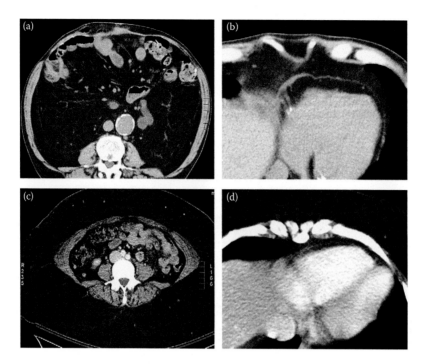

FIGURE 11.6   CT sections of the fat distribution of individuals with and without metabolic syndrome. (a) Large amount of visceral and (b) corresponding epicardial fat. Note the paucity of subcutaneous fat as is typically observed in patients with metabolic syndrome. (c) A large amount of subcutaneous fat without increased visceral fat, and corresponding (d) paucity of epicardial fat in a different patient with a favorable metabolic profile. (Reprinted from *Acad Radiol*, 17(10), Meng, K., Lee, C. H., and Saremi, F., Metabolic syndrome and ectopic fat deposition: What can CT and MR provide? 1302–12, Copyright 2010, with permission from Elsevier.)

11.7 Conclusion

In this chapter, we discussed the imaging strategies that are used in detecting and measuring insulin secretion, β-cell mass, and ectopic fat deposition. Although many of the strategies cannot be applied in a clinical setting, the information provided by these techniques should ultimately lead to the development and eventual implementation of noninvasive, or minimally invasive, molecular diagnostic strategies, better selection of metabolic disease therapies, and improved assessment of the efficacy of a treatment.

References

Aaen, K., Rygaard, J., Josefsen, K. et al. 1990. Dependence of antigen expression on functional state of beta-cells. *Diabetes*, 39, 697–701.

Antkowiak, P. F., Tersey, S. A., Carter, J. D. et al. 2009. Noninvasive assessment of pancreatic beta-cell function *in vivo* with manganese-enhanced magnetic resonance imaging. *Am J Physiol Endocrinol Metab*, 296, E573–8.

Biancone, L., Crich, S. G., Cantaluppi, V. et al. 2007. Magnetic resonance imaging of gadolinium-labeled pancreatic islets for experimental transplantation. *NMR Biomed*, 20, 40–8.

Brom, M., Andralojc, K., Oyen, W. J., Boerman, O. C., and Gotthardt, M. 2010. Development of radiotracers for the determination of the beta-cell mass *in vivo*. *Curr Pharm Des*, 16, 1561–7.

Cali, A. M. and Caprio, S. 2009. Ectopic fat deposition and the metabolic syndrome in obese children and adolescents. *Horm Res*, 71 Suppl 1, 2–7.

Cassidy, P. J. and Radda, G. K. 2005. Molecular imaging perspectives. *J R Soc Interface,* 2, 133–44.

Christ, E., Wild, D., Forrer, F. et al. 2009. Glucagon-like peptide-1 receptor imaging for localization of insulinomas. *J Clin Endocrinol Metab,* 94, 4398–405.

Cohen, B., Dafni, H., Meir, G., Harmelin, A., and Neeman, M. 2005. Ferritin as an endogenous MRI reporter for noninvasive imaging of gene expression in C6 glioma tumors. *Neoplasia,* 7, 109–17.

Cohen, B., Ziv, K., Plaks, V. et al. 2007. MRI detection of transcriptional regulation of gene expression in transgenic mice. *Nat Med,* 13, 498–503.

Culver, J., Akers, W., and Achilefu, S. 2008. Multimodality molecular imaging with combined optical and SPECT/PET modalities. *J Nucl Med,* 49, 169–72.

de Lonlay, P., Simon-Carre, A., Ribeiro, M. J. et al. 2006. Congenital hyperinsulinism: Pancreatic [18F] fluoro-L-dihydroxyphenylalanine (DOPA) positron emission tomography and immunohistochemistry study of DOPA decarboxylase and insulin secretion. *J Clin Endocrinol Metab,* 91, 933–40.

Delporte, C., Virreira, M., Crutzen, R. et al. 2009. Functional role of aquaglyceroporin 7 expression in the pancreatic beta-cell line BRIN-BD11. *J Cell Physiol,* 221, 424–9.

Evgenov, N. V., Medarova, Z., Dai, G., Bonner-Weir, S., and Moore, A. 2006. *In vivo* imaging of islet transplantation. *Nat Med,* 12, 144–8.

Fowler, M., Virostko, J., Chen, Z. et al. 2005. Assessment of pancreatic islet mass after islet transplantation using *in vivo* bioluminescence imaging. *Transplantation,* 79, 768–76.

Genove, G., Demarco, U., Xu, H., Goins, W. F., and Ahrens, E. T. 2005. A new transgene reporter for *in vivo* magnetic resonance imaging. *Nat Med,* 11, 450–4.

Gilad, A. A., Van Laarhoven, H. W., Mcmahon, M. T. et al. 2009. Feasibility of concurrent dual contrast enhancement using CEST contrast agents and superparamagnetic iron oxide particles. *Magn Reson Med,* 61, 970–4.

Gimi, B., Leoni, L., Oberholzer, J. et al. 2006. Functional MR microimaging of pancreatic beta-cell activation. *Cell Transplant,* 15, 195–203.

Goldhawk, D. E., Lemaire, C., Mccreary, C. R. et al. 2009. Magnetic resonance imaging of cells overexpressing MagA, an endogenous contrast agent for live cell imaging. *Mol Imaging,* 8, 129–39.

Gotthardt, M., Lalyko, G., Van Eerd-Vismale, J. et al. 2006. A new technique for *in vivo* imaging of specific GLP-1 binding sites: First results in small rodents. *Regul Pept,* 137, 162–7.

Gustavsson, N. and Han, W. 2009. Calcium-sensing beyond neurotransmitters: Functions of synaptotagmins in neuroendocrine and endocrine secretion. *Biosci Rep,* 29, 245–59.

Gustavsson, N., Lao, Y., Maximov, A. et al. 2008. Impaired insulin secretion and glucose intolerance in synaptotagmin-7 null mutant mice. *Proc Natl Acad Sci USA,* 105, 3992–7.

Hampe, C. S., Wallen, A. R., Schlosser, M., Ziegler, M., and Sweet, I. R. 2005. Quantitative evaluation of a monoclonal antibody and its fragment as potential markers for pancreatic beta cell mass. *Exp Clin Endocrinol Diabetes,* 113, 381–7.

Han, W., Chuang, K. H., Chang, Y. T. et al. 2010. Imaging metabolic syndrome. *EMBO Mol Med,* 2, 196–210.

Hara, M., Wang, X., Kawamura, T. et al. 2003. Transgenic mice with green fluorescent protein-labeled pancreatic beta-cells. *Am J Physiol Endocrinol Metab,* 284, E177–83.

Harris, P. E., Ferrara, C., Barba, P., Polito, T., Freeby, M., and Maffei, A. 2008. VMAT2 gene expression and function as it applies to imaging beta-cell mass. *J Mol Med,* 86, 5–16.

Inabnet, W. B., Milone, L., Harris, P. et al. 2010. The utility of [(11)C] dihydrotetrabenazine positron emission tomography scanning in assessing beta-cell performance after sleeve gastrectomy and duodenal-jejunal bypass. *Surgery,* 147, 303–9.

Jaffer, F. A. and Weissleder, R. 2005. Molecular imaging in the clinical arena. *JAMA,* 293, 855–62.

Jirak, D., Kriz, J., Herynek, V. et al. 2004. MRI of transplanted pancreatic islets. *Magn Reson Med,* 52, 1228–33.

Kawai, Y., Aoki, I., Umeda, M. et al. 2010. *In vivo* visualization of reactive gliosis using manganese-enhanced magnetic resonance imaging. *Neuroimage,* 49, 3122–31.

Keren, S., Zavaleta, C., Cheng, Z., De La Zerda, A., Gheysens, O., and Gambhir, S. S. 2008. Noninvasive molecular imaging of small living subjects using Raman spectroscopy. *Proc Natl Acad Sci USA,* 105, 5844–9.

Kim, H. S., Choi, Y., Song, I. C., and Moon, W. K. 2009. Magnetic resonance imaging and biological properties of pancreatic islets labeled with iron oxide nanoparticles. *NMR Biomed,* 22, 852–6.

Kim, J. H., Kim, J. S., Choi, H. et al. 2006. Nanoparticle probes with surface enhanced Raman spectroscopic tags for cellular cancer targeting. *Anal Chem,* 78, 6967–73.

Koretsky, A. P. and Silva, A. C. 2004. Manganese-enhanced magnetic resonance imaging (MEMRI). *NMR Biomed,* 17, 527–31.

Kung, M. P., Hou, C., Lieberman, B. P. et al. 2008b. *In vivo* imaging of beta-cell mass in rats using 18F-FP-(+)-DTBZ: A potential PET ligand for studying diabetes mellitus. *J Nucl Med,* 49, 1171–6.

Kung, H. F., Lieberman, B. P., Zhuang, Z. P. et al. 2008a. *In vivo* imaging of vesicular monoamine transporter 2 in pancreas using an (18)F epoxide derivative of tetrabenazine. *Nucl Med Biol,* 35, 825–37.

Ladriere, L., Malaisse-Lagae, F., Alejandro, R., and Malaisse, W. J. 2001. Pancreatic fate of a (125)I-labelled mouse monoclonal antibody directed against pancreatic B-cell surface ganglioside(s) in control and diabetic rats. *Cell Biochem Funct,* 19, 107–15.

Lee, J. S., Kang, N. Y., Kim, Y. K. et al. 2009a. Synthesis of a BODIPY library and its application to the development of live cell glucagon imaging probe. *J Am Chem Soc,* 131, 10077–82.

Lee, J. S., Kim, Y. K., Vendrell, M., and Chang, Y. T. 2009b. Diversity-oriented fluorescence library approach for the discovery of sensors and probes. *Mol Biosyst,* 5, 411–21.

Leoni, L. and Roman, B. B. 2010. MR imaging of pancreatic islets: Tracking isolation, transplantation and function. *Curr Pharm Des,* 16, 1582–94.

Leoni, L., Serai, S. D., Magin, R. L., and Roman, B. B. 2010. Functional MRI characterization of isolated human islet activation. *NMR Biomed,* 23, 1158–65.

Lin, M., Lubag, A., McGuire, M. J. et al. 2008. Advances in molecular imaging of pancreatic beta cells. *Front Biosci,* 13, 4558–75.

Lu, Y., Dang, H., Middleton, B. et al. 2004. Bioluminescent monitoring of islet graft survival after transplantation. *Mol Ther,* 9, 428–35.

Lu, Y., Dang, H., Middleton, B. et al. 2006a. Long-term monitoring of transplanted islets using positron emission tomography. *Mol Ther,* 14, 851–6.

Lu, Y., Dang, H., Middleton, B. et al. 2006b. Noninvasive imaging of islet grafts using positron-emission tomography. *Proc Natl Acad Sci USA,* 103, 11294–9.

Lu, J., Gustavsson, N., Li, Q., Radda, G. K., Sudhof, T. C., and Han, W. 2009. Generation of transgenic mice for *in vivo* detection of insulin-containing granule exocytosis and quantification of insulin secretion. *J Innov Opt Health Sci,* 2, 397–405.

Ma, X., Holalkere, N. S., Kambadakone, R. A., Mino-Kenudson, M., Hahn, P. F., and Sahani, D. V. 2009. Imaging-based quantification of hepatic fat: Methods and clinical applications. *Radiographics,* 29, 1253–77.

Malaisse, W. J. 2005. Non-invasive imaging of the endocrine pancreas (review). *Int J Mol Med,* 15, 243–6.

Malaisse, W. J., Damhaut, P., Ladriere, L., and Goldman, S. 2000a. Fate of 2-deoxy-2-[18F]fluoro-D-glucose in hyperglycemic rats. *Int J Mol Med,* 6, 549–52.

Malaisse, W. J., Damhaut, P., Malaisse-Lagae, F., Ladriere, L., Olivares, E., and Goldman, S. 2000b. Fate of 2-deoxy-2-[18F]fluoro-D-glucose in control and diabetic rats. *Int J Mol Med,* 5, 525–32.

Malaisse, W. J., Doherty, M., Kadiata, M. M., Ladriere, L., and Malaisse-Lagae, F. 2001a. Pancreatic fate of D-[3H] mannoheptulose. *Cell Biochem Funct,* 19, 171–9.

Malaisse, W. J., Doherty, M., Ladriere, L., and Malaisse-Lagae, F. 2001b. Pancreatic uptake of [2-(14)C] alloxan. *Int J Mol Med,* 7, 311–5.

Malaisse, W. J., Greco, A. V., and Mingrone, G. 2000c. Effects of aliphatic dioic acids and glycerol-1,2,3-tris(dodecanedioate) on D-glucose-stimulated insulin release in rat pancreatic islets. *Br J Nutr,* 84, 733–6.

Malaisse, W. J. and Ladriere, L. 2001. Assessment of B-cell mass in isolated islets exposed to D-[3H]man-noheptulose. *Int J Mol Med,* 7, 405–6.

Malaisse, W. J., Ladriere, L., Kadiata, M. M., and Malaisse-Lagae, F. 2000d. Pancreatic fate of 14C-labelled hexoses. *Cell Biochem Funct,* 18, 281–91.

Malaisse, W. J., Ladriere, L., and Malaisse-Lagae, F. 2000e. Pancreatic fate of 6-deoxy-6-[125I]iodo-D-glucose: *In vivo* experiments. *Endocrine,* 13, 95–101.

Malaisse, W. J., Louchami, K., and Sener, A. 2009. Noninvasive imaging of pancreatic beta cells. *Nat Rev Endocrinol,* 5, 394–400.

Malaisse, W. J., Olivares, E., Laghmich, A., Ladriere, L., Sener, A., and Scott, F. W. 2000f. Feeding a protective hydrolysed casein diet to young diabetic-prone BB rats affects oxidation of L[U-14C]glutamine in islets and Peyer's patches, reduces abnormally high mitotic activity in mesenteric lymph nodes, enhances islet insulin and tends to normalize NO production. *Int J Exp Diabetes Res,* 1, 121–30.

Marzola, P., Longoni, B., Szilagyi, E. et al. 2009. *In vivo* visualization of transplanted pancreatic islets by MRI: Comparison between *in vivo*, histological and electron microscopy findings. *Contrast Media Mol Imaging,* 4, 135–42.

Massaad, C. A. and Pautler, R. G. 2011. Manganese-enhanced magnetic resonance imaging (MEMRI). *Methods Mol Biol,* 711, 145–74.

Medarova, Z., Bonner-Weir, S., Lipes, M., and Moore, A. 2005. Imaging beta-cell death with a near-infrared probe. *Diabetes,* 54, 1780–8.

Medarova, Z. and Moore, A. 2009. MRI as a tool to monitor islet transplantation. *Nat Rev Endocrinol,* 5, 444–52.

Meng, K., Lee, C. H., and Saremi, F. 2010. Metabolic syndrome and ectopic fat deposition: What can CT and MR provide? *Acad Radiol,* 17(10), 1302–12.

Moore, A. 2009. Advances in beta-cell imaging. *Eur J Radiol,* 70, 254–7.

Moore, A., Bonner-Weir, S., and Weissleder, R. 2001. Noninvasive *in vivo* measurement of beta-cell mass in mouse model of diabetes. *Diabetes,* 50, 2231–6.

Muoio, D. M. and Newgard, C. B. 2006. Obesity-related derangements in metabolic regulation. *Annu Rev Biochem,* 75, 367–401.

Murthy, R., Harris, P., Simpson, N. et al. 2008. Whole body [11C]-dihydrotetrabenazine imaging of baboons: Biodistribution and human radiation dosimetry estimates. *Eur J Nucl Med Mol Imaging,* 35, 790–7.

Otonkoski, T., Nanto-Salonen, K., Seppanen, M. et al. 2006. Noninvasive diagnosis of focal hyperinsulinism of infancy with [18F]-DOPA positron emission tomography. *Diabetes,* 55, 13–8.

Park, S. Y., Wang, X., Chen, Z. et al. 2005. Optical imaging of pancreatic beta cells in living mice expressing a mouse insulin I promoter-firefly luciferase transgene. *Genesis,* 43, 80–6.

Porte, D., Jr. 1991. Banting lecture 1990. Beta-cells in type II diabetes mellitus. *Diabetes,* 40, 166–80.

Qian, X., Peng, X. H., Ansari, D. O. et al. 2008. *In vivo* tumor targeting and spectroscopic detection with surface-enhanced Raman nanoparticle tags. *Nat Biotechnol,* 26, 83–90.

Ran, C., Pantazopoulos, P., Medarova, Z., and Moore, A. 2007. Synthesis and testing of beta-cell-specific streptozotocin-derived near-infrared imaging probes. *Angew Chem Int Ed Engl,* 46, 8998–9001.

Ren, J., Trokowski, R., Zhang, S., Malloy, C. R., and Sherry, A. D. 2008. Imaging the tissue distribution of glucose in livers using a PARACEST sensor. *Magn Reson Med,* 60, 1047–55.

Rutter, G. A. 2004. Visualising insulin secretion. The Minkowski Lecture 2004. *Diabetologia,* 47, 1861–72.

Samli, K. N., McGuire, M. J., Newgard, C. B., Johnston, S. A., and Brown, K. C. 2005. Peptide-mediated targeting of the islets of Langerhans. *Diabetes,* 54, 2103–8.

Schneider, S., Feilen, P. J., Schreckenberger, M. et al. 2005. *In vitro* and *in vivo* evaluation of novel glibenclamide derivatives as imaging agents for the non-invasive assessment of the pancreatic islet cell mass in animals and humans. *Exp Clin Endocrinol Diabetes,* 113, 388–95.

Schneider, S., Ueberberg, S., Korobeynikov, A. et al. 2007. Synthesis and evaluation of a glibenclamide glucose-conjugate: A potential new lead compound for substituted glibenclamide derivatives as islet imaging agents. *Regul Pept,* 139, 122–7.

Sener, A., Bessieres, B., Courtois, P. et al. 2001. Uptake of 1-deoxy-1-[125I]iodo-D-mannoheptulose by different cell types: *In vitro* and *in vivo* experiments. *Int J Mol Med,* 7, 495–500.

Shi, C., Zhang, Y., Gu, C. et al. 2009. Molecular fiber sensors based on surface enhanced Raman scattering (SERS). *J Nanosci Nanotechnol,* 9, 2234–46.

Siegel, M. J., Hildebolt, C. F., Bae, K. T., Hong, C., and White, N. H. 2007. Total and intraabdominal fat distribution in preadolescents and adolescents: Measurement with MR imaging. *Radiology,* 242, 846–56.

Silva, A. C., Lee, J. H., Aoki, I., and Koretsky, A. P. 2004. Manganese-enhanced magnetic resonance imaging (MEMRI): Methodological and practical considerations. *NMR Biomed,* 17, 532–43.

Simpson, N. R., Souza, F., Witkowski, P. et al. 2006. Visualizing pancreatic beta-cell mass with [11C] DTBZ. *Nucl Med Biol,* 33, 855–64.

Souza, F., Simpson, N., Raffo, A. et al. 2006. Longitudinal noninvasive PET-based beta cell mass estimates in a spontaneous diabetes rat model. *J Clin Invest,* 116, 1506–13.

Speier, S., Nyqvist, D., Cabrera, O. et al. 2008. Noninvasive *in vivo* imaging of pancreatic islet cell biology. *Nat Med,* 14, 574–8.

Tai, J. H., Foster, P., Rosales, A. et al. 2006. Imaging islets labeled with magnetic nanoparticles at 1.5 Tesla. *Diabetes,* 55, 2931–8.

Tai, J. H., Nguyen, B., Wells, R. G. et al. 2008. Imaging of gene expression in live pancreatic islet cell lines using dual-isotope SPECT. *J Nucl Med,* 49, 94–102.

Tansey, D. K., Aly, Z., and Sheppard, M. N. 2005. Fat in the right ventricle of the normal heart. *Histopathology,* 46, 98–104.

van Zijl, P. C., Jones, C. K., Ren, J., Malloy, C. R., and Sherry, A. D. 2007. MRI detection of glycogen *in vivo* by using chemical exchange saturation transfer imaging (glycoCEST). *Proc Natl Acad Sci USA,* 104, 4359–64.

Virostko, J., Chen, Z., Fowler, M., Poffenberger, G., Powers, A. C., and Jansen, E. D. 2004. Factors influencing quantification of *in vivo* bioluminescence imaging: Application to assessment of pancreatic islet transplants. *Mol Imaging,* 3, 333–42.

von Maltzahn, G., Centrone, A., Park, J. H. et al. 2009. SERS-coded gold nanorods as a multifunctional platform for densely multiplexed near-infrared imaging and photothermal heating. *Adv Mater,* 21, 3175–3180.

Wang, C., Gao, X., and Su, X. 2010. *In vitro* and *in vivo* imaging with quantum dots. *Anal Bioanal Chem,* 397, 1397–415.

Wangler, B., Beck, C., Shiue, C. Y. et al. 2004a. Synthesis and *in vitro* evaluation of (S)-2-([11C]methoxy)-4-[3-methyl-1-(2-piperidine-1-yl-phenyl)-butyl-carbam oyl]-benzoic acid ([11C]methoxy-repaglinide): A potential beta-cell imaging agent. *Bioorg Med Chem Lett,* 14, 5205–9.

Wangler, B., Schneider, S., Thews, O. et al. 2004b. Synthesis and evaluation of (S)-2-(2-[18F]fluoroethoxy)-4-([3-methyl-1-(2-piperidin-1-yl-phenyl)-butyl-carbamoyl]-methyl)-benzoic acid ([18F]repaglinide): A promising radioligand for quantification of pancreatic beta-cell mass with positron emission tomography (PET). *Nucl Med Biol,* 31, 639–47.

Weir, G. C., Bonner-Weir, S., and Leahy, J. L. 1990. Islet mass and function in diabetes and transplantation. *Diabetes,* 39, 401–5.

Wicki, A., Wild, D., Storch, D. et al. 2007. [Lys40(Ahx-DTPA-111In)NH2]-Exendin-4 is a highly efficient radiotherapeutic for glucagon-like peptide-1 receptor-targeted therapy for insulinoma. *Clin Cancer Res,* 13, 3696–705.

Wild, D., Behe, M., Wicki, A. et al. 2006. [Lys40(Ahx-DTPA-111In)NH2]exendin-4, a very promising ligand for glucagon-like peptide-1 (GLP-1) receptor targeting. *J Nucl Med,* 47, 2025–33.

Woods, M., Zhang, S., and Sherry, A. D. 2004. Toward the design of MR agents for imaging beta-cell function. *Curr Med Chem Immunol Endocr Metab Agents,* 4, 349–369.

Xiao, M., Nyagilo, J., Arora, V. et al. 2009. Gold nanotags for combined multi-colored Raman spectroscopy and x-ray computed tomography. *Nanotechnology,* 21, 035101.

Yoshizumi, T., Nakamura, T., Yamane, M. et al. 1999. Abdominal fat: Standardized technique for measurement at CT. *Radiology,* 211, 283–6.

Zabow, G., Dodd, S., Moreland, J., and Koretsky, A. 2008. Micro-engineered local field control for high-sensitivity multispectral MRI. *Nature,* 453, 1058–63.

Zavaleta, C. L., Smith, B. R., Walton, I. et al. 2009. Multiplexed imaging of surface enhanced Raman scattering nanotags in living mice using noninvasive Raman spectroscopy. *Proc Natl Acad Sci USA,* 106, 13511–6.

Zhang, Y., Shi, C., Gu, C., Seballos, L., and Zhang, J. Z. 2007. Liquid core photonic crystal fiber sensor based on surface enhanced Raman scattering. *Appl Phys Lett*, 90, 193504–6.

Zhang, X., Yin, H., Cooper, J. M., and Haswell, S. J. 2008. Characterization of cellular chemical dynamics using combined microfluidic and Raman techniques. *Anal Bioanal Chem,* 390, 833–40.

Zurkiya, O., Chan, A. W., and Hu, X. 2008. MagA is sufficient for producing magnetic nanoparticles in mammalian cells, making it an MRI reporter. *Magn Reson Med,* 59, 1225–31.

12

High-Resolution X-Ray Cone-Beam Microtomography

Xiaochun Xu
National University of Singapore

Ping-Chin Cheng
Institute of Plant and Microbial Biology

State University of New York

12.1 Introduction

X-ray imaging fulfills a domain where other microscopy cannot attend, namely optically opaque/heterogenic specimen in the size range of millimeter to centimeter. The internal structures are inherently difficult to study in these types of specimens. Because of the technological complexity in focusing x-rays until the mid-twentieth century, microradiography was the only practical method of microscopy by x-rays. With respect to the shadow projection x-ray microscope of the 1950s, without the benefit of electronic recording or computer, high-quality three-dimensional (3D) image pairs could be easily generated (Newberry, 1987). High-resolution projection x-ray microscopy has been done by using micro-x-ray sources generated in a specially modified scanning electron microscope (Yada and Takahashi, 1989; Johnson et al., 1990, 1992; Johnson, 1993; Horikoshi et al., 1995; Yoshimura et al., 1997), laser-produced plasma x-ray source (Cheng et al., 1992; Kim et al., 1992), or by using a microslit in a synchrotron radiation beam line (Haddad et al., 1994), 3D reconstruction in terms of x-ray tomography has been studied for more than 30 years (Smith, 1982, 1985; Cheng et al., 1991, 1998 ; Xu 2010). Because of the high-penetration capability of x-rays, microtomography is a powerful tool for nondestructive analysis and visualization of 3D structures in opaque specimens. It avoids the need to cut thin sections of specimens; thus, visualization of thin layers of a living specimen is possible without fixation and physical sectioning.

12.2 Cone-Beam Microtomography

There are a number of imaging geometries used in x-ray tomography, namely parallel-beam, fan-beam, and cone-beam geometries. The use of cone-beam geometry provides numerous advantages over conventional parallel- or fan-beam geometry. Cone-beam geometry permits premagnification for matching the resolution to the detector pixel size, faster data acquisition, and minimal radiation dose to specimens. Figure 12.1 shows the imaging coordinate system used in the current microtomographic system.

The x-ray source and scintillation screen are placed in a fixed geometry, that is, fixed D_d; therefore, magnification change is achieved by moving the specimen in between the source and detector. This

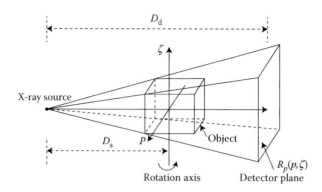

FIGURE 12.1 Geometrical representative of a cone-beam tomography setup. The reconstructed voxel $g(x,y,z)$ are within the center cube. Samples are placed between the x-ray source and the detector and are rotated along the rotation axis. The ratio of the source–detector distance (D_d) to the source sample (D_s) gives the magnification factor of the projection system.

setup has added advantage that the background x-ray intensity on the scintillation screen remains constant despite the magnification change. The position of the specimen is encoded by a high-resolution linear encoder. Figure 12.2 shows a schematic diagram of the x-ray tomographic imaging system. In this system, a 5 μm (in diameter) spot-size micro-focused x-ray tube (operating at 45–90 keV with a current of 0.18–0.09 mA, respectively) was used as the x-ray source. Samples were attached to the rotational axis of a specimen stage and were placed between the x-ray source and a fiber–optical plate-supported scintillation screen was used (Hamamatsu Co.). The x-ray projection image was formed on a high-resolution scintillation screen. The scintillated visible light image was then captured by a complementary metal–oxide–semiconductor (CMOS) camera (Lumenera Infinity2-2M) via a relay lens system. Figure 12.3 shows an example of an x-ray projection image of living maize male spikelets.

Control of the image acquisition is achieved via an integrated computer–user interface (Figure 12.4). The large window displays the current image while the small window below displays the previous captured images. The window on the lower right displays the visible image of the specimen under study. 3D icon activates a separate anaglyph window displaying two projection images obtained at two different

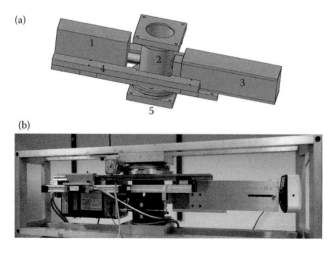

FIGURE 12.2 (a) A schematic representation of a cone-beam x-ray microtomographic imaging system (only the essential parts are shown). (1) X-ray source, (2) specimen chamber, (3) detector assembly, (4) magnification change slider, and (5) rotation assembly. (b) An experimental setup.

FIGURE 12.3 X-ray projection images of a maize spikelet.

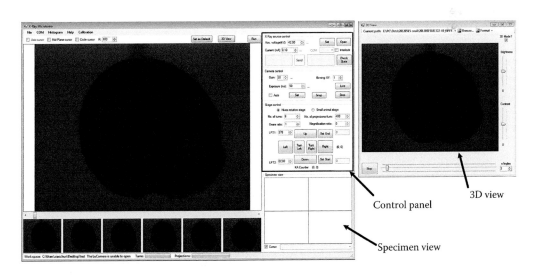

FIGURE 12.4 The user interface of the control software. The control panel provides control of x-ray source, camera parameter, and stage movements. The main display window shows the current real-time x-ray image and the six thumbnail images are previously captured as x-ray images. The specimen view provides real-time visible image of the specimen. 3D view provides real-time anaglyph stereo image of the x-ray projections.

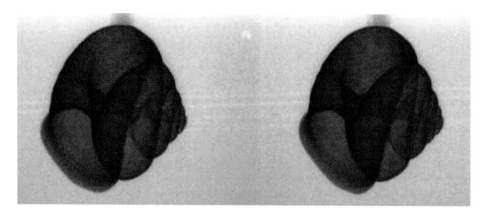

FIGURE 12.5   A stereo-pair of a fresh-water snail used to generate the anaglyph display.

viewing angles. Therefore, a 3D view of the specimen can be viewed during the data acquisition. Frame-sequenced 3D display, anaglyph (Figure 12.5), or lenticular liquid crystal display (LCD) screen can also be used. The user interface provides control to the acceleration voltage and beam current of the x-ray source, x-ray safety interlock, CMOS camera control, magnification, and specimen stage control.

The design philosophy separates the image acquisition from image reconstruction operation. Image reconstruction requires significantly advanced computational resource and is time consuming. On the other hand, image acquisition requires minimum computer resource as the speed of image acquisition is limited by the x-ray integration time.

12.3 Tomographic Reconstruction

Tomographic reconstruction algorithms can be summarized into two major categories, namely iterative and analytic methods. Analytic methods are represented by the filtered back projection (FBP) and Feldkamp-Davis-Kress (FDK); on the other hand, examples of iterative methods are algebraic reconstruction technique (ART) and order subset expectation maximization (OSEM). Recently, the introduction of compressed sensing can amazingly optimize the iterative reconstruction and reduce the required number of projections for a given reconstruction quality. This will significantly reduce the radiation dosage. This method may be introduced to our future work. All in all, iterative reconstruction requires significantly higher computational resource and much more time consuming than analytic reconstruction. Therefore, compromised with the goal of reconstruction accuracy, computational efficiency, and hardware implementation, a generalized Feldkamp algorithm (1984) for cone-beam x-ray microtomography was used in the system described in this chapter (Wang et al., 1992a,b, 1993). The algorithm was initially implemented by Pan et al. (1998) in a laboratory test system. Our current implementation was coded in C and utilizes multiple graphics processing units (GPUs) for fast reconstruction operation. The following steps describe the current implementation.

Step 1 : Preweighted convolution filtering

$$Q_\beta(p,\zeta) = \left(\frac{D_S}{\sqrt{D_S^2 + p^2 + \zeta^2}} R_\beta(p,\zeta) \right) * h(p)$$

where D_{SO} is the distance between the x-ray source and the origin of the reconstruction coordinate system, and $R_\beta(p,\zeta)$ is the projection data.

Step 2 : Weighted back projection

$$g(x,y,z) = \int_0^{2\pi} \frac{D_S^2}{U(x,y,\beta)^2} Q_\beta[a(x,y,\beta),b(x,y,\beta)]\mathrm{d}\beta$$

where

$$U(x,y,\beta) = D_S + x\cos\beta + y\sin\beta$$

$$a(x,y,\beta) = D_S \frac{x\cos\beta - y\sin\beta}{D_S + x\cos\beta + y\sin\beta}$$

$$b(x,y,\beta) = D_S \frac{z}{D_S + x\cos\beta + y\sin\beta}$$

The quality of the reconstructed image depends on the quality of the two-dimensional projection images and the number of such images, as shown in Figure 12.6a–d; the quality of the reconstructed images increases as the number of projections increases. The point-spread function of the cone-beam reconstruction has been discussed by Wang et al. (1992b).

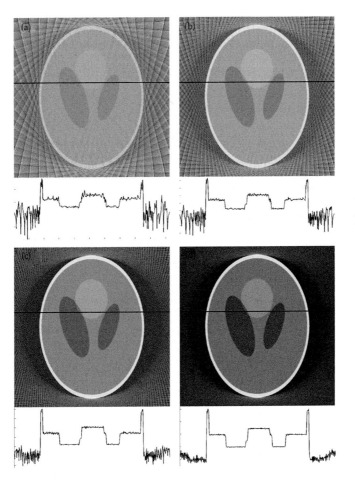

FIGURE 12.6 Comparison of tomographic reconstruction using (a) 50 (7.2°), (b) 100 (3.6°), (c) 200 (1.8°), and (d) 400 (0.9°) equispatial projections from a test phantom. The intensity plots reveal the edge response and noise level of the reconstructions.

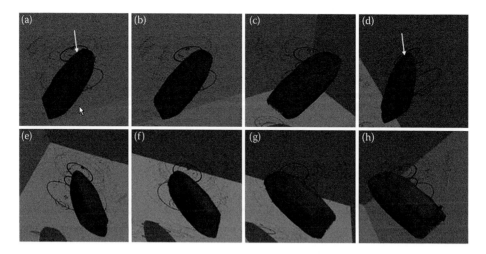

FIGURE 12.7 Surface-rendered image of a rice grain (*Oryza sativa*). The arrow indicates the location of the embryo (removed). Source size: 5 μm, 40 kV at 40 mA. The volume of the rice grain (purple volume) is 0.8415 mm³.

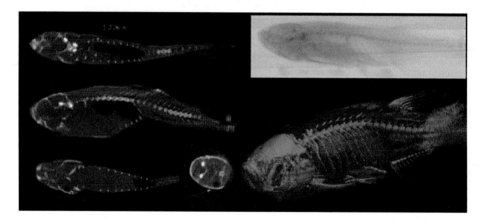

FIGURE 12.8 3D reconstruction of a small tropical fish. (Inset: one of the x-ray projection images.)

12.4 Discussion

The use of a cone beam for tomography permits the specimen to be much closer to the source. Since the intensity falls off the square of the source distance, this gain of intensity makes the simple thermionic cathode sources adequate for tomography, as demonstrated in this work. We have demonstrated that this technology is suitable for studies of internal 3D structures in an optically opaque specimen in a variety of applications (Figures 12.7 and 12.8). The image contrast, which is determined by the x-ray energy and the elemental composition of the specimen, can be manipulated by changing the suitable target for different applications (Pan et al. 1997).

Acknowledgments

This project was supported by a grant from Academia Sinica (Taipei, Taiwan, Republic of China) to PCC for constructing an x-ray microtomographic imaging system for botanical specimens. Special thanks to Song Tao Lin for his wonderful machining work.

References

Cheng, P. C., T. H. Lin, D. M. Shinozaki, and S. P. Newbernr. 1991. Projection microscopy and microto-mography using x-rays. *J Scan Microsc*, 13, 10–12.

Cheng, P. C., S. I. Pan, A. Shih, K. S. Kim, W. S. Liou, and M. S. Park. 1998. High efficient upconverters for multi-photon fluorescence microscopy. *J Microsc*, 189, 199–212.

Cheng, P. C., D. M. Shinozaki, T. H. Lin, S. P. Newberry, R. Sridhar, W. Tarng, M. T. Chen, and L. H. Chen. 1992. X-ray shadow projection microscopy and microtomography. In: *X-ray Microscopy -III*, eds. A. Michette, G. R. Morries and C. J. Buckley, Springer-Verlag, Berlin, pp.184–9.

Feldkamp Lzta Davis, L. C. and J. W. Kress. 1984. Practical cone-beam altubes. The phosphor, however, must have a very short decay algorithm. *J Opt Soc Am*, 1(A), 612–19.

Haddad, W. S., I. McNulty, L. E. Trebes, E. H. Anderson, P. A. Levesque, and L. Yang. 1994. Ultrahigh-resolution x-ray tomography. *Science*, 266, 1213–5.

Horikoshi, T., H. Chiba, K. Takahashi, W. Hiraoka, T. Mitsui, K. Yada, and S. J. Pan et al. 1995. Biological application of a projection x-ray microscope. *Zool Stud*, 34(Suppl), 207–8.

Johnson, R. H. 1993. 3D microanalysis of tissue volumes using dual-energy conebeam x-ray microtomo-graphy. *Proceedings of the 51st Annual Meeting of the Microscopy Society of America*, 652–3.

Johnson, R. H., A. C. Nelson, and D. H. Burns. 1990. Instrument design and image reconstruction for a laboratory x-ray microtomograph. *Proc XII ICEM*, l, 518–9.

Johnson, R. H., R. M. Fisher, and A. C. Nelson. 1992. 3D x-ray microscopy with a storage phosphor plate detector. *Proceedings of the 50th Annual Meeting of the Electron Microscopy Society of America*, 584–5.

Kim, H., B. Yaakobi, J. M. Soures, and P. C. Cheng. 1992. Laser-produced plasma as a source for x-ray microscopy. In: *X-ray Microscopy -III*, eds. A. Michette, G. R. Morries and C. J. Buckley, Springer-Verlag, Berlin, pp. 47–53.

Newberry, S. P. 1987. The shadow projection type of x-ray microscopy. In: *X-ray Microscopy: Instrumentation and Biological Applications*, eds. P. C. Cheng and G. J. Jan, Springer-Verlag, Berlin, pp. 126–141.

Pan, S. J., A. Shih, W. S. Liou, M. S. Park, G. Wang, S. P. Newberry, H. Kim, and P. C. Cheng. 1997. Cone-beam x-ray microtomography of human inner ear. *J Anal Morphism*, 4, 216–7.

Pan, S. J., W. S. Liou, A. Shih, M. S. Park, G. Wang, S. P. Newberry, H. Kim, D. M. Shinozaki, and P. C. Cheng. 1998. Experimental system for x-ray cone-beam microtomography. *Microsc Microanal* 4, 56–62.

Smith, H. D. 1985. Image reconstruction from cone-beam projections: Necessary and sufficient condi-tions and reconstruction methods. *IEEE Trans Mod Imag*, 5114, 14–28.

Smith, K. T. 1982. Reconstruction formulas in computed tomography. *Carps Proc Symp Appl Math*, 27, 7–23.

Wang, G., T. H. Lin, P. C. Cheng, and D. M. Shinozaki. 1992a. Cone-beam x-ray reconstruction of plate-like specimens. *J Scan Microsc*, 14, 350–4.

Wang, G., T. H. Lin, P. C. Cheng, and D. M. Sizinozaki. 1992b. Point spread function of the general cone-beam x-ray reconstruction formula. *J Scan Microsc*, l 4(4), 187–93.

Wang, G., T. H. Lin, P. C. Cheng, and D. M. Shinozaki. 1993. A general cone-beam reconstruction algo-rithm. *IEEE Tracts Med Imag*, 12, 486–96.

Xu Xiaochun. 2010. Acceleration of x-ray computed tomography reconstruction for rice tiller. MEng dis-sertation, Huazhung University of Science and Technology, Wuhan, China.

Yada, K. and S. Takahashi. 1989. Target materials suitable for projection x-ray microscope observation of biological samples. *J Electron Microsc*, 38, 321–3l.

Yoshimura, H., S. Kumagai, D. Sitoutsu, Y. Sekiya, and T. Mitsui. 1997. Phase contrast in a projection x-ray microscopy and time resolved imaging. *J Arlfl/Morphol*, 4, 210–11.

IV

Data Analysis

13

Image Analysis for Cellular and Tissue Engineering

Shuoyu Xu
Singapore-MIT Alliance

Piyushkumar
A. Mundra
Nanyang Technological University

Huipeng Li
Singapore-MIT Alliance

Shiwen Zhu
Singapore-MIT Alliance

Roy E. Welsch
Massachusetts Institute of Technology

Jagath C. Rajapakse
Nanyang Technological University

13.1 Introduction

Cellular and tissue engineering is a developing research field that aims to regenerate tissues and restore organ functions, which focus on the precise controls of structure and function at cellular resolution via the extracellular microenvironments (Langer and Vacanti, 1993). Both the top-down and bottom-up approaches (Ananthanarayanan et al., 2011) that are widely adopted in the field to construct tissues require the understanding of cell and tissue responses to biomaterials or tissue-engineered constructs at different levels. Imaging techniques play an important role to serve this purpose by medical imaging modalities at the tissue and organ level *in vivo* and by microscopic imaging modalities at the cell level *in vitro*. While the details of these imaging modalities are elaborated in this book, image analysis methods to interpret the images from these imaging modalities are equally important. The common approach of image analysis through manual visualization is qualitative and has several limitations, including low throughput, high observer variations, and difficulties in evaluating high-dimensional images (Peng, 2008). Hence, image analysis methods that incorporate both automated image processing and pattern recognition tools are attracting increasing interest in cellular and tissue engineering

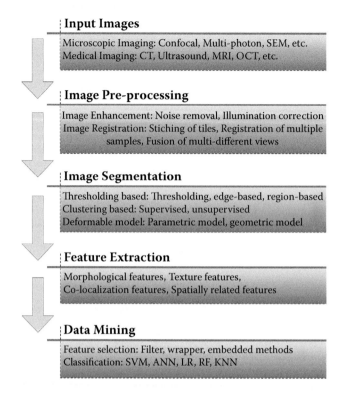

Input Images

Microscopic Imaging: Confocal, Multi-photon, SEM, etc.
Medical Imaging: CT, Ultrasound, MRI, OCT, etc.

Image Pre-processing

Image Enhancement: Noise removal, Illumination correction
Image Registration: Stiching of tiles, Registration of multiple
          samples, Fusion of multi-different views

Image Segmentation

Thresholding based: Thresholding, edge-based, region-based
Clustering based: Supervised, unsupervised
Deformable model: Parametric model, geometric model

Feature Extraction

Morphological features, Texture features,
Co-localization features, Spatially related features

Data Mining

Feature selection: Filter, wrapper, embedded methods
Classification: SVM, ANN, LR, RF, KNN

FIGURE 13.1   Flowchart of a typical image analysis system.

applications. These automated analysis methods are capable of providing quantitative characterization of cells and tissues with no discrepancies at high speed, which are helpful for tissue engineers to rationally design biomaterial scaffolds, bioreactors, microenvironments, and tissue constructs at the desirable levels of precisions and scale.

A typical flowchart of an image analysis system is illustrated in Figure 13.1, which contains several procedures, such as image preprocessing to remove noise and enhance image quality and image segmentation to identify objects of interest, followed by the feature extraction to generate quantitative measurements from the recognized objects and classification to recognize a pattern from the extracted features. In this chapter, we will introduce typical segmentation methods and useful features that are applicable in cellular and tissue engineering applications. Widely used feature selection and classification tools will also be covered. Several examples will be reviewed that contain the usage of image analysis methods to characterize cell, scaffold, vascular network, and tissue constructs. We will also briefly introduce the available image analysis software for the researchers in the field.

13.2 Image Segmentation

Segmentation is the process of partitioning an image into multiple regions by assigning the same label to the pixels with common features. The function of image segmentation in an image analysis system for cellular and tissue engineering is to identify biologically meaningful objects such as cells and functional tissue regions for further quantitative characterization. So far, the segmentations are achieved manually or with the assistance of user-interactive tools available in microscope software. However, these manual or semiautomatic segmentation methods are often labor intensive and are affected by high inter- and intra observer variations, which limit their application for high-dimensional or high-content images.

TABLE 13.1 Summary of Image Segmentation Methods

Category	Method	Advantage	Disadvantage
Thresholding-based	Global thresholding: Otsu, histogram Edge-based: Canny, Sobel, Laplacian Region-based: region growing, split-and-merge Hybrid-based: watershed	Computationally efficient, easy to implement, available in commercialized software	Sensitive to noise and nonuniformity of intensity

Clustering-based	Supervised classification: SVM, ANN,
	   KNN
	Unsupervised clustering: K-means,
	   fuzzy-C-means, Gaussian mixture model
		Computationally efficient, incorporate learning procedures	Sensitive to noise, training sample size and initialization setting of clusters

Deformable models-based	Parametric model: snake Geometric model: level set	Insensitive to noise, subpixel accuracy of object boundary, easy to incorporate prior knowledge	Sensitive to the parameters of initial contours, time-consuming

Therefore, the development of reliable automated image segmentation methods remains important for the characterization of cellular and tissue images. We next present a brief overview of automated image segmentation methods that are suitable for applications in cellular and tissue engineering research. There are different ways to classify the types of image segmentation methods (Somasundaram and Alli, 2012); we adopt Ma's approach (Ma et al., 2010) to introduce the image segmentation methods of three main types: the ones based on threshold, the ones based on clustering techniques, and the one based on deformable models (Table 13.1).

13.2.1 Thresholding-Based Methods

The representative thresholding method is to recognize the pixels with intensities in the range defined by a certain threshold. The threshold can be chosen manually according to prior knowledge or experiments, as well as be defined automatically based on certain criteria. The Otsu method (Otsu, 1975) is a widely used threshold algorithm that aims to find the best threshold automatically from the image histogram to minimize intraclass and maximize interclass variances. Another effective method is to identify proper threshold from the shape and distribution of the intensity histogram (Sezan, 1990, Guo and Pandit, 1998). Other than intensity, the threshold value can come from any distinctive feature of the pixel such as color and gradient magnitude (Kulkarni, 2012). The thresholding method is useful in edge-based and region-based segmentation methods as well. The principle of edge-based methods is to locate the edge points on the boundaries and segment the objects inside the closed region boundaries. Several well-studied edge detectors such as Canny (Canny, 1986), Sobel, and Laplacian (Gonzales and Woods, 2002) calculate the gradient magnitude or second-order derivative of pixel intensities and require further hysteresis thresholding (Medina-Carnicer et al., 2010). Region-growing methods (Adams and Bischof, 1994) are the typical region-based segmentation methods with the assumption of homogeneity inside the object. The region expands from a seed point by merging the neighbor pixels, which have the intensity differences smaller than the threshold. The watershed method (Roerdink and Meijster, 2000) combines both edge-based and region-based methods and begins with local minima in gradient images and forms the regions in a region-growing way until the boundaries located at high gradient points are reached.

13.2.2 Clustering-Based Methods

Clustering-based methods apply classification tools to image segmentation by treating different structures in the image as different classes. The features used for the classification can be any distinctive measures of pixels or regions such as intensity, color, and texture (Thilagamani, 2011). According to the requirement of learning procedures, these segmentation methods can be divided into supervised classification methods and unsupervised classification methods. In supervised classification methods, a certain classifier is trained from pixels in the training images with the extracted features and assigned class labels, and is applied to unlabeled pixels in the new images (Egmont-Petersen et al., 2002, Kotropoulos and Pitas, 2003, Vrooman et al., 2007). Frequently used classifiers include support vector machines (SVM) (Lloyd, 1982), artificial neural networks (ANN) (Bishop, 1996), and k-nearest neighbor (KNN) (Cover and Hart, 1967) classifiers. On the other hand, unsupervised classification methods identify pixels or regions with similar features using clustering techniques (K-means clustering (Hartigan and Wong, 1979), fuzzy-C-mean clustering (Bezdek and Ehrlich, 1984), Gaussian mixture models (Day, 1969), etc.) without the need for training. The K-means algorithm is one of the most popular iterative clustering methods that aims to partition n observations (pixels or regions) into K clusters in which each observation belongs to the cluster with the nearest mean (Pappas, 1992). Unlike K-means clustering, in fuzzy-C-means clustering, each point has a degree of belonging to clusters rather than just belonging completely to one cluster and associated with each point is a set of membership levels (Chen and Zhang, 2004). These indicate the strength of the association between that data element and a particular cluster. Another way of clustering is the model-based approach, which uses certain models for clusters and attempts to optimize the fit between the data and the model. For example, we can assume the histogram of the intensities of pixels as the mixture of several Gaussian distributions, one for each class of the objects. Clustering is performed by estimating the best values of the parameters of all Gaussian distributions using the expectation-maximization (EM) method (Dempster et al., 1977).

13.2.3 Deformable Models-Based Methods

Deformable models have been extensively studied in microscopic and medical image segmentation. These methods model the curve or surface evolution from the initial contour to the final boundary of the structure. The contours move under both the internal forces defined by the curve or surface and the external forces computed from the image. The internal forces smooth the contour and the external forces attract the contour toward the structure boundary. The deformable models have two main types: parametric deformable models (McInerney and Terzopoulos, 1996) and geometric deformable models (Niessen et al., 1998). The typical parametric deformable models derive from a snake model (Kass et al., 1988) representing the contours in their parametric forms. Tension and rigidity of the contour can be used as the internal force and the external force is computed by the integral of a locally computed edge map along the contour.

On the other hand, geometric deformable models based on level sets (Chan and Vese, 2001, Osher and Sethian, 1988) represent contours as a level set of a signed function and compute image energy terms using intensity variances inside and outside the contours.

13.2.4 Advantages and Disadvantages of Segmentation Methods

The performances of segmentation methods are affected by low signal-to-noise ratios, large variability of objects, and partial volume average effects (Soret et al., 2007) in the images. Noise coming from the sensor of the digital camera or photon detector is inevitable in the imaging process and can randomly alter the intensity of a pixel. Segmentation is very challenging when the noise level is comparable to the useful signals. Nonuniformity of intensity can degrade image quality significantly as the pixel intensities vary gradually over the entire extent of the image. Nonuniformity could be attributed to out-of-flatness of the tissue slice for optical imaging or patient anatomy inside and outside the field of view in medical imaging. The partial volume effect often occurs in medical imaging where the pixel resolution is limited so that individual pixel may represent more than one tissue structure. The advantages and disadvantages of the three types of segmentation methods introduced are discussed next regarding their abilities to handle these difficulties.

The thresholding-based segmentation methods are effective when there is a clear difference of certain features such as intensity between the objects to be segmented and other parts, and are usually easy to implement and computationally efficient. However, these methods are sensitive to noise and nonuniformity of intensity due to the difficulty of identifying a proper threshold when the pixel intensities are altered in such situations. The clustering-based methods engage the power of pattern classification tools and can achieve good segmentation results if the patterns of different structures in the image are properly modeled. However, these methods also require small intensity variations inside the objects or structures and are influenced by noise and nonuniformity of intensity. Moreover, the choice of certain parameters such as the size of training samples in supervised classification methods or the number of clusters in unsupervised classification methods affects the segmentation accuracy significantly. The deformable model-based methods are noise insensitive and capable of segmenting complex structures at subpixel accuracy. However, owing to the complexity of such methods, the selection of parameters to initialize the algorithm is not a trivial task to obtain a promising result. Furthermore, it is usually time-consuming for the initial contours of the object boundaries deforming to their exact locations.

The choice of a proper segmentation method is application driven in tissue engineering research, which depends on the imaging modality used and the properties of the biological objects to be studied. For example, bone is a kind of high-density structure with high Hounsfield unit values in computed tomography (CT) images but no distinctive features in magnetic resonance (MR) images. Hence, the thresholding-based segmentation methods are appropriated for bone segmentation in CT images (Zhang et al., 2010) but more advanced techniques are required for MR images (Schmid and Magnenat-Thalmann, 2008). In general, thresholding-based segmentation methods are suitable when the intensities inside the structures are homogeneous or there is a need for fast processing. Clustering-based methods can improve the segmentation results under similar conditions if enough training samples are provided or the initialization of the clusters is performed well. Deformable methods are preferred when the image is noisy and structures have complex topology. Moreover, applications in tissue engineering research usually require the identification of each single structure from a group of touching structures. In such cases, thresholding- and clustering-based methods are usually used with the watershed method. The deformable model-based methods can segment and separate all the objects simultaneously. The parametric deformable model-based methods are suitable for tubular objects with open boundaries such as fiber (Smith et al., 2010) and vessel (Al-Diri et al., 2009). The geometric deformable model-based methods can handle structures with large shape variety and are suitable for globular objects such as cell and nuclei (Dufour et al., 2005). In summary, no segmentation method applies to all the applications. The segmentation method needs to be carefully selected according to its advantages, disadvantages, and applicability to certain imaging conditions.

13.3 Feature Extraction

The quantitative assessment of tissue and organ function relies on capturing proper features to characterize cellular and tissue structures. To measure the deviations at the cellular and tissue level, morphological, texture, colocalization, and spatially related features are widely used (Table 13.2). Morphological and spatially related features require the segmentation of objects of interest in advance, while the other features are applicable to both segmented and original images.

13.3.1 Morphological Features

Morphological features characterize the size and shape of an object of interest in the image, which, for instance, can be nuclei, cell, fiber, pore, or functional organ area depending on the application in cellular and tissue engineering. Morphological features are purely object-based, which are sensitive to the object

TABLE 13.2   Summary of Useful Features in Image Analysis

Category	Useful Features
Morphological features	Globular objects: area, radius, perimeter, convex area, eccentricity, Euler number, extent, filled area, length of the major and minor axes, orientation, solidity, compactness, smoothness, symmetry and concavity, etc.
	Tubular objects: length, width, curvature, number of branch points, cross-link spaces and cross-link density, etc.
Texture features	GLCM: contrast, correlation, average, variance and entropy etc. from the occurrence matrix
	Run-length matrix: short run emphasis, long run emphasis, gray-level nonuniformity, run-length nonuniformity and run percentage, etc.
	Transform-based: Fourier, Gabor, and wavelet transformation
	Fractal dimension: spatial correlation estimator, box counting and surface estimator, etc.
Colocalization features	Global statistic-based: Pearson's coefficient, Mander's overlap coefficient, Costes approach, and intensity correlation quotient
	Object-based: distance of the geometrical or intensity-weighted centroids, percentage of overlapping objects
Spatially related features	Delaunay triangulation: number of nodes, edge length, number of edges, and number of triangles
	Voronoi diagram: the area and shape of the polygons
	Minimum spinning tree: total tree length

segmentation and require accurate characterization of object boundaries. For globular objects, the size measurements include area, radius, and perimeter, while the shape is described by features such as convex area, eccentricity, Euler number, extent, filled area, length of the major and minor axes, orientation, solidity, compactness, smoothness, symmetry, and concavity. The detailed mathematical definition of these features can be found in Chen et al. (2012) and Demir and Yener (2005). The size and shape of tubular structures are expressed by length, width, curvature, number of branch points, cross-link spaces, and cross-link density (Stein et al., 2008) from the centerline of the structure extracted. Furthermore, the ratio of the same morphological feature from different biological structures, for example, nuclear area/cytoplasm area ratio, can be used as a feature to reflect the relative morphological change between different objects. When the number of objects in the image is big, the statistical measures such as mean, median, standard deviation, and skewness (Groeneveld and Meeden, 1984) of the features described above can be calculated and used as new features to represent the morphology of the population.

13.3.2 Texture Features

Image texture is defined as the repeated pattern of the spatial arrangement of pixel intensities. Texture features can be extracted from objects of interest after segmentation, as well as from the entire image. A complete survey of texture analysis can be found in several good reviews (Castellano et al., 2004, Tuceryan and Jain, 1993, Zhang and Tan, 2002) and is beyond the scope of this chapter. We will introduce gray-level co-occurrence matrix (GLCM) features (Haralick et al., 1973), run-length matrix features (Galloway, 1975), Fourier and Gabor wavelet transformed features (Manjunath and Ma, 1996, Azencott et al., 1997), and fractal dimension features (Chaudhuri and Sarkar, 1995) that are popular in cellular and tissue engineering applications. GLCM features are based on the occurrence of pairs of pixel intensities for a given displacement at different directions that indicate the spatial distribution of the gray levels. A number of features such as contrast, correlation, average, variance, and entropy are proposed by Haralick et al. (1973) for calculation from the occurrence matrix. The run-length matrix measures the number of consecutive pixels in a specific direction that has the same intensity, which reveals the coarseness of a certain texture. Various features such as short run emphasis, long run emphasis, gray-level nonuniformity, run-length nonuniformity, and run percentage (Galloway, 1975) are derived from the matrix. Transform-based methods are widely used for texture analysis as well, which represent the image in the frequency (Fourier transform) or spatial (Gabor filters) domain to characterize the texture. The values and the statistical measures of the transformed coefficients are exploited as the features. The fractal dimension features characterize the self-similarity level of the object to reveal its regularity and complexity. These features are relatively insensitive to the image scaling and hence suitable to describe objects with irregular shapes. Measures as spatial correlation estimator (Chen et al., 1989), box counting (Gagnepain and Roques-Carmes, 1986), and surface and blanket estimator (Peleg et al., 1984) are the examples of such features. The parameters in texture analysis such as the displacement and the number of coefficients are critical but no method of selecting these parameters has been well established. The available approach is to generate various texture features with different settings of the parameters and to adopt feature selection methods to find the most relevant features.

13.3.3 Colocalization Features

Colocalization features are used to quantify the co-occupying or the overlapping of the physical distribution of two or more markers, which can be grouped as global statistic features and object-based features. The global statistic features are mainly based on intensity correlation coefficient analysis, including Pearson's coefficient, Manders' overlap coefficient (Manders et al., 1993), colocalization coefficient based on Costes' approach (Costes et al., 2004), and the intensity correlation quotient (Li et al., 2004). The Pearson's coefficient measures the linear relationship between the intensities of two signals. Manders' overlap coefficient is developed from the expression of Pearson's coefficient to reveal the correlation for different conditions of each channel. The colocalization coefficient represents the proportion of segmented overlapped pixels over total pixels of every particular channel. The intensity correlation quotient is an interpretable colocalization based on a nonparametric sign test of the product of the differences from the mean. Pixel-based global features measure well the strength of the linear relationship between two channels; however, the estimation is for the whole image or region of interest but not for a unique structure. These features are sensitive to noise and therefore background subtraction is crucial for the quantification of results. On the other hand, objected-based colocalization features incorporate knowledge of the objects of interest. One of the object-based features is the distance of the geometrical centroids or intensity-weighted centroids of the structures of interest in two channels (Zhu et al., 2010). The percentage of colocalized objects out of all objects will be defined as a specific feature to measure the degree of the colocalization. Another object-based feature is modified from the intensity correlation quotient, by replacing the product of the differences from the mean with the normalized mean deviation product (Jaskolski et al., 2005), comparing the values from one set of segmented object images to another.

13.3.4 Spatially Related Features

The spatial relationship between tissue structures provides important information to characterize tissue functions. The common approach is to locate each structure as the node and create graph or network accordingly. The identification of each structure depends on the segmentation result; however, the location information such as the centriod of each structure is more important than the exact boundary to calculate spatially related features. Hence, the segmentation methods selected do not necessarily require an accurate representation of the boundaries. Among all the methods to establish edges between the nodes, the most widely used ones are Delaunay triangulation and minimum spanning tree (De Berg et al., 2008). The Delaunay triangulation connects the nodes by creating the graph of a set of triangles where no node is inside the circumcircle of any triangle. The centers of the circumcircles of the triangles are connected to produce the Voronoi diagram (Okabe et al., 1992) so that, for a particular node, every pixel in its polygon is closer to itself than to another node in the graph. In addition, a minimum spinning tree can also be generated from the Delaunay triangulation. The possible spatially related features to extract include number of nodes, edge length, number of edges, and number of triangles from Delaunay triangulation, the area and shape of the polygons in the Voronoi diagram, and total tree length from minimum spinning tree. The complete introduction of spatially related features is provided in Gurcan et al. (2009). If properly modeled, the spatially related features have great power to characterize spatial organization and structural relationships especially for the analysis of complicated tissue environment involving different types of structures.

13.4 Data Mining

Identification of objects in optical and medical images largely depends on the features that characterize the individual objects. As discussed in the previous section, an exhaustive set of features could be generated using various strategies. However, all the features may not be equally important for the highly accurate automatic identification and it is necessary to either reduce the dimensionality or determine the most relevant features that give the best performance to object classification. These features can be classified into two categories: (i) relevant and (ii) irrelevant, with respect to the classification of samples. The relevant features can be further classified into strongly relevant and weakly relevant (Kohavi and John, 1997, Yu and Liu, 2004). The strongly relevant features are absolutely necessary in the optimal feature subset while weakly relevant features are conditionally important. However, weakly relevant features may also be redundant. Highly correlated features are generally referred to as redundant features since they do not provide any additional information for building a classifier. Hence, it becomes imperative to develop strategies to determine the best feature groups, individual features, or dimensionality reduction techniques that need to be developed.

In general, feature selection approaches can be classified into (1) filter, (2) wrapper, and (3) embedded methods (Guyon and Elisseeff, 2003, Inza et al., 2004). Similarly, based on how the feature space is searched, methods are classified into forward and/or backward methods, sequential floating search, and optimization-based methods such as evolutionary search methods (Wahde and Szallasi, 2006).

Filter methods use a ranking criteria based on the relationship between features and class labels without taking into account the functionality of the classifier. Typical filter criteria such as statistical tests have been used in feature selection of breast cancer images or breast masses in mammograms (Akay, 2009). Other popular filter methods are T-score (Inza et al., 2004), F-score (Dudoit et al., 2002), Kruskal Wallis (KW)-score (Chen et al., 2005), Wilcoxon rank sum score (Zhang and Deng, 2007), and the minimum redundancy maximum relevancy (MRMR) criterion (Peng et al., 2005). Wrapper methods refer to algorithms that take into account the classifier characteristics for feature selection. The feature selection algorithms conduct a search for the optimal subset of features, incorporating the classifier characteristics as part of the evaluation function. Popular wrapper methods are SVM-recursive feature elimination (SVM-RFE) (Guyon et al., 2002, Duan et al., 2005, Mundra and Rajapakse, 2010). Embedded methods refer to a set of methods/algorithms in which feature selection is embedded within the construction of the classifier. Therefore, the structure of a classifier and discriminant function play a crucial role in embedded methods of feature selection. Embedded methods include feature selection in the classification such as the LASSO. A recent review on various penalized feature selection methods can be found in Fan and Lv (2010). In general, filter methods have been attractive because of their simplicity and speed, but may result in poor classification performance. Wrappers and embedded methods are complex but result in improved classification performance. In the following, we discuss two widely used feature selection methods, support vector machine-recursive feature elimination (SVM-RFE) and random forest-recursive feature elimination (RF-RFE).

13.4.1 SVM-RFE

SVM weights represent the importance of the corresponding input or feature for the classification. Using such weights and starting from a full set of features, SVM-RFE eliminates features recursively in a backward elimination manner. The smallest such weight corresponds to the least significant feature for classification and that feature is removed. SVM is learned again and the least significant feature is removed and so on until all the features are ranked. The standard SVM-RFE algorithm was originally proposed for two-class classification problem. Zhou and Tuck (2007) recently proposed an extension of the SVM-RFE algorithm for multiclass problems. They proposed to sum SVM weights obtained from different class-wise decision functions for gene ranking. The optimization of the SVM objective function is possible with any one approach to multiclass problem: one-versus-all, one-versus-one, and all-at-one. These weights were then used for the computation of weights of each gene.

13.4.2 RF-RFE

Random forest (Breiman, 2001) is an ensemble of classification trees, which uses both bagging (bootstrap aggregation) and random feature selection for tree building. Díaz-Uriarte and de Andrés (2006) studied a random forest-based gene importance method. For each tree, the candidate set of features is a random subset of features from a subset of samples. The feature importance measure is based on the decrease of classification accuracy when values of a feature in a node of a tree are permuted randomly. Like SVM-RFE, starting with all features, a forest was built and feature importance was measured using the permutation-based criterion. The importance of a feature is based on the decrease of classification accuracy when values of a feature in a node of a tree are permuted randomly. The feature with smallest permutation-based importance is discarded and a new forest was constructed. The procedure is continued until the minimum number of features is achieved, which have an out-of-bag error within the one standard error of minimum error.

Apart from selecting individual features, the dimensionality of the problem could be reduced by employing popular feature dimension reduction techniques like principal component analysis (PCA), partial least squares (PLS), and so on. These methods find the linear combination of all the features that explain the variance in the data. Generally, very few components are needed to achieve good classification performance. However, these methods lack the interpretability of individual features.

Similarly, instead of selection individual features, a set of best-performing feature groups (e.g., a combination of morphology, texture, colocalization, and/or spatially related features) could be identified. For example, Meier et al. (2008) have proposed a group lasso technique for logistic regression (LR) that could identify the best-performing feature groups at once, eliminating the need of heuristic or brute force search.

13.4.3 Classifier

The choice of a classifier also plays an important role in object identification. Supervised classifiers like SVM (Wang et al., 2008), ANN (Boland and Murphy 2001), LR, decision trees, random forests (RF), kNN, Ada-boost, Bayesian techniques, and so on, and unsupervised techniques like K-means, EM, self-organizing feature map (SOFM), and so on, have been used for object identifications. Though ANN is very widely used in image analysis, RF and SVM have gained lots of attention recently for object classification. As discussed earlier, the advantages of RF over other methods are that it has an embedded property of selection of relevant features and is robust and unaffected by high dimensions and irrelevant features. This is an ensemble-based training algorithm, which has very few parameters to tune and is applicable to both two-class and multiclass problems. On the other hand, SVM has become popular for several reasons: the dependence of its solution on a number of samples rather than their dimensionality, inherent ability to select data points important for classification, and good generalization capabilities (Vapnik, 2000). Since the performance of a classifier depends on many factors, a few classifiers should be tested on the same dataset to choose the best for a given application.

Most of these classifiers are now available as "R" packages or in WEKA software. For example, the "RandomForest" (Liaw and Wiener, 2002) package in R is hugely popular while LibSVM is a very widely used SVM toolbox due to its availability in a large number of programming languages (Chang and Lin, 2011).

13.5 Applications

In this section, we summarize several applications of image analysis methods for cellular and tissue engineering research. These applications include, but are not limited to, the characterization of cell behaviors, vessel networks, scaffold architectures, and bone and cartilage tissue construct structures.

13.5.1 Characterization of Cell Behaviors

Assessment of the appropriate cellular behaviors is critical in determining the performance of tissue scaffolds and medical implants. Image analysis-based approaches are capable of providing quantitative tools to characterize migration, distribution, and growth of cells throughout the tissue construct. The applications of these image analysis methods include the evaluation of biocompatibility and cell–substrate contacts of a scaffold (Cuijpers et al., 2011, Shaw et al., 2012, Sommerhage et al., 2008, Thevenot et al., 2008), the prediction of cell quality in tissue-engineered products (Kagami et al., 2011), the examination of the functional effectiveness of the engineered microenvironments (Xu et al., 2011, Xylas et al., 2010), and the assessment of stem cell therapy in regenerative medicine (Rabbani and Javanmard 2011, Polzer et al., 2010).

Cell segmentation in most of these applications was performed by thresholding-based methods in scanning electron microscopy (SEM), phase contrast, confocal, and multiphoton microscopy images. The threshold was either manually selected (Kagami et al., 2011, Polzer et al., 2010, Sommerhage et al., 2008, Thevenot et al., 2008, Xylas et al., 2010) or automatically decided by Canny edge detector (Shaw et al., 2012) and Sauvola's method (Xu et al., 2011). More sophisticated segmentation methods such as level sets and fast marching method (Meijering et al., 2009) were applied to the time-lapse fluorescence microscopy images due to the demand for accurate demonstration of single-cell boundary from the touching cells for the cell tracking.

Different cell features were extracted for different application purposes. Morphological features were adopted for the quantification of cell–substrate interactions. These features were, but not limited to, cell density, cell footprint (the maximum cross-sectional area), cell thickness, and volume (Cuijpers et al., 2011, Shaw et al., 2012, Thevenot et al., 2008). Their results showed that cell spreading was well characterized by these morphological features, which were good indicators of biocompatibility. Sommerhage et al. (2008) further analyzed the change of cell surface area at different cell heights to create a membrane allocation profile reflecting the influences of substrates on cell attachment. Cell morphological features were also shown to be useful for cell quality prediction. 120 morphological features were extracted from segmented cell images and three of them were selected using multiple regression analysis for cell yield prediction (Kagami et al., 2011). A similar study of the prediction of the level of osteogenic differentiation utilized 740 morphological features of human stem cells and selected the 10 most important parameters among them (Kagami et al., 2011). Coupled with classification tools such as regression, the image analysis system would have the prediction power to perform not only "in-sample analysis" but "out-of-sample analysis" as well. To assess the cell microenvironment in tissue constructs, cell alignment quantified from the distribution of cell orientations (Xu et al., 2011) was useful, as well as the intensity-based features (Xylas et al., 2010), which were shown to be able to characterize cell migration. For the evaluation of stem cell-based therapies in regenerative medicine, the fluorescence intensity features and cell number were applied to assess the cell fate (Polzer et al., 2010). The spatial-temporal features were required for the tracking of stem cells *in vitro* (Liu et al., 2010) but were difficult to quantify *in vivo* due to the limited image resolution. Hence, the bioluminescence signal intensity was used as an alternative feature that was evaluated in detail in van der Bogt et al. (2008). A good review of image analysis method for *in vivo* stem cell tracking is provided in Rabbani and Javanmard (2011).

13.5.2 Characterization of Angiogenesis

To maintain the long-term function of tissue constructs after implantation, sufficient vascularization is required to support the growth of newly formed tissue. Various *in vitro* and *in vivo* angiogenesis assays have been developed that lead to the need for quantitative analysis of the growth of new blood vessels. Manual tracing of the vessels can be done in microscopic images using commercialized image analysis software but the procedure is tedious and time-consuming.

Several semiautomated and automated approaches have been reported to quantify angiogenesis both *in vitro* (Khoo et al., 2011, Niemisto et al., 2005, Rytlewski et al., 2012) and *in vivo* (Hegen et al., 2011, Kagadis et al., 2008, Seaman et al., 2011). Thresholding-based methods were widely used in these studies for vessel segmentation as clear intensity differences were observed between the vessels and the background in microscopic (Hegen et al., 2011, Rytlewski et al., 2012, Seaman et al., 2011), phase-contrast (Khoo et al., 2011), and digital subtraction angiography (DSA) (Kagadis et al., 2008) images. While the global thresholding with a fixed threshold worked fine in some studies (Hegen et al., 2011, Khoo et al., 2011, Seaman et al., 2011), the hysteresis thresholding method based on two thresholds that were generated automatically (Niemisto et al., 2005) and the histogram-based thresholding on multiscale structure tensors enhanced images (Kagadis et al., 2008) were found useful in other reports. A more complicated segmentation method was presented for 3D images (Rytlewski et al., 2012) using the level set approach initialized with manually selected seeds. The surface of the segmented vessels was further smoothed by geodesic active contour evolution to construct the 3D vessel network.

The centerlines or skeletons of the vessels were generated after segmentation for feature extraction. The common features to characterize the vessel network include the number of the vessel, the diameter of the vessel, the length of the vessel, the number of junctions, and the area/volume of the vessels, although named differently in various studies (Hegen et al., 2011, Kagadis et al., 2008, Khoo et al., 2011, Niemisto et al., 2005, Rytlewski et al., 2012, Seaman et al., 2011). More features were available for certain application particularly, for instance, the number of loops, the mean perimeter loop, and the number of nets (Khoo et al., 2011) as well as the fractal dimension (Seaman et al., 2011). These features were found to characterize vessel networks accurately and proved to be useful vascular parameters for the evaluation of bioengineered tissues. It is worth mentioning the differences between 2D features with their 3D counterparts indicated in a recent study by Rytlewski et al. (2012), which revealed the importance of 3D morphological quantification when 3D imaging was feasible.

13.5.3 Characterization of Scaffold Architecture

The use of scaffolds plays an important role in tissue engineering, which acts as a template for attachment and migration of the cell, delivery of biochemical factors, and maintenance of mechanical influences for cell proliferation and differentiation. The quantitative characterization of scaffold architecture is receiving increased attention to better understand its effects on cell morphology, tissue growth, and material mechanical behavior both *in vitro* and *in vivo*.

The first step of characterization of scaffold architecture is the accurate segmentation of images into solid and porous spaces. Twelve thresholding-based segmentation methods were compared by Rajagopalan et al. (2005) for micro-CT images of the porous scaffold. Their results revealed that the local thresholding methods were more robust than global thresholding techniques and the further quantification of features such as porosity was sensitive to the segmentation methods. They suggested that the indicator kriging method (Oh and Lindquist, 1999) and the Mardia-Hainsworth technique (Mardia and Hainsworth, 1988) performed best by visual assessment. A similar study was reported by Brun et al. (2011), where seven segmentation methods were evaluated on micro-CT images from bone tissue engineering scaffolds and the Kittler and Illingworth thresholding (Kittler and Illingworth, 1986) method was found the best. More studies simply adopted a global thresholding method with a fixed threshold for micro-CT (Chiang et al., 2006, 2008, Jones et al., 2009, Mather et al., 2008, Rajagopalan et al., 2005), SEM (Guarino et al., 2010,

Mather et al., 2008), and multiphoton (Liu et al., 2007) images. Additional watershed methods might be required to separate the touching porous spaces (Mather et al., 2008). For fiber segmentation in electrospun fibrous scaffolds, both global thresholding methods (Hu et al., 2008, D'Amore et al., 2010) and edge-based methods (Facco and Tomba, 2010, Vatankhah et al., 2012) were used.

The most widely used features to characterize porous spaces are pore size/diameter, porosity, surface area, and the density of the pores (Brun et al., 2011, Chiang et al., 2008, D'Amore et al., 2010, Guarino et al., 2010, Jones et al., 2009, Liu et al., 2007, Mather et al., 2008, Rajagopalan et al., 2005, Vatankhah et al., 2012). Pore size/diameter can be computed as the average area/diameter of all porous spaces and porosity is the fraction of total porous space area within the image. These features are critical to characterize the ability of cell mass delivery in the scaffold for tissue repair. Surface area is defined as the ratio of the total number of boundary pixels/voxels of the porous spaces to the total image size/volume, which is the important feature to reflect the capability of a scaffold to enable cell attachment and growth. Owing to the irregular shape of the porous spaces, more features can be investigated from the skeleton representation of these areas. These features include interconnectivity (Brun et al., 2011, D'Amore et al., 2010, Jones et al., 2009), which is given as the number of skeleton branch points per porous space and the cross-sectional area at the branch points (throat size) (Brun et al., 2011). A set of structural anisotropy measurements was proposed in Chiang et al. (2006) on unsegmented images to characterize the local variation of the orientation in the highly porous media using the mean intercept length (MIL) method (Harrigan and Mann, 1984). The fractal dimension of the porous space was also adopted to characterize the interconnection degree of scaffolds based on the box counting method (Guarino et al., 2010). As for the fiber characterization in fibrous media, the diameter and orientation of the fibers were well studied (D'Amore et al., 2010, Facco and Tomba, 2010, Hu et al., 2008, Vatankhah et al., 2012). Other useful fiber features are the number and density of the fiber intersections (D'Amore et al., 2010) as well as texture measurements such as energy, entropy, standard deviation, skewness, and kurtosis of the coefficients after wavelet transform of the images (Facco and Tomba, 2010).

13.5.4 Characterization of Tissue Constructs

Evaluation of the structural properties before and after implantation is the key step to develop a useful engineered tissue construct. Image analysis methods can greatly contribute to such an assessment, by providing quantitative measures with no observer discrepancies. Examples of these methods are shown in the following, which cover the applications in cartilage and bone tissue engineering.

13.5.5 Cartilage Tissue

Magnetic resonance imaging (MRI) is the most widely used imaging modality in cartilage tissue engineering, especially for *in vivo* applications. A wide range of segmentation methods have been applied to MRI images of cartilage tissue. Thresholding-based methods were found useful in several studies (Cashman et al., 2002, Cohen et al., 1999, Shim et al., 2009) but the deformable model-based methods (Tejos et al., 2004) attracted more attention recently because these methods are able to incorporate prior knowledge of cartilage shape easily and provide more accurate contour of the tissue. A detailed review of these segmentation methods is provided (Swamy and Holi, 2012). The engineered cartilage tissue construct can also be imaged by optical microscopy (Hagiwara et al., 2011, Stok and M Ller, 2009) for *in vitro* studies and thresholding-based methods were used to segment cartilage tissue in most cases. A list of important structural features of articular cartilage is provided in Hunziker et al. (2002), which can be potentially applied in image-based feature extraction. The commonly used features among the list are the thickness of the cartilage layer and the cartilage volume (Swamy and Holi, 2012, Stok et al., 2010, Stok and M Ller, 2009). The intensity-based features, such as the autofluorescence intensity (Hagiwara et al., 2011) and the fluorescence lifetime and decay constants (Sun et al., 2012) were reported to be useful to correlate with extracellular matrix (ECM) composition and tissue microstructure.

13.5.6 Bone Tissue

X-ray and micro-CT imaging techniques are most suitable image modalities to investigate the structure of tissue-engineered bones (Mathieu et al., 2006, Mauney et al., 2007, Zakaria et al., 2012). Global thresholding remains the most commonly used method to identify bone phase and marrow phase from the image (Hofmann et al., 2007, Lopez-Heredia et al., 2012, Stok et al., 2010, van Lenthe et al., 2007). However, as reported by Hara et al. (2002), the segmentation results were highly sensitive to the threshold while the variation of bone volume quantified could be as high as 15%. To improve the segmentation, several approaches have been proposed, such as the dual threshold method (Buie et al., 2007), the edge-based method (Polak et al., 2011), and deformable models (Demenegas et al., 2011). Thresholding-based segmentation methods were also used in SEM (Lopez-Heredia et al., 2012) or confocal microscopy images (Stok et al., 2010) of the cyrosection of the engineered bone tissue. Volume, volume density, and surface density of newly formed bone are important parameters to characterize bone tissue growth, which can be quantified from both 2D and 3D images (Mathieu et al., 2006, van Lenthe et al., 2007, Zakaria et al., 2012). With the development of 3D image processing techniques, more features such as trabecular thickness, trabecular separation, and trabecular number (Hildebrand et al., 1999, Odgaard, 1997) were used to characterize the microstructure of engineered bone tissues. Besides, the Eular numbers in the 3D image were able to reveal the connectivity of the tissue (Odgaard, 1997) and the topological features could be generated from the 3D skeleton of the tissue (Pothuaud et al., 2000).

13.6 Image Analysis Software

In biomedical research, the functions of image analysis software includes cell/tissue segmentation, cell/particle tracking, 3D reconstruction, and visualization of cellular/subcellular/tissue-level structures, image preprocessing (image stitching, image registration, deconvolution), and image postprocessing (feature extraction, and statistics and patter recognition). This section introduces some of the popular image analysis software and their main characteristics.

Some of the commercialized image analysis software integrate image acquisition functions and are compatible with certain types of image acquisition devices. This software can do imaging system control, image acquisition, and image analysis together. This software include MetaMorph (Molecular Devices, California, USA), SlideBook (Intelligent Imaging Innovations, Gottingen, Germany), Image-Pro (MediaCybernetics, Rockville, USA), and Volocity (Perkin-Elmer, Massachusetts, USA). Imaging device companies also develop their own image analysis software, such as AxioVision (Ziess, Jena, Germany), NIS-Elements (Nikon, Melville, USA), and cellSens (Olympus, Tokyo, Japan). This software provides its own image acquisition, processing, and analysis functions and has great compatibility with its imaging instruments. However, an important shortcoming of this software is that there are very limited opportunities for users to customize or extend its function.

Another group of image analysis software is particularly designed for specific application. Huygens Software (Scientific Volume Imaging, Hilversum, the Netherlands) and Imaris (Bitplane Scientific Software, Zurich, Switzerland) are two typical examples. Huygens is famous for its deconvolution ability. With fast algorithms and smart memory allocation schemes, Huygens can do powerful deconvolution for microscopic images. Imaris is well known for its powerful live cell imaging ability. One of its components, ImarisTrack, applies many sophisticated manual and automatic tracking algorithms to capture and analyze biological signals in a dynamic cellular environment.

Among medical image analysis software, OsiriX (Rosset et al., 2004) and 3Dslicer (Pieper et al., 2004) are widely used. OsiriX is an open-source image processing software dedicated to images produced by medical imaging devices such as MRI, CT, PET, PET-CT, SPECT-CT, and ultrasounds. Its multidimensional visualization module offers all types of modern rendering modes. 3D Slicer is another open-source software package offering multimodality imaging functions, including MRI, CT, and microscopy images. It provides imaging analysis modules for multiple organs.

TABLE 13.3 Summary of Image Analysis Software Used in Cellular and Tissue Engineering Applications

Application	Commercialized/Open-Source Software	Customized Software
Angiogenesis	ImageJ (Rytlewski et al., 2012), 3D Slicer (Rytlewski et al., 2012), Photoshop (Khoo et al., 2011), Angiosys (Khoo et al., 2011), Wimasis (Khoo et al., 2011), Imaris (Hegen et al., 2011)	MATLAB (Seaman et al., 2011, Kagadis et al., 2008)
Cell	ImageJ (Thevenot et al., 2008, Xu et al., 2011, Polzer et al., 2010)	MATLAB (Sommerhage et al., 2008)
Scaffold	ImageJ (Guarino et al., 2010, Liu et al., 2007)	MATLAB (Mather et al., 2008, Hu et al., 2008, D'Amore et al., 2010, Vatankhah et al., 2012)
Bone	Leica Qwin (Lopez-Heredia et al., 2012), Imaris (Hofmann et al., 2007), ImagePro (Panseri et al., 2012), CtAnalyser (Demenegas et al., 2011)	C+ (Buie et al., 2007)
Cartilage	ImageJ (Spangenberg et al., 2002)	MATLAB (Stok and M Ller, 2009)
Skin	Photoshop (McMullen et al., 2010), ImagePro (McMullen et al., 2010)	
Lung	Scion Image (Agarwal et al., 2001)	
Kidney	ImageJ (Frohlich et al., 2012), Cell Profiler (Frohlich et al., 2012)	

Besides this commercialized software, open-source software is being developed by different research groups. This software is maintained and updated frequently, and contains state-of-the-art image analysis algorithms. They are designed either for general image analysis purposes, such as ImageJ (Rasband, 1997), BioImageXD (Kankaanpää et al., 2012), and CellProfiler (Carpenter et al., 2006), or to analyze certain types of tissue or cell images. More information can be found at their respective websites.

We summarize the usage of image analysis software in cellular and tissue engineering applications in Table 13.3. ImageJ is most popular in different applications due to its abundant plug-ins that are well developed for all kinds of applications. It also provides the user certain freedom to customize and extend image analysis functions if needed. Programming software is available to build fully customized image analysis functions, among which MATLAB® (MathWorks, MA, USA) is most widely used.

13.7 Case Study

In this section, we show an example of using image analysis methods to quantify collagen network changes along the fibrosis progression from second harmonic generation (SHG) microscopy images of liver capsule at the surface. Although liver biopsy remains the gold standard for fibrosis assessment, it has several disadvantages such as being invasive and having high sampling error. We hypothesize that the less invasive liver surface imaging by endoscope is potentially an alternative way to monitor fibrosis progression and to evaluate drug efficacy. Hence, it is important to demonstrate whether nonlinear optical imaging of the liver surface capsule would yield sufficient information to allow quantitative staging of fibrosis. To achieve this, we have established a capsule index based on significant features extracted from the capsule images through various image analysis methods, including segmentation, feature extraction, feature selection, and LR, which are elaborated in the previous sections.

The flowchart of the proposed image analysis system is shown in Figure 13.2. As we see, the original SHG image suffered uneven illumination due to out-of-flatness of the sample surface. Therefore, preprocessing is necessary to restore the image quality. A Gaussian mixture model-based segmentation method, which was shown to be more accurate than other methods such as global thresholding and clustering methods for collagen segmentation (Xu et al., 2010), was applied next to identify collagen areas. The centerline of each collagen segment after segmentation was traced after skeletonization (Stein et al., 2008). Sixteen morphological features and 108 texture features were extracted as listed in Figure 13.2. Eleven features were selected using a SVM-RFE method. Each of these features was selected at least

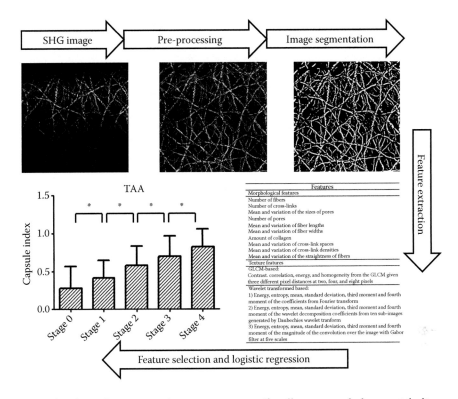

FIGURE 13.2 Flowchart of an image analysis system to quantify collagen network changes at the liver surface.

75 times from 100 bootstrap samples. We finally trained a multinomial LR model based on selected features to estimate the probability of each stage for certain samples and normalized the probabilities into the capsule index. The results indicated that capsule index was able to differentiate between different fibrosis stages for the thioacetomide (TAA)-induced rat fibrosis model. By incorporating capsule index quantification with SHG endoscopes, we see the feasibility of liver fibrosis diagnosis from the liver surface, which would therefore give us the potential to image a larger sampling area than from biopsy and extract enough information over a long period of time without the complications brought forth by the invasive biopsy.

13.8 Conclusion and Discussion

So far the majority of the image analysis applications in cellular and tissue engineering need user interactions such as highlighting the contour of the objects and deciding the threshold for segmentation. The main reason is that most of the research in cellular and tissue engineering thus far is low throughput, either with a limited number of images or with a limited number of objects in one image to be analyzed. However, with the demand to better understand the tissue dynamics and to integrate tissue-engineered constructs in living hosts at higher spatial and temporal resolution, high throughput and high content imaging modalities are becoming more and more important in the field to bridge the size and depth limits of the current imaging modalities. Besides, 3D and temporal images are generated to provide more accurate illustrations of the cell and tissue structures than conventional 2D images. These factors significantly raise the complexity of analyzing the images and require analysis consistency among large numbers of images, which makes manual or semiautomated image analysis methods inappropriate.

As we have illustrated in this chapter, automated image analysis methods have been applied in different studies for cellular and tissue engineering research, from the characterization of cell behaviors, scaffold architectures, to tissue construct structures. The research in these studies has benefited from quantitative measurements generated from image analysis. However, the challenges in developing more accurate, robust, and fast image analysis methods in the more complicated cell and tissue context still exist. For example, thresholding-based methods are most widely adopted in these applications for segmentation purposes. Although such methods are easy to implement and are provided in most of the commercial software, they are highly sensitive to intensity variations due to imaging conditions and are incapable of recognizing single objects in the intensive environment. Deformable models-based methods have the advantage of being able to handle these issues; however, they are difficult to perform and are not available in most commercial image analysis software. Customized programming may be needed for these methods, which is hard for tissue engineers who are not professionals in the image analysis field. This requires more communication between researchers in both fields. Image analysis algorithm development is application driven for cellular and tissue engineering, which needs the objects and features to be well defined by tissue engineers in advance so that computer scientists can develop image analysis methods properly. As seen above, several studies have reported their own image analysis software using MATLAB for certain cellular and tissue engineering applications, and it might also be important to encourage the sharing of this software between the research groups in the field.

References

Adams, R. and Bischof, L. 1994. Seeded region growing. *IEEE Transactions on Pattern Analysis and Machine Intelligence,* 16, 641–647.

Agarwal, A., Coleno, M. L., Wallace, V. P. et al. 2001. Two-photon laser scanning microscopy of epithelial cell-modulated collagen density in engineered human lung tissue. *Tissue Engineering,* 7, 191–202.

Akay, M. F. 2009. Support vector machines combined with feature selection for breast cancer diagnosis. *Expert Systems with Applications,* 36, 3240–3247.

AL-Diri, B., Hunter, A. and Steel, D. 2009. An active contour model for segmenting and measuring retinal vessels. *IEEE Transactions on Medical Imaging,* 28, 1488–1497.

Ananthanarayanan, A., Narmada, B. C., Mo, X., Mcmillian, M. and Yu, H. 2011. Purpose-driven biomaterials research in liver-tissue engineering. *Trends in Biotechnology,* 29, 110–118.

Azencott, R., Wang, J. P. and Younes, L. 1997. Texture classification using windowed Fourier filters. *IEEE Transactions on Pattern Analysis and Machine Intelligence,* 19, 148–153.

Bezdek, J. C. and Ehrlich, R. 1984. FCM: The fuzzy c-means clustering algorithm. *Computers & Geosciences,* 10, 191–203.

Bishop, C. M. 1996. *Neural Networks for Pattern Recognition,* Oxford University Press, New York, NY.

Boland, M. V. and Murphy, R. F. 2001. A neural network classifier capable of recognizing the patterns of all major subcellular structures in fluorescence microscope images of HeLa cells. *Bioinformatics,* 17, 1213–1223.

Breiman, L. 2001. Random forests. *Machine Learning,* 45, 5–32.

Brun, F., Turco, G., Accardo, A. and Paoletti, S. 2011. Automated quantitative characterization of alginate/hydroxyapatite bone tissue engineering scaffolds by means of micro-CT image analysis. *Journal of Materials Science: Materials in Medicine,* 22(12), 2617–2629.

Buie, H. R., Campbell, G. M., Klinck, R. J., Macneil, J. A. and Boyd, S. K. 2007. Automatic segmentation of cortical and trabecular compartments based on a dual threshold technique for *in vivo* micro-CT bone analysis. *Bone,* 41, 505–515.

Canny, J. 1986. A computational approach to edge detection. *IEEE Transactions on Pattern Analysis and Machine Intelligence,* 6(8), 679–698.

Carpenter, A. E., Jones, T. R., Lamprecht, M. R. et al. 2006. CellProfiler: Image analysis software for identifying and quantifying cell phenotypes. *Genome Biology,* 7, R100.

Cashman, P. M. M., Kitney, R. I., Gariba, M. A. and Carter, M. E. 2002. Automated techniques for visualization and mapping of articular cartilage in MR images of the osteoarthritic knee: A base technique for the assessment of microdamage and submicro damage. *IEEE Transactions on NanoBioscience*, 1, 42–51.

Castellano, G., Bonilha, L., Li, L. and Cendes, F. 2004. Texture analysis of medical images. *Clinical Radiology*, 59, 1061–1069.

Chan, T. F. and Vese, L. A. 2001. Active contours without edges. *IEEE Transactions on Image Processing*, 10, 266–277.

Chang, C. C. and Lin, C. J. 2011. LIBSVM: A library for support vector machines. *ACM Transactions on Intelligent Systems and Technology (TIST)*, 2, 27.

Chaudhuri, B. and Sarkar, N. 1995. Texture segmentation using fractal dimension. *IEEE Transactions on Pattern Analysis and Machine Intelligence*, 17, 72–77.

Chen, C. C., Daponte, J. S. and Fox, M. D. 1989. Fractal feature analysis and classification in medical imaging. *IEEE Transactions on Medical Imaging*, 8, 133–142.

Chen, D., Liu, Z., Ma, X. and Hua, D. 2005. Selecting genes by test statistics. *Journal of Biomedicine and Biotechnology*, 2, 132.

Chen, S. and Zhang, D. 2004. Robust image segmentation using FCM with spatial constraints based on new kernel-induced distance measure. *IEEE Transactions on Systems, Man, and Cybernetics, Part B: Cybernetics*, 34, 1907–1916.

Chen, S., Zhao, M., Wu, G., Yao, C. and Zhang, J. 2012. Recent advances in morphological cell image analysis. *Computational and Mathematical Methods in Medicine*, 2012, 101536.

Chiang, M. Y. M., Landis, F. A., Wang, X. et al. 2008. Local thickness and anisotropy approaches to characterize pore size distribution of three-dimensional porous networks. *Tissue Engineering Part C: Methods*, 15, 65–76.

Chiang, M. Y. M., Wang, X., Landis, F. A., Dunkers, J. and Snyder, C. R. 2006. Quantifying the directional parameter of structural anisotropy in porous media. *Tissue Engineering*, 12, 1597–1606.

Cohen, Z. A., Mccarthy, D. M., Kwak, S. D. et al. 1999. Knee cartilage topography, thickness, and contact areas from MRI: In-vitro calibration and in-vivo measurements. *Osteoarthritis and Cartilage*, 7, 95–109.

Costes, S. V., Daelemans, D., Cho, E. H., Dobbin, Z., Pavlakis, G. and Lockett, S. 2004. Automatic and quantitative measurement of protein-protein colocalization in live cells. *Biophysical Journal*, 86, 3993–4003.

Cover, T. and Hart, P. 1967. Nearest neighbor pattern classification. *IEEE Transactions on Information Theory*, 13, 21–27.

Cuijpers, V. M. J. I., Walboomers, X. F. and Jansen, J. A. 2011. Scanning electron microscopy stereoimaging for three-dimensional visualization and analysis of cells in tissue-engineered constructs: Technical note. *Tissue Engineering Part C: Methods*, 17, 663–668.

Díaz-Uriarte, R. and de Andres, S. A. 2006. Gene selection and classification of microarray data using random forest. *BMC Bioinformatics*, 7, 3.

D'Amore, A., Stella, J. A., Wagner, W. R. and Sacks, M. S. 2010. Characterization of the complete fiber network topology of planar fibrous tissues and scaffolds. *Biomaterials*, 31, 5345–5354.

Day, N. E. 1969. Estimating the components of a mixture of normal distributions. *Biometrika*, 56, 463–474.

De Berg, M., Cheong, O. and Van Kreveld, M. 2008. *Computational Geometry: Algorithms and Applications*, Springer-Verlag, New York.

Demenegas, F., Tassani, S. and Matsopoulos, G. K. 2011. Active segmentation of micro-CT trabecular bone images. Biomedical Engineering, 10th International Workshop on, IEEE, 1–4.

Demir, C. and Yener, B. 2005. Automated cancer diagnosis based on histopathological images: A systematic survey. Technical Report, Rensselaer Polytechnic Institute.

Dempster, A. P., Laird, N. M. and Rubin, D. B. 1977. Maximum likelihood from incomplete data via the EM algorithm. *Journal of the Royal Statistical Society. Series B (Methodological)*, 39(1), 1–38.

Duan, K. B., Rajapakse, J. C., Wang, H. and Azuaje, F. 2005. Multiple Svm-Rfe for gene selection in cancer classification with expression data. *IEEE Transactions on NanoBioscience*, 4, 228–234.

Dudoit, S., Fridlyand, J. and Speed, T. P. 2002. Comparison of discrimination methods for the classification of tumors using gene expression data. *Journal of the American Statistical Association*, 97, 77–87.

Dufour, A., Shinin, V., Tajbakhsh, S., Guill N-Aghion, N., Olivo-Marin, J. C. and Zimmer, C. 2005. Segmenting and tracking fluorescent cells in dynamic 3-D microscopy with coupled active surfaces. *IEEE Transactions on Image Processing*, 14, 1396–1410.

Egmont-Petersen, M., De Ridder, D. and Handels, H. 2002. Image processing with neural networks—A review. *Pattern Recognition*, 35, 2279–2301.

Facco, P. and Tomba, E. 2010. Soft sensor for the characterization of fibrous materials through multivariate and multiresolution image analysis. Sensors, IEEE, 1319–1324.

Fan, J. and Lv, J. 2010. A selective overview of variable selection in high dimensional feature space. *Statistica Sinica*, 20, 101.

Frohlich, E. M., Zhang, X. and Charest, J. L. 2012. The use of controlled surface topography and flow-induced shear stress to influence renal epithelial cell function. *Integrative Biology*, 4, 75–83.

Gagnepain, J. and Roques-Carmes, C. 1986. Fractal approach to two-dimensional and three-dimensional surface roughness. *Wear*, 109, 119–126.

Galloway, M. M. 1975. Texture analysis using gray level run lengths. *Computer Graphics and Image Processing*, 4, 172–179.

Gonzales, R. C. and Woods, R. E. 2002. *Digital Image Processing*, Prentice Hall Press, Upper Saddle River, NJ, ISBN: 0-201-18075-8.

Groeneveld, R. A. and Meeden, G. 1984. Measuring skewness and kurtosis. *Journal of the Royal Statistical Society. Series D (The Statistician)*, 33(4), 391–399.

Guarino, V., Guaccio, A., Netti, P. A. and Ambrosio, L. 2010. Image processing and fractal box counting: User-assisted method for multi-scale porous scaffold characterization. *Journal of Materials Science: Materials in Medicine*, 21, 3109–3118.

Guo, R. and Pandit, S. 1998. Automatic threshold selection based on histogram modes and a discriminant criterion. *Machine Vision and Applications*, 10, 331–338.

Gurcan, M. N., Boucheron, L. E., Can, A., Madabhushi, A., Rajpoot, N. M. and Yener, B. 2009. Histopathological image analysis: A review. *Biomedical Engineering, IEEE Reviews in*, 2, 147–171.

Guyon, I. and Elisseeff, A. 2003. An introduction to variable and feature selection. *The Journal of Machine Learning Research*, 3, 1157–1182.

Guyon, I., Weston, J., Barnhill, S. and Vapnik, V. 2002. Gene selection for cancer classification using support vector machines. *Machine Learning*, 46, 389–422.

Hagiwara, Y., Hattori, K., Aoki, T., Ohgushi, H. and Ito, H. 2011. Autofluorescence assessment of extracellular matrices of a cartilage-like tissue construct using a fluorescent image analyser. *Journal of Tissue Engineering and Regenerative Medicine*, 5, 163–168.

Hara, T., Tanck, E., Homminga, J. and Huiskes, R. 2002. The influence of microcomputed tomography threshold variations on the assessment of structural and mechanical trabecular bone properties. *Bone*, 31, 107–109.

Haralick, R. M., Shanmugam, K. and Dinstein, I. H. 1973. Textural features for image classification. *IEEE Transactions on Systems, Man and Cybernetics*, 3, 610–621.

Harrigan, T. and Mann, R. 1984. Characterization of microstructural anisotropy in orthotropic materials using a second rank tensor. *Journal of Materials Science*, 19, 761–767.

Hartigan, J. A. and Wong, M. A. 1979. Algorithm AS 136: A k-means clustering algorithm. *Journal of the Royal Statistical Society. Series C (Applied Statistics)*, 28, 100–108.

Hegen, A., Blois, A., Tiron, C. E. et al. 2011. Efficient *in vivo* vascularization of tissue-engineering scaffolds. *Journal of Tissue Engineering and Regenerative Medicine*, 5, e52–e62.

Hildebrand, T., Laib, A., M Ller, R., Dequeker, J. and R Egsegger, P. 1999. Direct three-dimensional morphometric analysis of human cancellous bone: Microstructural data from spine, femur, iliac crest, and calcaneus. *Journal of Bone and Mineral Research,* 14, 1167–1174.

Hofmann, S., Hagenm Ller, H., Koch, A. M. et al. 2007. Control of *in vitro* tissue-engineered bone-like structures using human mesenchymal stem cells and porous silk scaffolds. *Biomaterials,* 28, 1152–1162.

Hu, J. J., Humphrey, J. D. and Yeh, A. T. 2008. Characterization of engineered tissue development under biaxial stretch using nonlinear optical microscopy. *Tissue Engineering Part A,* 15, 1553–1564.

Hunziker, E. B., Quinn, T. M. and H Uselmann, H. J. 2002. Quantitative structural organization of normal adult human articular cartilage. *Osteoarthritis and Cartilage,* 10, 564–572.

Inza, I., Larranaga, P., Blanco, R. and Cerrolaza, A. J. 2004. Filter versus wrapper gene selection approaches in DNA microarray domains. *Artificial Intelligence in Medicine,* 31, 91–103.

Jaskolski, F., Mulle, C. and Manzoni, O. J. 2005. An automated method to quantify and visualize colocalized fluorescent signals. *Journal of Neuroscience Methods,* 146, 42–49.

Jones, J. R., Atwood, R. C., Poologasundarampillai, G., Yue, S. and Lee, P. D. 2009. Quantifying the 3D macrostructure of tissue scaffolds. *Journal of Materials Science: Materials in Medicine,* 20, 463–471.

Kagadis, G. C., Spyridonos, P., Karnabatidis, D. et al. 2008. Computerized analysis of digital subtraction angiography: A tool for quantitative in-vivo vascular imaging. *Journal of Digital Imaging,* 21, 433–445.

Kagami, H., Agata, H., Kato, R., Matsuoka, F. and Tojo, A. 2011. Fundamental technological developments required for increased availability of tissue engineering. In: *Regenerative Medicine and Tissue Engineering—Cells and Biomaterials,* Daniel Eberli (Ed.), InTech, ISBN: 978-953-307-663-8, InTech, DOI: 10.5772/21137.

Kankaanpää, P., Paavolainen, L., Tiitta, S. et al. 2012. BioImageXD: An open, general-purpose and high-throughput image-processing platform. *Nature Methods,* 9, 683–689.

Kass, M., Witkin, A. and Terzopoulos, D. 1988. Snakes: Active contour models. *International Journal of Computer Vision,* 1, 321–331.

Khoo, C. P., Micklem, K. and Watt, S. M. 2011. A comparison of methods for quantifying angiogenesis in the matrigel assay in vitro. *Tissue Engineering Part C: Methods,* 17, 895–906.

Kittler, J. and Illingworth, J. 1986. Minimum error thresholding. *Pattern Recognition,* 19, 41–47.

Kohavi, R. and John, G. H. 1997. Wrappers for feature subset selection. *Artificial Intelligence,* 97, 273–324.

Kotropoulos, C. and Pitas, I. 2003. Segmentation of ultrasonic images using support vector machines. *Pattern Recognition Letters,* 24, 715–727.

Kulkarni, N. 2012. Color thresholding method for image segmentation of natural images. *International Journal of Image, Graphics and Signal Processing,* 4, 28–34.

Langer, R. and Vacanti, J. P. 1993. Tissue engineering. *Science,* 260, 920–926.

Li, Q., Lau, A., Morris, T. J., Guo, L., Fordyce, C. B. and Stanley, E. F. 2004. A syntaxin 1, Gαo, and N-type calcium channel complex at a presynaptic nerve terminal: Analysis by quantitative immunocolocalization. *The Journal of Neuroscience,* 24, 4070–4081.

Liaw, A. and Wiener, M. 2002. Classification and regression by RandomForest. *R News,* 2, 18–22.

Liu, E., Treiser, M. D., Johnson, P. A. et al. 2007. Quantitative biorelevant profiling of material microstructure within 3D porous scaffolds via multiphoton fluorescence microscopy. *Journal of Biomedical Materials Research Part B: Applied Biomaterials,* 82, 284–297.

Liu, M., Yadav, R. K., Roy-Chowdhury, A. and Reddy, G. V. 2010. Automated tracking of stem cell lineages of Arabidopsis shoot apex using local graph matching. *The Plant Journal,* 62, 135–147.

Lloyd, S. 1982. Least squares quantization in PCM. *IEEE Transactions on Information Theory,* 28, 129–137.

Lopez-Heredia, M. A., Bongio, M., Cuijpers, V. M. J. I. et al. 2012. Bone formation analysis: Effect of quantification procedures on the study outcome. *Tissue Engineering Part C: Methods,* 18(5), 369–73.

Ma, Z., Tavares, J. M., Jorge, R. N. and Mascarenhas, T. 2010. A review of algorithms for medical image segmentation and their applications to the female pelvic cavity. *Computer Methods in Biomechanics and Biomedical Engineering,* 13, 235–246.

Manders, E., Verbeek, F. and Aten, J. 1993. Measurement of co-localization of objects in dual-colour confocal images. *Journal of Microscopy,* 169, 375–382.

Manjunath, B. S. and Ma, W. Y. 1996. Texture features for browsing and retrieval of image data. *IEEE Transactions on Pattern Analysis and Machine Intelligence,* 18, 837–842.

Mardia, K. V. and Hainsworth, T. 1988. A spatial thresholding method for image segmentation. *IEEE Transactions on Pattern Analysis and Machine Intelligence,* 10, 919–927.

Mather, M. L., Morgan, S. P., White, L. J. et al. 2008. Image-based characterization of foamed polymeric tissue scaffolds. *Biomedical Materials,* 3, 015011.

Mathieu, L. M., Mueller, T. L., Bourban, P. E., Pioletti, D. P., M Ller, R. and M Nson, J. A. E. 2006. Architecture and properties of anisotropic polymer composite scaffolds for bone tissue engineering. *Biomaterials,* 27, 905–916.

Mauney, J. R., Nguyen, T., Gillen, K., Kirker-Head, C., Gimble, J. M. and Kaplan, D. L. 2007. Engineering adipose-like tissue *in vitro* and *in vivo* utilizing human bone marrow and adipose-derived mesenchymal stem cells with silk fibroin 3D scaffolds. *Biomaterials,* 28, 5280–5290.

Mcmullen, R., Bauza, E., Gondran, C. et al. 2010. Image analysis to quantify histological and immunofluorescent staining of ex vivo skin and skin cell cultures. *International Journal of Cosmetic Science,* 32, 143–154.

Mcinerney, T. and Terzopoulos, D. 1996. Deformable models in medical image analysis: A survey. *Medical Image Analysis,* 1, 91–108.

Medina-Carnicer, R., Carmona-Poyato, A., Mu Oz-Salinas, R. and Madrid-Cuevas, F. 2010. Determining hysteresis thresholds for edge detection by combining the advantages and disadvantages of thresholding methods. *IEEE Transactions on Image Processing,* 19, 165–173.

Meier, L., Van De Geer, S. and B Hlmann, P. 2008. The group lasso for logistic regression. *Journal of the Royal Statistical Society: Series B (Statistical Methodology),* 70, 53–71.

Meijering, E., Dzyubachyk, O., Smal, I. and Van Cappellen, W. A. 2009. Tracking in cell and developmental biology. *Seminars in Cell & Developmental Biology,* 20, 894–902.

Mundra, P. A. and Rajapakse, J. C. 2010. SVM-RFE with MRMR filter for gene selection. *IEEE Transactions on NanoBioscience,* 9, 31–37.

Niemisto, A., Dunmire, V., Yli-Harja, O., Zhang, W. and Shmulevich, I. 2005. Robust quantification of *in vitro* angiogenesis through image analysis. *IEEE Transactions on Medical Imaging,* 24, 549–553.

Niessen, W. J., Romeny, B. and Viergever, M. A. 1998. Geodesic deformable models for medical image analysis. *IEEE Transactions on Medical Imaging,* 17, 634–641.

Odgaard, A. 1997. Three-dimensional methods for quantification of cancellous bone architecture. *Bone,* 20, 315–328.

Oh, W. and Lindquist, B. 1999. Image thresholding by indicator kriging. *IEEE Transactions on Pattern Analysis and Machine Intelligence,* 21, 590–602.

Okabe, A., Boots, B. N., Sugihara, K. and Chiu, S. N. 1992. *Spatial Tessellations: Concepts and Applications of Voronoi Diagrams,* Wiley & Sons, Chichester.

Osher, S. and Sethian, J. A. 1988. Fronts propagating with curvature-dependent speed: Algorithms based on Hamilton-Jacobi formulations. *Journal of Computational Physics,* 79, 12–49.

Otsu, N. 1975. A threshold selection method from gray-level histograms. *Automatica,* 11, 285–296.

Panseri, S., Cunha, C., D'Alessandro, T. et al. 2012. Magnetic hydroxyapatite bone substitutes to enhance tissue regeneration: Evaluation *in vitro* using osteoblast-like cells and *in vivo* in a bone defect. *PloS One,* 7, e38710.

Pappas, T. N. 1992. An adaptive clustering algorithm for image segmentation. *IEEE Transactions on Signal Processing,* 40, 901–914.

Peleg, S., Naor, J., Hartley, R. and Avnir, D. 1984. Multiple resolution texture analysis and classification. *IEEE Transactions on Pattern Analysis and Machine Intelligence,* 518–523.

Peng, H. 2008. Bioimage informatics: A new area of engineering biology. *Bioinformatics,* 24, 1827–1836.

Peng, H., Long, F. and Ding, C. 2005. Feature selection based on mutual information criteria of max-dependency, max-relevance, and min-redundancy. *IEEE Transactions on Pattern Analysis and Machine Intelligence,* 27, 1226–1238.

Pieper, S., Halle, M., and Kikinis, R 2004. 3D Slicer. 1st IEEE International Symposium on Biomedical Imaging: From Nano to Macro, 632–635.

Polak, S. J., Candido, S., Levengood, S. K. L. and Wagoner Johnson, A. J. 2011. Automated segmentation of micro-CT images of bone formation in calcium phosphate scaffolds. *Computerized Medical Imaging and Graphics.*

Polzer, H., Haasters, F., Prall, W. C. et al. 2010. Quantification of fluorescence intensity of labeled human mesenchymal stem cells and cell counting of unlabeled cells in phase-contrast imaging: An open-source-based algorithm. *Tissue Engineering Part C: Methods,* 16, 1277–1285.

Pothuaud, L., Porion, P., Lespessailles, E., Benhamou, C. and Levitz, P. 2000. A new method for three-dimensional skeleton graph analysis of porous media: Application to trabecular bone microarchitecture. *Journal of Microscopy,* 199, 149–161.

Rabbani, H. and Javanmard, S. H. 2011. Image analysis in *in vivo* stem cell tracking. *Annual Review & Research in Biology,* 1, 123–142.

Rajagopalan, S., Lu, L., Yaszemski, M. J. and Robb, R. A. 2005. Optimal segmentation of microcomputed tomographic images of porous tissue-engineering scaffolds. *Journal of Biomedical Materials Research Part A,* 75, 877–887.

Rasband, W. S. 1997. ImageJ, US National Institutes of Health, Bethesda, Maryland, USA.

Roerdink, J. B. T. M. and Meijster, A. 2000. The watershed transform: Definitions, algorithms and parallelization strategies. *Fundamenta Informaticae,* 41, 187–228.

Rosset, A., Spadola, L. and Ratib, O. 2004. OsiriX: An open-source software for navigating in multidimensional DICOM images. *Journal of Digital Imaging,* 17, 205–216.

Rytlewski, J. A., Geuss, L. R., Anyaeji, C. I., Lewis, E. W. and Suggs, L. J. 2012. Three-dimensional image quantification as a new morphometry method for tissue engineering. *Tissue Engineering Part C: Methods,* 18(7), 507–516.

Schmid, J. and Magnenat-Thalmann, N. 2008. MRI bone segmentation using deformable models and shape priors. *Medical Image Computing and Computer-Assisted Intervention–MICCAI 2008,* 119–126.

Seaman, M. E., Peirce, S. M. and Kelly, K. 2011. Rapid analysis of vessel elements (RAVE): A tool for studying physiologic, pathologic and tumor angiogenesis. *PloS One,* 6, e20807.

Sezan, M. I. 1990. A peak detection algorithm and its application to histogram-based image data reduction. *Computer Vision, Graphics, and Image Processing,* 49, 36–51.

Shaw, M., Faruqui, N., Gurdak, E. and Tomlins, P. 2012. Three-dimensional cell morphometry for the quantification of cell–substrate interactions. *Tissue Engineering Part C: Methods,* 19(1), 48–56.

Shim, H., Chang, S., Tao, C., Wang, J. H., Kwoh, C. K. and Bae, K. T. 2009. Knee cartilage: Efficient and reproducible segmentation on high-spatial-resolution MR images with the semiautomated graph-cut algorithm method 1. *Radiology,* 251, 548–556.

Smith, M. B., Li, H., Shen, T., Huang, X., Yusuf, E. and Vavylonis, D. 2010. Segmentation and tracking of cytoskeletal filaments using open active contours. *Cytoskeleton,* 67, 693–705.

Somasundaram, S. and Alli, P. 2012. A review on recent research and implementation methodologies on medical image segmentation. *Journal of Computer Science,* 8(1), 170–174.

Sommerhage, F., Helpenstein, R., Rauf, A., Wrobel, G., Offenh Usser, A. and Ingebrandt, S. 2008. Membrane allocation profiling: A method to characterize three-dimensional cell shape and attachment based on surface reconstruction. *Biomaterials,* 29, 3927–3935.

Soret, M., Bacharach, S. L. and Buvat, I. 2007. Partial-volume effect in PET tumor imaging. *Journal of Nuclear Medicine,* 48, 932–945.

Spangenberg, K. M., Peretti, G. M., Trahan, C. A., Randolph, M. A. and Bonassar, L. J. 2002. Histomorphometric analysis of a cell-based model of cartilage repair. *Tissue Engineering,* 8, 839–846.

Stein, A. M., Vader, D. A., Jawerth, L. M., Weitz, D. A. and Sander, L. M. 2008. An algorithm for extracting the network geometry of three-dimensional collagen gels. *Journal of microscopy,* 232, 463–475.

Stok, K. S. and M Ller, R. 2009. Morphometric characterization of murine articular cartilage—Novel application of confocal laser scanning microscopy. *Microscopy Research and Technique,* 72, 650–658.

Stok, K. S., No L, D., Apparailly, F. et al. 2010. Quantitative imaging of cartilage and bone for functional assessment of gene therapy approaches in experimental arthritis. *Journal of Tissue Engineering and Regenerative Medicine,* 4, 387–394.

Sun, Y., Responte, D., Xie, H. et al. 2012. Nondestructive evaluation of tissue engineered articular cartilage using time-resolved fluorescence spectroscopy and ultrasound backscatter microscopy. *Tissue Engineering Part C: Methods,* 215–226.

Swamy, M. S. M. and Holi, M. S. 2012. Knee joint articular cartilage segmentation, visualization and quantification using image processing techniques: A review. *International Journal of Computer Applications,* 42(19), 36–43.

Tejos, C., Hall, L. D. and Cardenas-Blanco, A. 2004. Segmentation of articular cartilage using active contours and prior knowledge. Engineering in Medicine and Biology Society, 2004. IEMBS '04. 26th Annual International Conference of the IEEE, IEEE, 1648–1651.

Thevenot, P., Nair, A., Dey, J., Yang, J. and Tang, L. 2008. Method to analyze three-dimensional cell distribution and infiltration in degradable scaffolds. *Tissue Engineering Part C: Methods,* 14, 319–331.

Thilagamani, S. 2011. A survey on image segmentation through clustering. *International Journal of Research and Reviews in Information Sciences,* 1(1), 14–17.

Tuceryan, M. and Jain, A. K. 1993. Texture analysis. In: *Handbook of Pattern Recognition and Computer Vision.* C. H. Chen, L. F. Pau, and P. S. P. Wang (Eds.), World Scientific Publishing Co., River Edge, NJ, pp. 235–276.

Van Der Bogt, K. E. A., Sheikh, A. Y., Schrepfer, S. et al., 2008. Comparison of different adult stem cell types for treatment of myocardial ischemia. *Circulation,* 118, S121–S129.

van Lenthe, G. H., Bohner, M., Hollister, S. J. and Meinel, L. 2007. Nondestructive micro-computed tomography for biological imaging and quantification of scaffold-bone interaction in vivo. *Biomaterials,* 28, 2479–2490.

Vapnik, V. N. 2000. *The Nature of Statistical Learning Theory,* Springer-Verlag, New York.

Vatankhah, E., Semnani, D. and Tadayon, M. 2012. Structural characterization of electrospun scaffolds by image analysis techniques. International Symposium on Biomedical Imaging: From Nano to Macro, IEEE, 1381–1384.

Vrooman, H. A., Cocosco, C. A., Van Der Lijn, F. et al. 2007. Multi-spectral brain tissue segmentation using automatically trained k-nearest-neighbor classification. *Neuroimage,* 37, 71–81.

Wahde, M. and Szallasi, Z. 2006. Improving the prediction of the clinical outcome of breast cancer using evolutionary algorithms. *Soft Computing—A Fusion of Foundations, Methodologies and Applications,* 10, 338–345.

Wang, M., Zhou, X., Li, F., Huckins, J., King, R. W. and Wong, S. T. C. 2008. Novel cell segmentation and online SVM for cell cycle phase identification in automated microscopy. *Bioinformatics,* 24, 94–101.

Xu, F., Beyazoglu, T., Hefner, E., Gurkan, U. A. and Demirci, U. 2011. Automated and adaptable quantification of cellular alignment from microscopic images for tissue engineering applications. *Tissue Engineering Part C: Methods,* 17, 641–649.

Xu, S., Tai, D., So, P., Yu, H. and Rajapakse, J. 2010. Automated scoring system for liver fibrosis diagnosis with second harmonic generation microscopy. *Australian Journal of Intelligent Information Processing Systems,* 12(2).

Xylas, J., Alt-Holland, A., Garlick, J., Hunter, M. and Georgakoudi, I. 2010. Intrinsic optical biomarkers associated with the invasive potential of tumor cells in engineered tissue models. *Biomedical Optics Express,* 1, 1387–1400.

Yu, L. and Liu, H. 2004. Efficient feature selection via analysis of relevance and redundancy. *The Journal of Machine Learning Research,* 5, 1205–1224.

Zakaria, O., Madi, M. and Kasugai, S. 2012. A novel osteogenesis technique: The expansible guided bone regeneration. *Journal of Tissue Engineering*, 3(1), 2041731412441194.

Zhang, J. G. and Deng, H. W. 2007. Gene selection for classification of microarray data based on the Bayes error. *BMC Bioinformatics*, 8, 370.

Zhang, J. and Tan, T. 2002. Brief review of invariant texture analysis methods. *Pattern Recognition*, 35, 735–747.

Zhang, J., Yan, C. H., Chui, C. K. and Ong, S. H. 2010. Fast segmentation of bone in CT images using 3D adaptive thresholding. *Computers in Biology and Medicine*, 40, 231–236.

Zhou, X. and Tuck, D. P. 2007. MSVM-RFE: Extensions of SVM-RFE for multiclass gene selection on DNA microarray data. *Bioinformatics*, 23, 1106–1114.

Zhu, S., Matsudaira, P., Welsch, R. and Rajapakse, J. 2010. Quantification of cytoskeletal protein localization from high-content images. *Pattern Recognition in Bioinformatics*, 289–300.

Index